REVISE BTEC NATIONAL
Applied Science

REVISION GUIDE

Series Consultant: Harry Smith

Authors: David Brentnall, Ann Fullick, Karlee Lees, Chris Meunier, and Carol Usher

A note from the publisher

While the publishers have made every attempt to ensure that advice on the qualification and its assessment is accurate, the official specification and associated assessment guidance materials are the only authoritative source of information and should always be referred to for definitive guidance.

This qualification is reviewed on a regular basis and may be updated in the future. Any such updates that affect the content of this Revision Guide will be outlined at **www.pearsonfe.co.uk/BTECchanges**. The eBook version of this Revision Guide will also be updated to reflect the latest guidance as soon as possible.

Introduction

Which units should you revise?

This Revision Guide has been designed to support you in preparing for the externally assessed units of your course. Remember that you won't necessarily be studying all the units included here – it will depend on the qualification you are taking.

BTEC National Qualification	Externally assessed units
Certificate	1 Principles and Applications of Science I
For both: Extended certificate Foundation diploma	1 Principles and Applications of Science I 3 Science Investigation Skills
Diploma	1 Principles and Applications of Science I 3 Science Investigation Skills 5 Principles and Applications of Science II
Extended diploma	1 Principles and Applications of Science I 3 Science Investigation Skills 5 Principles and Applications of Science II 7 Contemporary Issues in Science

Your Revision Guide

Each unit in this Revision Guide contains two types of pages, shown below.

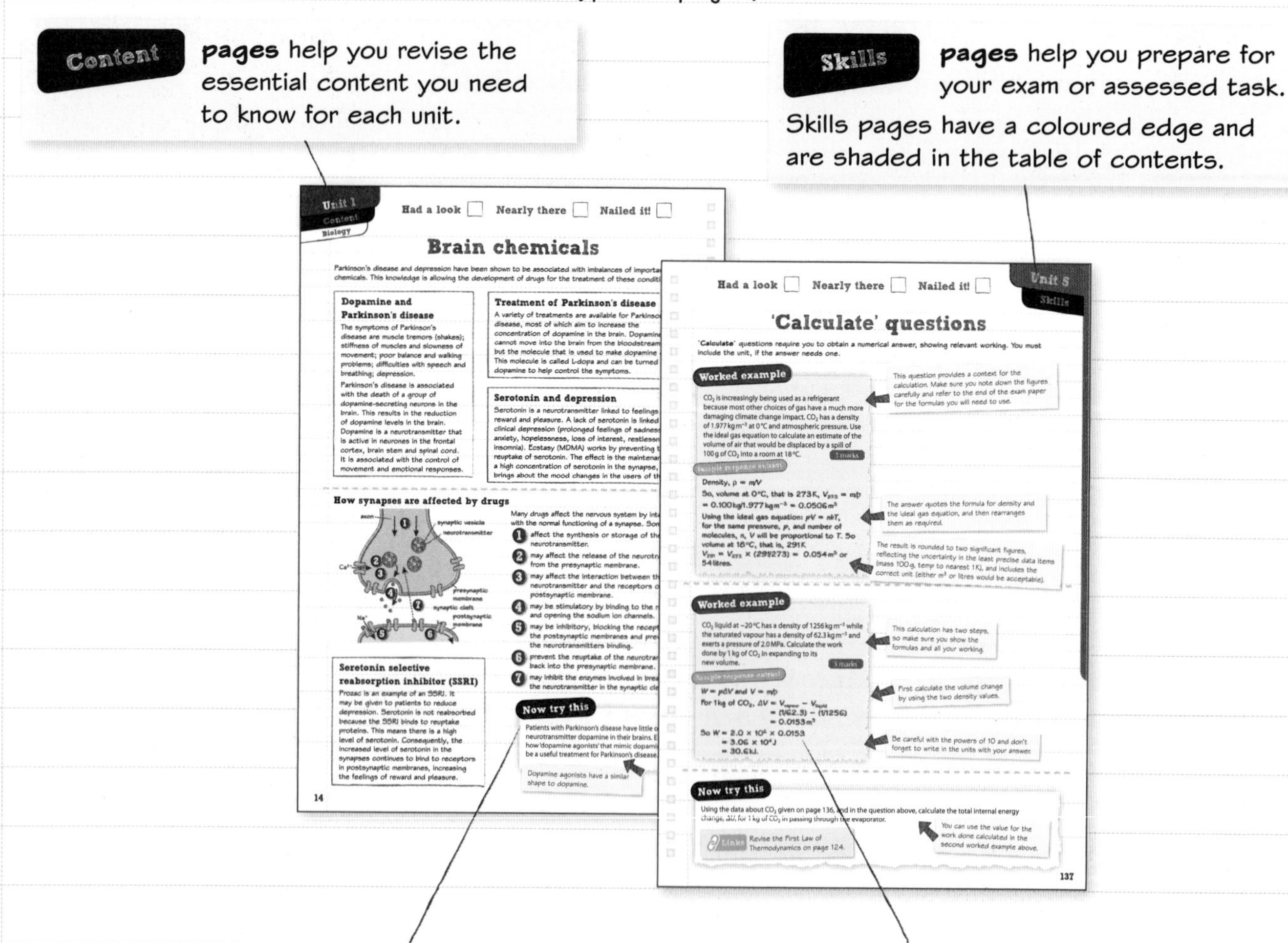

Use the **Now try this** activities on every page to help you test your knowledge and practise the relevant skills.

Look out for the **sample response extracts** to revision questions or tasks on the skills pages. Post-its will explain their strengths and weaknesses.

Contents

Unit 1 Principles and Applications of Science I

Unit 3 Science Investigation Skills

Unit 5 Principles and Applications of Science II

Unit 7 Contemporary Issues in Science

A small bit of small print
Pearson publishes Sample Assessment Material (SAMs) and the Specification on its website. This is the official content and this book should be used in conjunction with it. The questions in *Now try this* have been written to help you test your knowledge and skills. Remember: the real assessment may not look like this.

Cells and microscopy

All living things are made of cells. Cells are the basic units of life and all cells come from other cells. **Microscopes** magnify images so it is easier to see cells and their structures more clearly.

The light microscope

Magnification is an important property of microscopes. The magnification is calculated as follows:

Total magnification = magnification of eyepiece lens × magnification of objective lens.

The eyepiece lens usually has a ×10 magnification and the greatest magnification of the objective lens is usually ×100. So the greatest total magnification is usually ×1000.

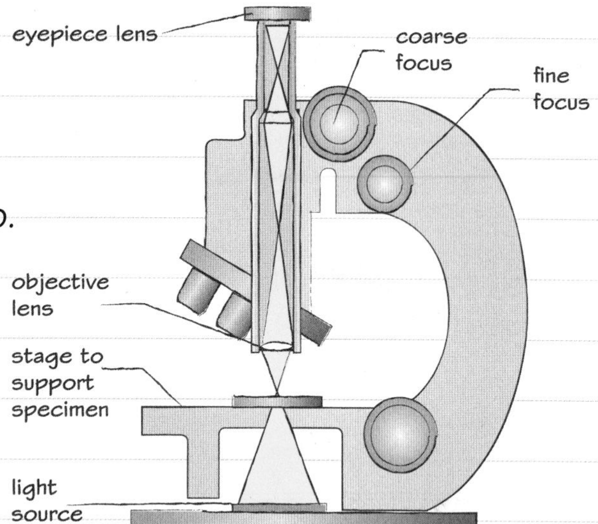

The specimen

The material to be viewed on the slide or specimen needs to be **thin**, so that light or an electron beam can pass through it. A **coverslip** is needed to protect the specimen and also the lens if they should touch. **Stains** can help to distinguish different features in the specimen.

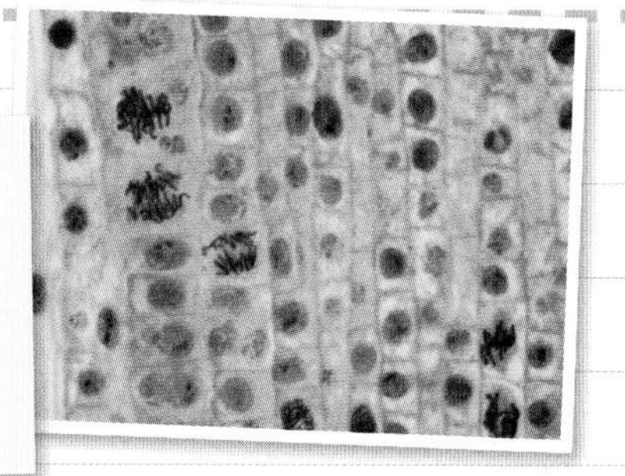

Light micrograph of onion root tip cells, undergoing mitosis (nuclear division). Magnification ×300. Compare the detail of this image to the one below taken with an electron microscope.

Electron microscopes

Electron microscopes were invented in the middle of the twentieth century and have a far greater magnification. However, you can only examine dead material with an electron microscope.

Worked example

The photograph shows an electron micrograph of part of a cell.

Calculate the width of P at its widest part (between A and B). **3 marks**

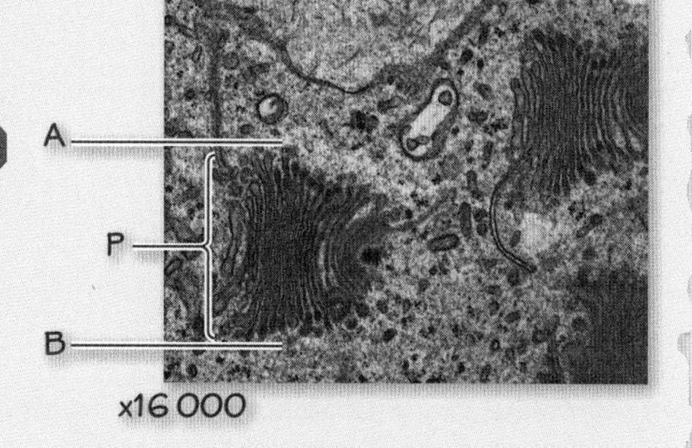

Sample response extract

$$\text{magnification} = \frac{\text{size of image}}{\text{size of real object}}$$

$$\text{so } 16\,000 = \frac{20\,\text{mm (1)}}{\text{size of real object}}$$

$$\text{rearrange, so width of P} = \frac{20\,\text{mm (1)}}{16\,000\,\text{mm (1)}}$$

$$= 0.001\,25\,\text{mm (1)}$$

Take care with the very small numbers. Remember $1\,\mu m = 0.000\,001\,m = 1 \times 10^{-6}\,m = 0.001\,mm = 1 \times 10^{-3}\,mm$. So the above answer is also $1.25\,\mu m$.

Now try this

A structure viewed under a light microscope with a magnification of ×400 is measured using a scale in the eyepiece. Each division in the scale is equal to 0.06 mm. The structure measures 7 divisions.

Calculate the real length of the structure.

Had a look ☐ Nearly there ☐ Nailed it! ☐

Cells

All living organisms are made of cells, which share some common features. They all contain **DNA**, **cytoplasm**, **ribosomes** and have **plasma**. However, some of these structures differ in prokaryote and eukaryote cells.

Cells are tiny!

It was not realised that living things are made from cells until Robert Hooke invented the microscope in 1665 and saw cells for the first time in tiny slivers of cork.

Light microscopes are very limited as to what they can allow us to see inside cells, because the wavelength of light is the limiting factor.

Electron beams have a much shorter wavelength than beams of light and allow much more detail to be seen through an electron microscope. However, electron microscopes can allow only dead material to be examined.

Structures seen with a light microscope

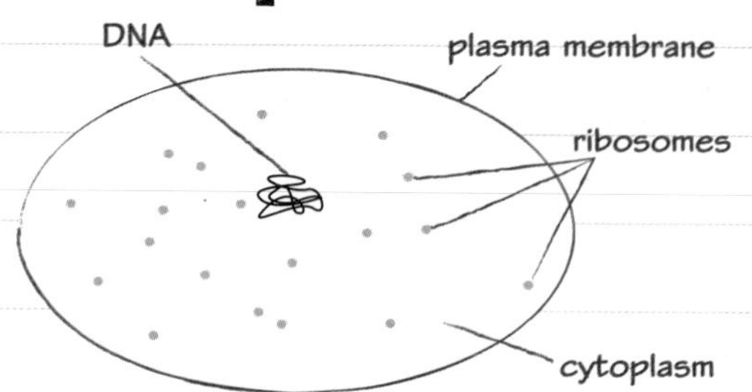

All cells have the four features shown, although they are not always the same, for example bacterial cells have 70S ribosomes, whilst plant and animal cells have 80S ribosomes.

Structures seen with an electron microscope (ultrastructure)

Eukaryotic cells contain **organelles**, which are structures in cells with specialised functions often bound by a membrane. Only the plasma membrane, nucleus and nucleolus are visible under the light microscope.

plasma membrane
- protects cell from its surroundings
- regulates movement of substances in and out of cells

vesicle
- small, membrane-bound sac
- transports and stores substances in the cell

rER (rough endoplasmic reticulum)
- a series of single, flattened sacs enclosed by a membrane
- has ribosomes on the surface
- proteins made here

nucleolus
- region of dense DNA and protein
- makes ribosomes

nucleus
- surrounded by a double membrane (envelope)
- pores (holes) in the nuclear envelope

centrioles
- two hollow cylinders
- arranged at right-angles to each other
- makes the spindle in cell division

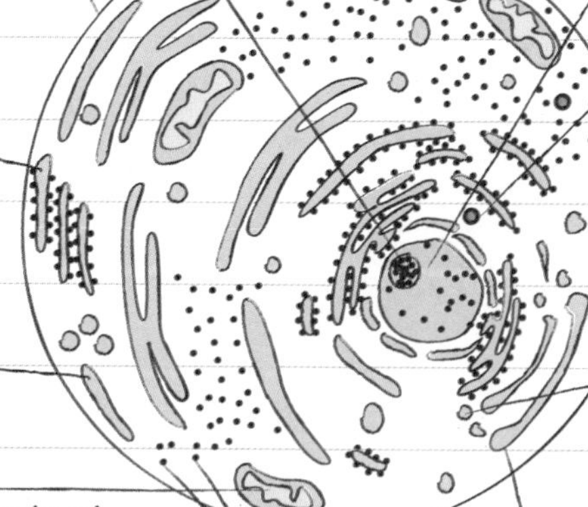

sER (smooth endoplasmic reticulum)
- a series of single, tubular sacs made of membrane
- lipids made here

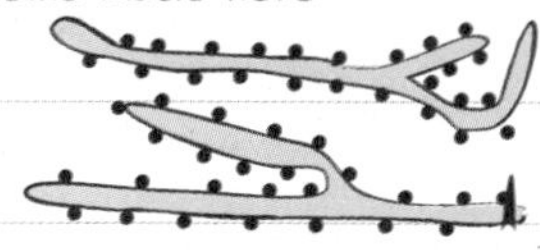

lysosome
- enclosed by a single membrane
- containing digestive enzymes
- destroys old organelles and pathogens

mitochondrion
- surrounded by a double membrane (envelope)
- inner membrane folded into finger-like projections called cristae (singular, crista)
- central area contains a jelly called the matrix
- containing 70S ribosomes and DNA
- site of respiration

80S ribosomes
site of protein synthesis, contrast with 70S in prokaryotes

Golgi apparatus
- a series of single, curved sacs enclosed by a membrane
- many vesicles cluster around the Golgi apparatus
- modifies proteins and packages them in vesicles for transport

cytoplasm
- fluid that fills a cell
- many molecules dissolved in solution (enzymes, sugars, amino acids, fatty acids)
- site of many metabolic processes

Now try this

Make a table listing animal cell organelles not surrounded by a membrane, those surrounded by a single membrane and those surrounded by a double membrane.

Prokaryotes

Bacteria are **prokaryotes**, made of a single cell with no membrane-bound organelles.

Bacteria cell structure

Plasmids
- double-stranded DNA in a circular structure
- often contain additional genes that aid the bacterium's survival, such as antibiotic resistance or toxin producing genes.

Ribosome
- makes proteins
- 70S (S is a Svedberg, a measure of size by rate of sedimentation)

Nucleoid
- region where single circular, length of DNA, is folded.
- DNA carries all essential information.

Capsule
- polysaccharide layer outside the cell wall
- protects cells from drying out, being engulfed by, for example, white blood cells
- helps cells to stick to surfaces.

Cell wall
- made of long-chained molecule made up of a sugar and amino acids called peptidoglycan.

0.5–5μm

Gram-negative and gram-positive bacteria

Gram-negative bacteria **do not** retain the gram stain (crystal violet) when washed with acetone and absolute alcohol because their cell wall has an outer layer.

Gram-positive bacteria **do** retain the gram stain because the thick peptidoglycan wall absorbs the stain and they do not have an outer wall.

Gram-negative bacteria are more resistant to antibiotics than gram-positive bacteria. This is because gram-negative bacteria have a cell wall with an outer membrane, which protects them from the antibiotic.

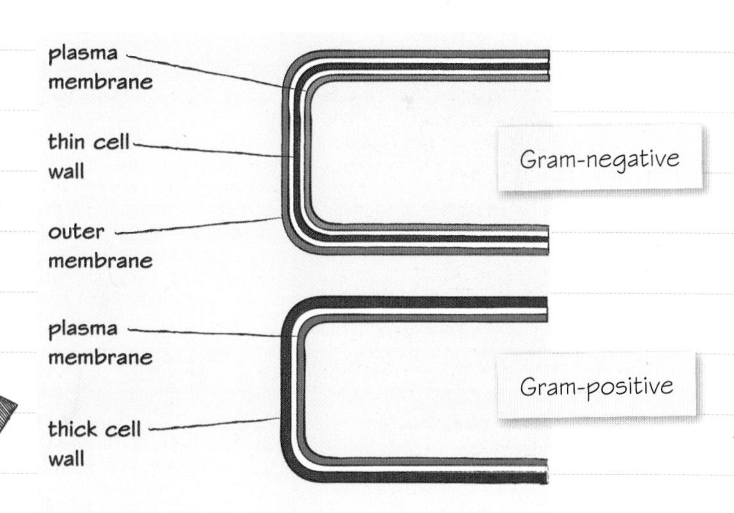

Now try this

The photograph shows a technician testing a swab from meat for bacteria.

1 Examine the photograph and state what the results show.

2 Explain why some bacteria stained purple.

Another stain, usually safranin, is also used at the same time. This stains the thin peptidoglycan cell wall red.

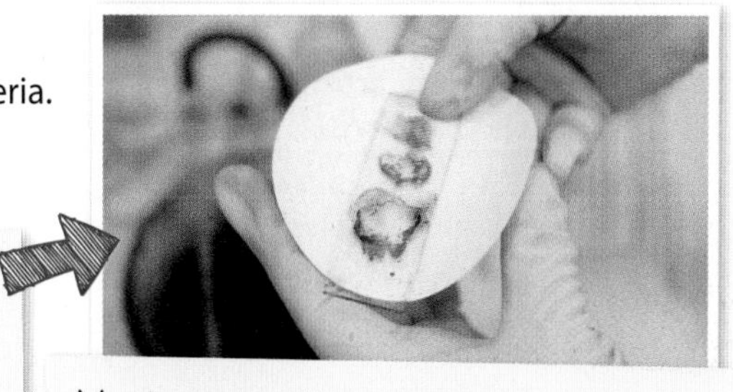

Most meat pathogens are gram-positive.

Plant cells

Plants are eukaryotes, but their cells differ from those of other eukaryotes, such as animals.

The more complex plant cell

Plant cells include all the structures (except centrioles) that are in animal cells, **as well as:**

- **chloroplasts** (for photosynthesis)
- a **vacuole** stores water and other substances
- a **tonoplast membrane** controls movement of molecules into and out of the vacuole
- a **cell wall** (for support and protection)
- **amyloplasts** to store starch
- **middle lamella** to stick cells together
- **plasmodesmata** and **pits** to allow communication between one cell and another.

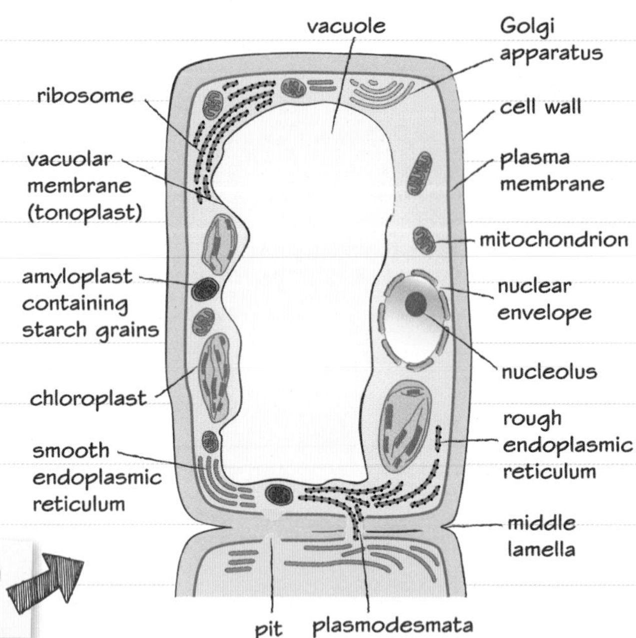

A generalised plant cell as seen through a low power electron microscope

Electron micrographs of the key organelles

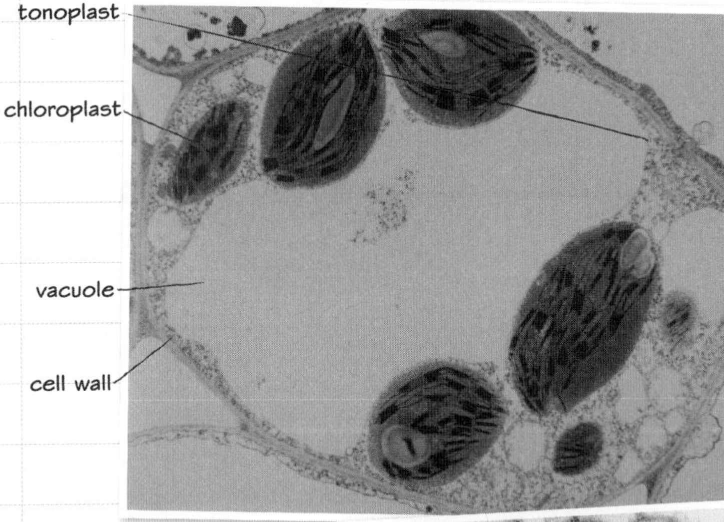

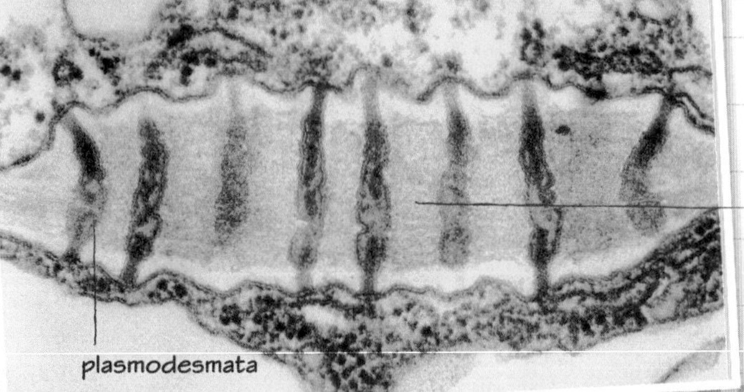

Now try this

Compare animal, plant and bacterial cells, making reference to cell wall, chloroplasts, nuclear membrane, cell membrane, ribosomes and centrioles.

A table would be a good way of doing this.

Specialised cells: Plant cells

The organs of the plant, for example, leaves and roots contain some cells that are specially **adapted** for their **function**.

Leaves

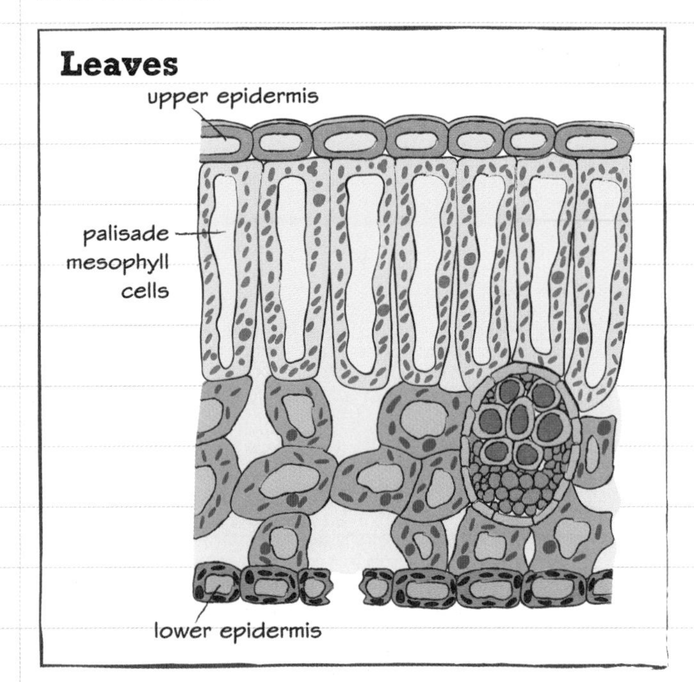

Palisade cell

Palisade cells are **cylindrical shaped**, so they pack tightly in the upper part of a leaf.They contain many **chloroplasts** to capture as much energy from light as possible for photosynthesis. They have a **large vacuole**, which helps to keep the cell and leaf structure rigid.

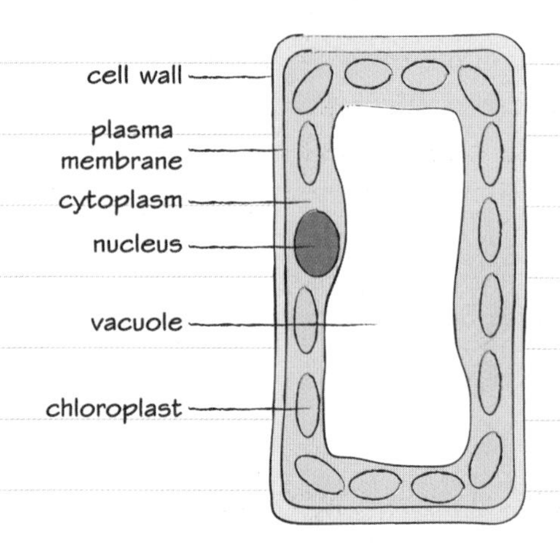

Root hair cell

Root hair cells are found in the **epithelium** (surface cells) near the root tip, where there is no thickening.

The hair is a fine protrusion from the cell out into the soil. This provides a **large surface area** to volume ratio to absorb water and dissolved minerals from the soil. The cells have a thin cell wall to make it easier for water absorption.

They contain many mitochondria to help supply energy for active transport of minerals from the soil into the cell.

Links See page 105 for information on active transport.

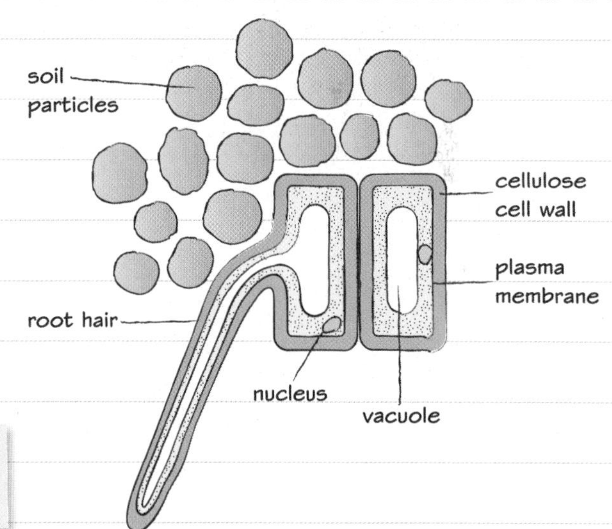

Now try this

1 Palisade mesophyll cells are found in leaf tissue. What is the main function of palisade mesophll cells?

A To capture energy transferred by light

B To provide water

C To provide carbon dioxide

D To make starch

2 Explain how root hair cells are adapted to their function.

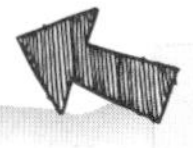

In this question, you need to state the function and the adaptation, then link them.

Specialised cells: Animal cells

Mammalian **gametes** (the sperm and egg), and red and white **blood cells** are specialised for their functions.

Structure and function in sperm and egg

Sperm

undulipodium for movement to swim to egg

haploid **nucleus** contains only one set of chromosomes so that full complement restored at fertilisation

acrosome containing enzymes to digest the outer layers of the egg

mid region with **mitochondria** to provide the energy (from respiration) for movement

Egg

haploid nucleus contains half the chromosomes of a body cell, so that when the nucleus from the sperm fuses with the egg cell nucleus the full number of chromosomes is restored at fertilisation

special **vesicles** (cortical granules) – these contain a substance that helps stop more than one sperm fertilising the egg

cytoplasm full of energy-rich material

zona pellucida (jelly layer) – to stop more than one sperm fertilising the egg

follicle cells (corona radiata) supplies vital proteins

White blood cells

These blood cells have a large nucleus, often with protrusions, and are made in the bone marrow and lymph nodes. There are different types of white blood cell and their function is to fight pathogens. They can squeeze through capillary walls and are found in tissues as well as the blood. So they are able to move to sites of infection easily.

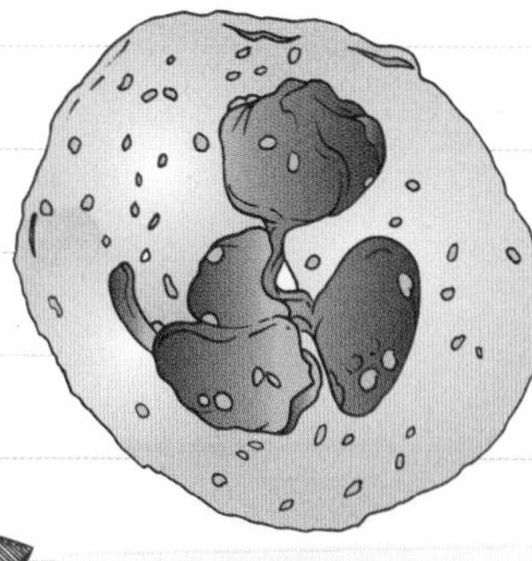

phagocyte (a neutrophil)

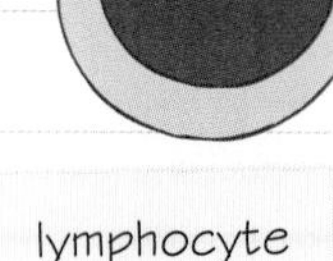

lymphocyte

Phagocytes and lymphocytes are types of white blood cell.

Red blood cells

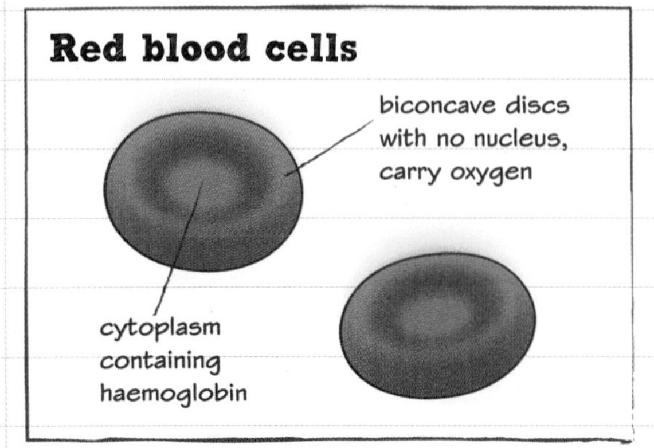

Mature red blood cells have no nucleus or other organelles, so there is room for more haemoglobin, which is the protein that carries oxygen.

They are small, round, biconcave discs. This optimises their **surface area to volume ratio**. This allows more oxygen and carbon dioxide to diffuse into the haemoglobin. Their size and shape also allow them to squeeze through narrow blood vessels.

Having no nucleus also means that they cannot divide, which is another difference between them and white blood cells. They are made in the bone marrow.

Now try this

Describe three structural differences between a human sperm cell and a human egg cell.

Epithelial tissue

Epithelial cells line any surface that is in contact with the external environment. They may also line the surface of internal organs, such as the lungs. In the lungs are two types of epithelial tissue: squamous epithelium and columnar epithelium.

Columnar epithelium

The upper airway (trachea and bronchi) is mainly lined with ciliated columnar epithelium cells. These cells have a lot of mitochondria. The cilia move mucus away from the lungs, preventing any inhaled particles causing infection. Goblet cells produce mucus.

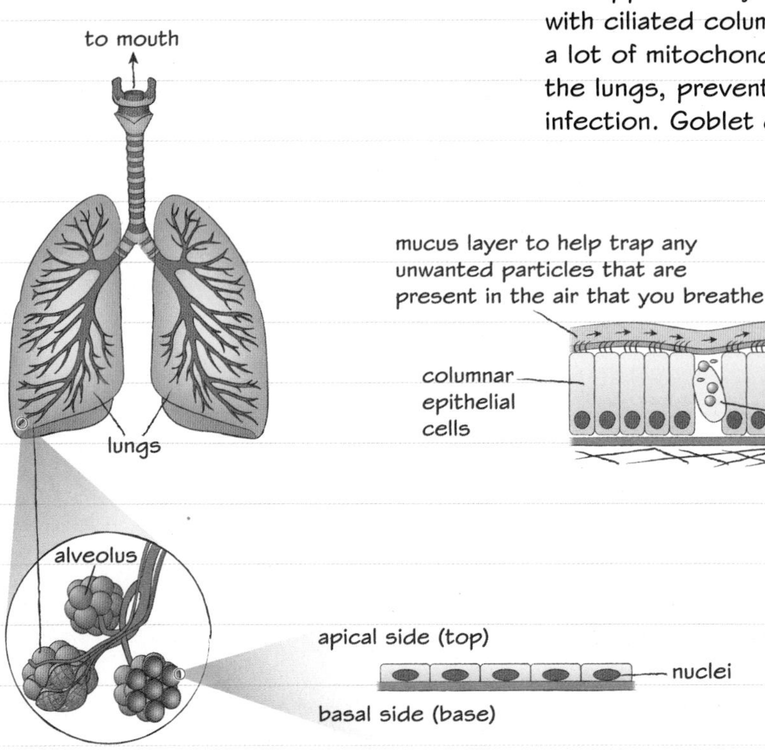

Squamous epithelium

Very flat and thin with egg-shaped nuclei, often only one cell thick and very good for surfaces where diffusion occurs as in the lungs.

Chronic obstructive pulmonary disease (COPD)

COPD includes several conditions and is more common in smokers than non-smokers, because substances in smoke damage the lungs.

Cigarette smoke causes the cilia of columnar epithelium to slow and stop beating and eventually die off so mucus builds up. This clogs the airways and this causes more coughing that ruptures the thin alveolar epithelial cells (destroying them), reducing the surface area for gas exchange. This also provides a good environment for pathogens to grow.

Now try this

Explain why ciliated columnar cells contain many mitochondria.

Remind yourself about the function of mitochondria on page 3.

Had a look ☐ Nearly there ☐ Nailed it! ☐

Blood vessels and atherosclerosis

Blood vessels are lined with **endothelial** tissue. Blood clots form when there is damage to tissue, if there is damage to an artery and a clot forms this can cause a number of health-related problems.

Endothelial tissue in blood vessels

Endothelial tissue lines the inside of blood vessels. It is made up of a single layer of flat, long cells, which are orientated lengthways in the direction of blood flow. Their function is to provide a smooth surface so that blood flows easily over them.

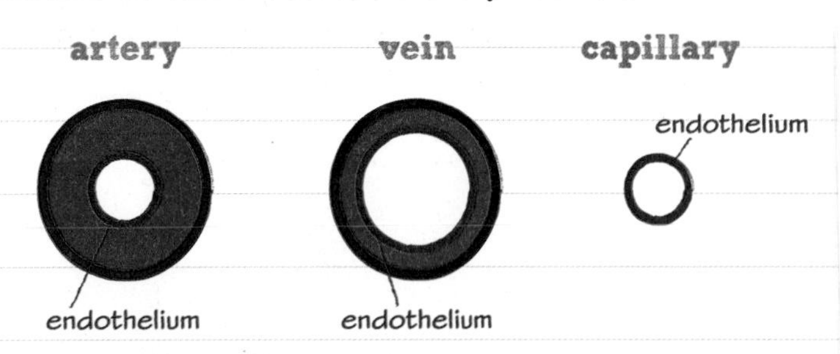

Arteries and veins are made of the same tissues but in different proportions. They both have an outer layer of connective and elastic tissue. This is thicker in veins to prevent collapse of the blood vessel. They also have a middle layer composed of smooth muscle, connective and elastic tissue. This is thicker in arteries to maintain blood pressure. The inner layer is made up of endothelial tissue. Capillaries only have an endothelium.

Atherosclerosis

Atherosclerosis is the disease process that leads to coronary disease and strokes (cardiovascular diseases). Fatty deposits (**atheroma**) can either block an artery directly or increase its chance of being blocked by a blood clot (**thrombosis**).

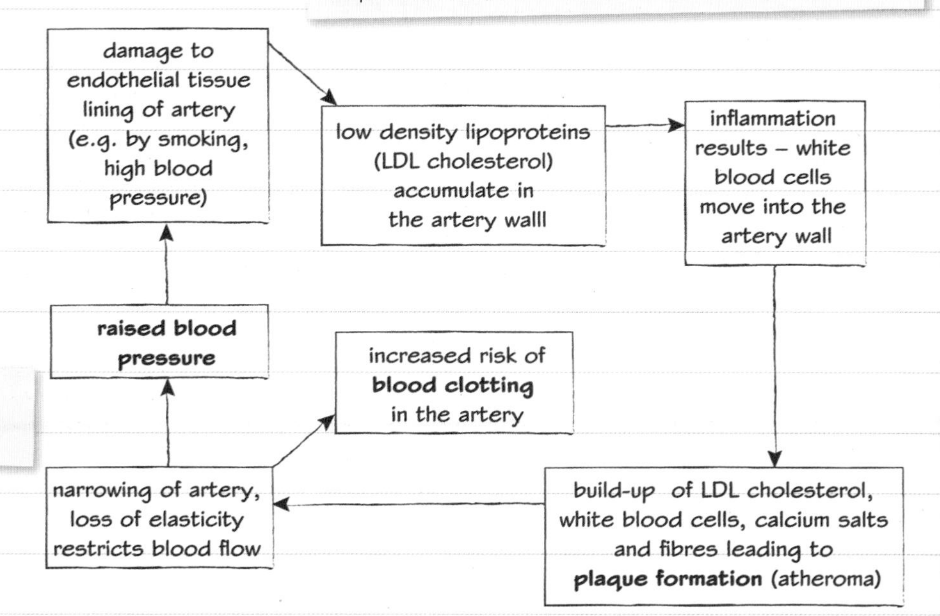

Development of atherosclerosis

Smoking and atherosclerosis

Cigarette smoke contains many toxic chemicals, which can lead to atherosclerosis. The thickness of the blood increases. The thick blood causes fatty deposits to build up on the walls of arteries and increases the risk of clotting. Smoking also increases blood pressure and heart rate, which can also cause damage to the endothelium.

Now try this

Describe endothelial tissue and explain how it is adapted to its function.

In this case, an annotated drawing would help to answer the question.

Fast and slow twitch muscle

Muscle is a soft tissue, which can contract to maintain a position in the body or produce movement.

Muscle structure

Muscle is made up of bundles of cells called muscle fibres (cells). These muscle fibres are made up of myofibrils.

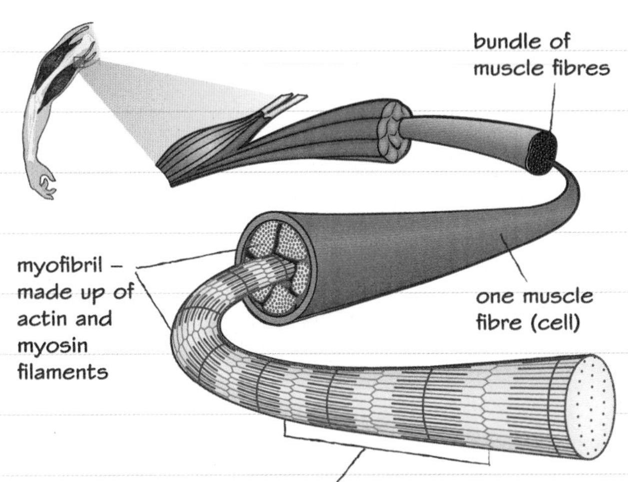

Muscle contraction

Muscles are made up of a thick protein filament composed of the protein myosin and a thin filament composed of actin. When muscle contracts, the thin actin filaments move between the thick myosin filaments, shortening the length of the sarcomere and the overall length of the muscle.

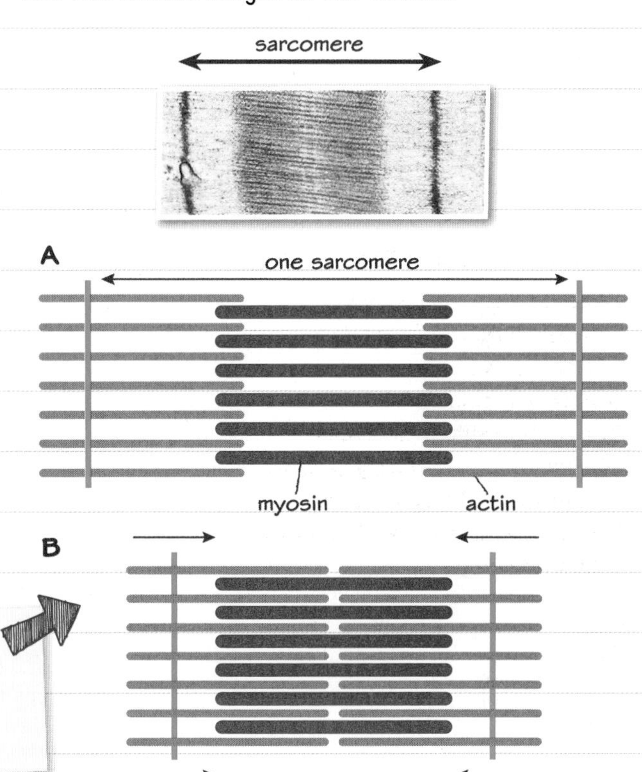

Look at the vertical blue lines to see the shortening of the sarcomere that has happened between A and B.

Fast and slow twitch muscles

There are two types of muscle fibres.

The properties of fast twitch muscle fibres suit short bursts of activity such as sprinting.

Slow twitch	Fast twitch
slow, sustained contraction for long periods of exercise	rapid intense contractions in short bursts
many mitochondria supply energy from aerobic respiration (requires oxygen)	few mitochondria – energy for contraction from anaerobic respiration (needs no oxygen)
lots of capillaries	few capillaries
does not tire easily	tires easily
large oxygen and glucose stores	little stored oxygen and glucose

Which sport?

Humans have differences in the proportions of slow and fast twitch muscles. People who have a higher proportion of fast twitch muscles can move quickly in short bursts, for example sprinters. This is because the muscles contract rapidly and strongly in short bursts using energy from anaerobic respiration. People who have a higher proportion of slow twitch muscles are better at endurance events like the marathon, as they use aerobic respiration and can work for a long time without getting tired.

Now try this

State the type of muscle cells that are an advantage to marathon runners and explain why.

Had a look ☐ Nearly there ☐ Nailed it! ☐

Nerve tissue

Nerves are made up of cells called **neurones**. Neurones carry messages in the form of electric signals, from one part of the body to another. These messages are also known as **nerve impulses**.

Structure of the nervous system

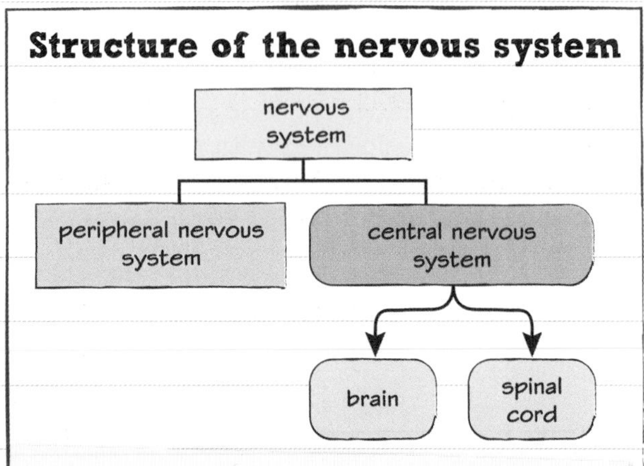

The nervous system of a mammal is divided into the **central nervous system** (CNS) and the **peripheral nervous system**. The CNS is made up of the brain and the spinal cord.

Structure of a neurone

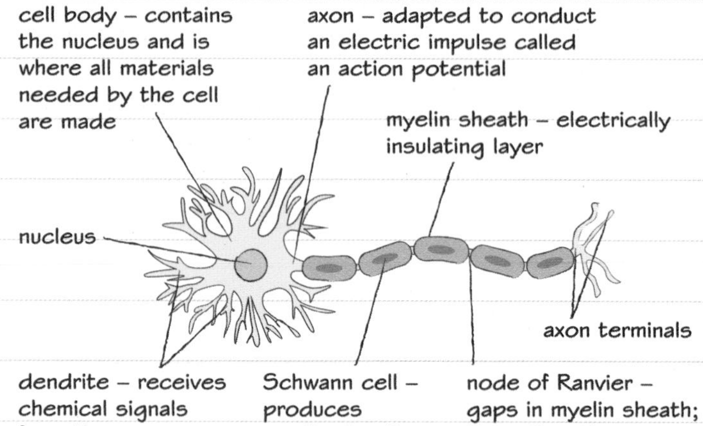

Myelin sheaths

Most nerve cells are myelinated. Those that are not are small diameter nerves usually responsible for transmitting pain such as aches and soreness rather than sharp pain. They also detect temperature changes.

This table compares non-myelinated nerve cells with myelinated nerve cells. It is a useful way of displaying the answer to a compare and contrast question.

Comparing non-myelinated and myelinated nerve cells

Non-myelinated	Myelinated
do not have a myelin sheath	have a myelin sheath
grey	white
transmit impulses slower	transmit impulses very fast
do not have nodes of Ranvier	do have nodes of Ranvier

Types of neurone

The mammalian nervous system is made up of cells called **neurones**. There are several types of neurone. The three most common types are shown here.

sensory neurone

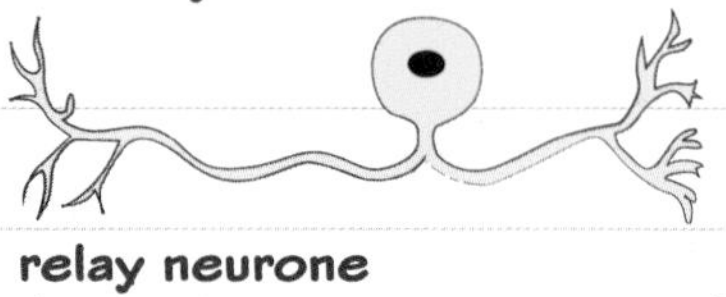

Connects sensory receptors to the central nervous systems

relay neurone

Found in the central nervous system

motor neurone

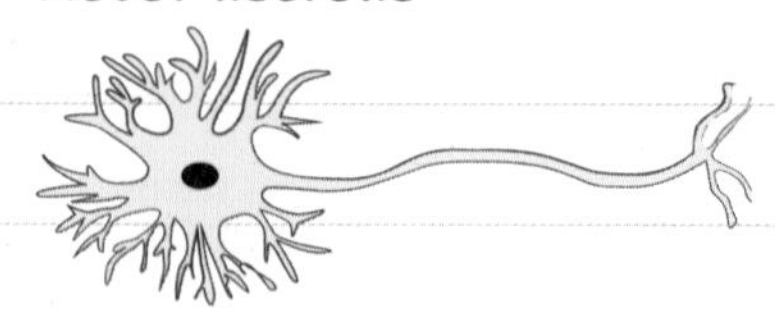

Communicates from central nervous system to effectors

Now try this

'Progressive' means 'increases over time' and 'de' before a word means 'removal of'.

Multiple sclerosis is known as a 'progressive de-myelinating disease' that affects the nervous system.

Suggest what 'progressive de-myelinating' means and how MS affects the transmission of nervous impulses.

Nerve impulse

Neurons send electric impulses. These are generated by changes in the concentrations of ions inside and outside the nerve cell causing a potential difference (PD), called an **action potential**, which transmits an electrical signal between nerve cells.

Changes in polarity

The **action potential** is triggered by the **depolarisation** of the nearby membrane changing the PD to the **threshold potential**. If the threshold potential is not reached, nothing happens.

A new action potential cannot be generated in the same section of membrane for about 5 milliseconds, which makes sure that the impulse travels in one direction along a nerve fibre. Between the action potentials the cell is at **resting potential** (between −60 and −70 mV, depending on the nerve cell).

1 Depolarisation

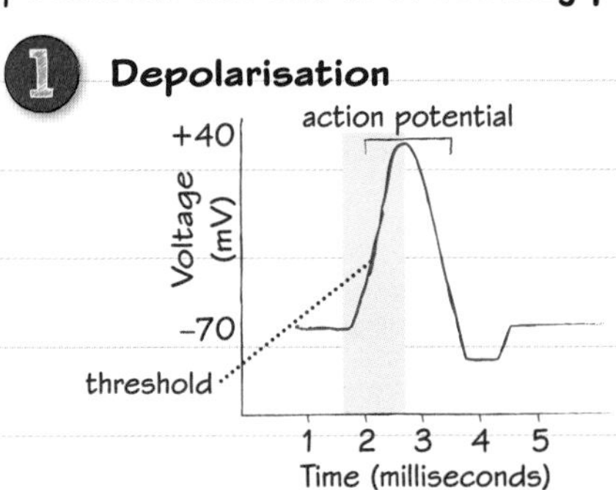

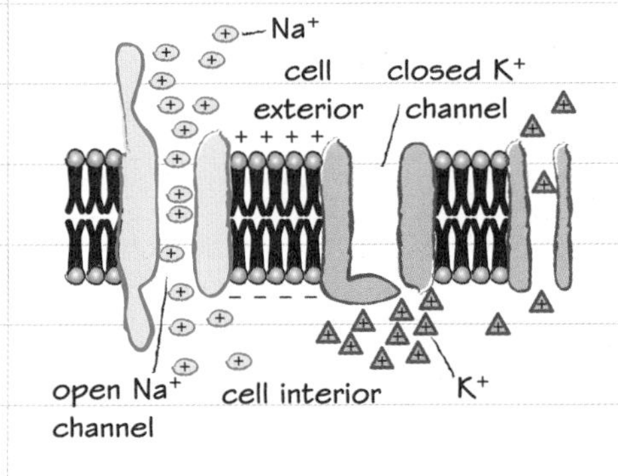

- Na⁺ gates open.
- Na⁺ diffuses into the cell carrying positive charge.
- Na⁺ gates close.

2 Repolarisation

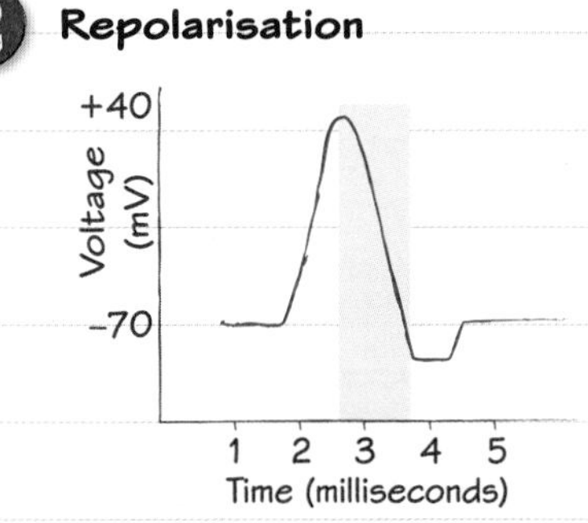

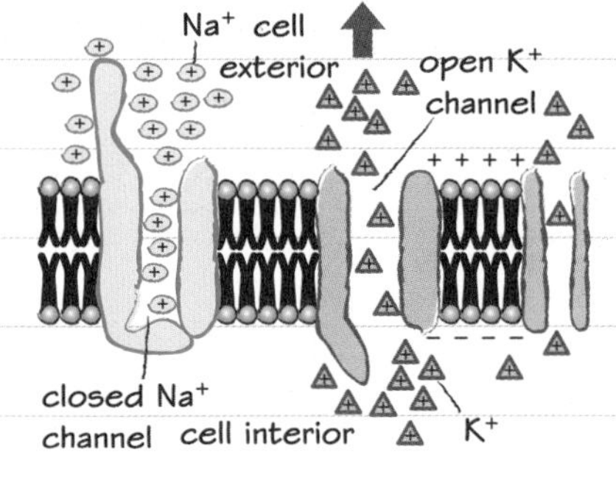

- K⁺ gates open.
- K⁺ diffuses out of the cell taking positive charge with it.
- K⁺ gates close.

3 Recovery

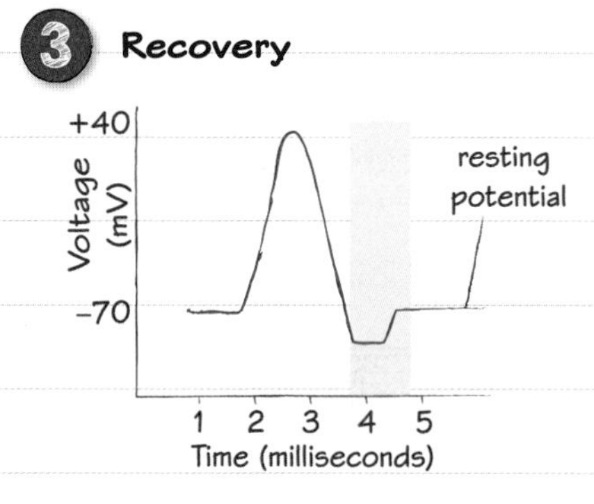

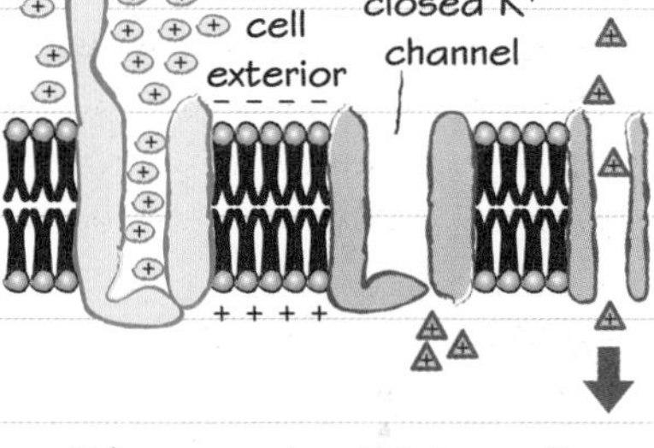

- K⁺ moves back into cell through non-voltage gated channels.
- Attracted by negative charge as **hyperpolarised** (that is, more negative than −60 to −70 mV).
- **Resting potential** equilibrium restored.

Saltatory (jumping) conduction

The only region of a myelinated nerve fibre that can depolarise is at the nodes of Ranvier where there is no myelin.

This means that nerve impulses can travel a much longer distance. They can also travel faster than without myelin, and the impulse 'jumps' from one node to the next.

Now try this

Describe and explain the movement of ions at point C in the diagram.

There are two different forces causing ions to move at this point.

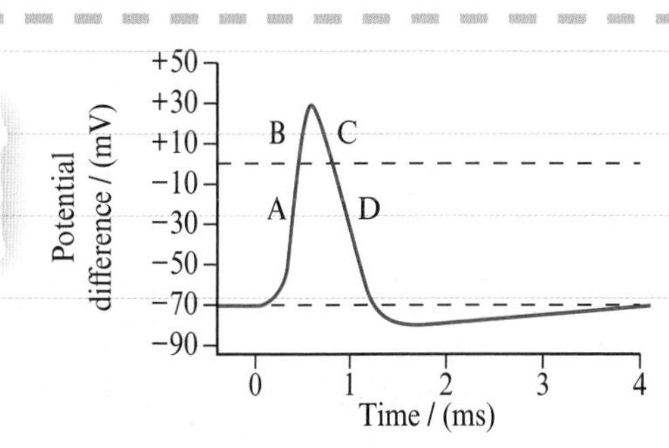

Electrocardiogram (ECG)

An electrocardiogram (ECG) measures the **action potentials** of the heart (see page 11). Electrodes are placed on different parts of the body to detect electrical impulses and a machine amplifies the impulses during each heart beat and records them.

Electrocardiogram trace

The electrical changes in the heart can be measured and presented as an Electrocardiogram (ECG).

If disease disrupts the heart's normal conduction pathways there is a disruption of the expected ECG pattern (which is 60 to 100 beats per minute at regular intervals). ECGs can therefore be used for diagnosis of cardiovascular disease.

Links See page 90 and 92 for more information on the heart and heart beat.

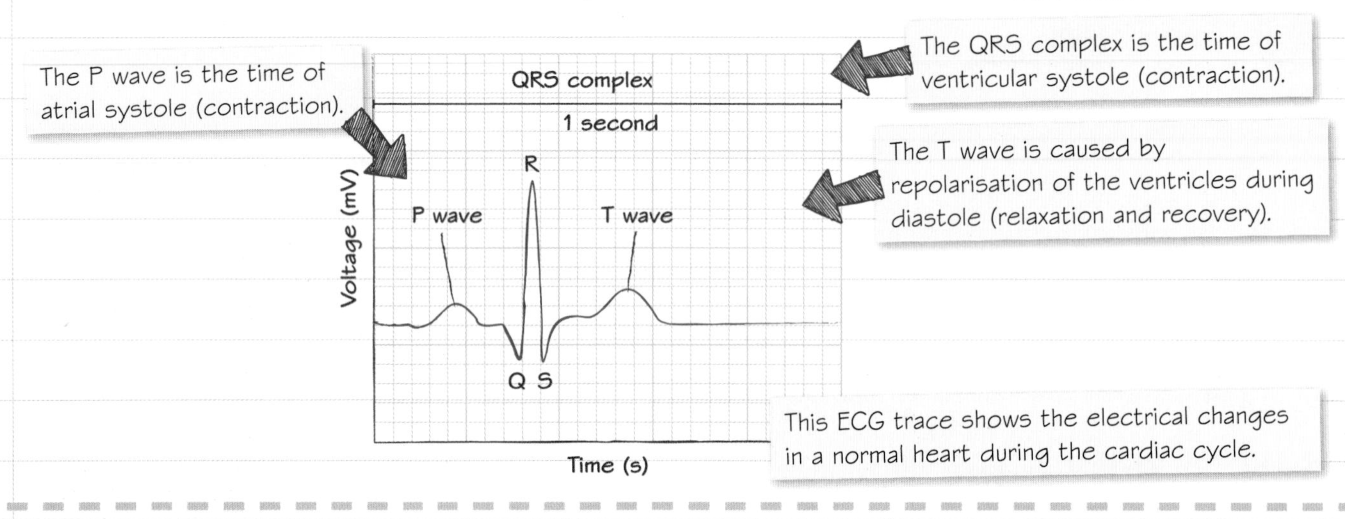

The heart beat

The heart contracts because a small cluster of cells (pacemaker) produces an electrical impulse, which causes the heart muscle to contract. The three key stages of the heart beat are shown below.

P wave

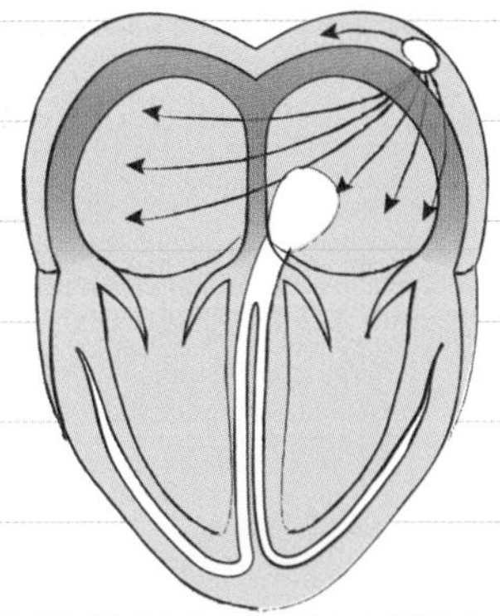

Depolarisation of the atria (upper heart chambers), so they contract.

QRS complex

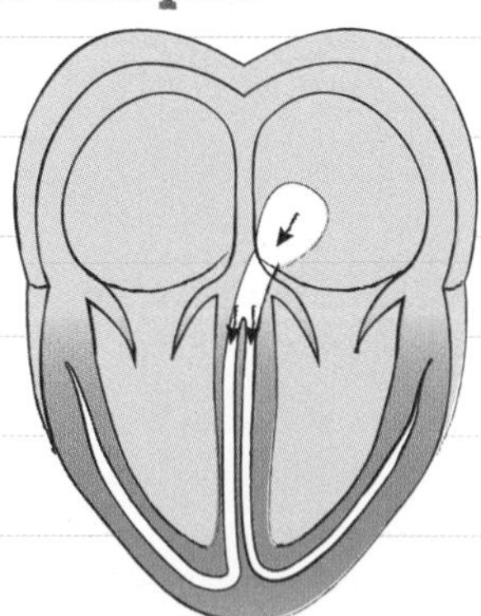

Depolarisation of the venticles (lower heart chamber), so they contract.

T wave

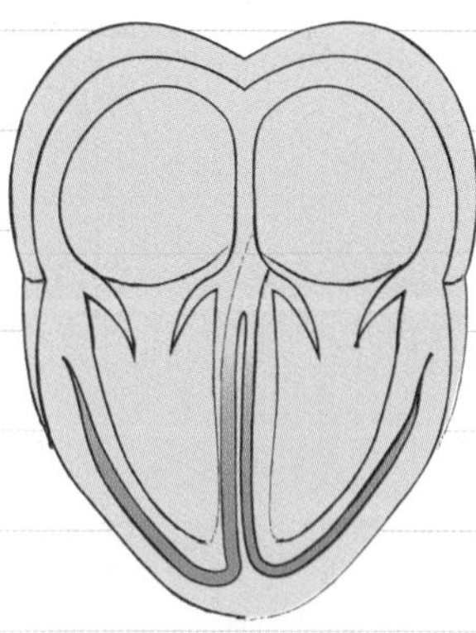

Repolarisation of the ventricles (recovery).

Heart muscle

The peripheral nervous system (PNS) (see page 10) connects the central nervous system (brain and spinal cord) to other organs of the body. This includes the **autonomic nerves**. These regulate automatic or involuntary functions of the body, for example contracting the heart muscle.

Now try this

Explain why the QRS complex is much larger than the P wave.

The atria move blood into the ventricles, whilst the ventricles move blood around the body. See page 90 to find out about the structure of the heart.

Synapses

Synapses are gaps between nerve cells. Action potentials cannot pass between nerve cells across the synapse, so transmitter substances are used.

Structure of a synapse

A synapse is the junction between two neurons. The **presynaptic membrane** (before the gap) allows the release of chemicals (**neurotransmitters**) when impulses are arriving to stimulate impulses in the cell after the gap (**postsynaptic cell**).

A typical synapse has a presynaptic knob containing vesicles filled with neurotransmitter. Receptor molecules on the postsynaptic membrane bind with neurotransmitter.

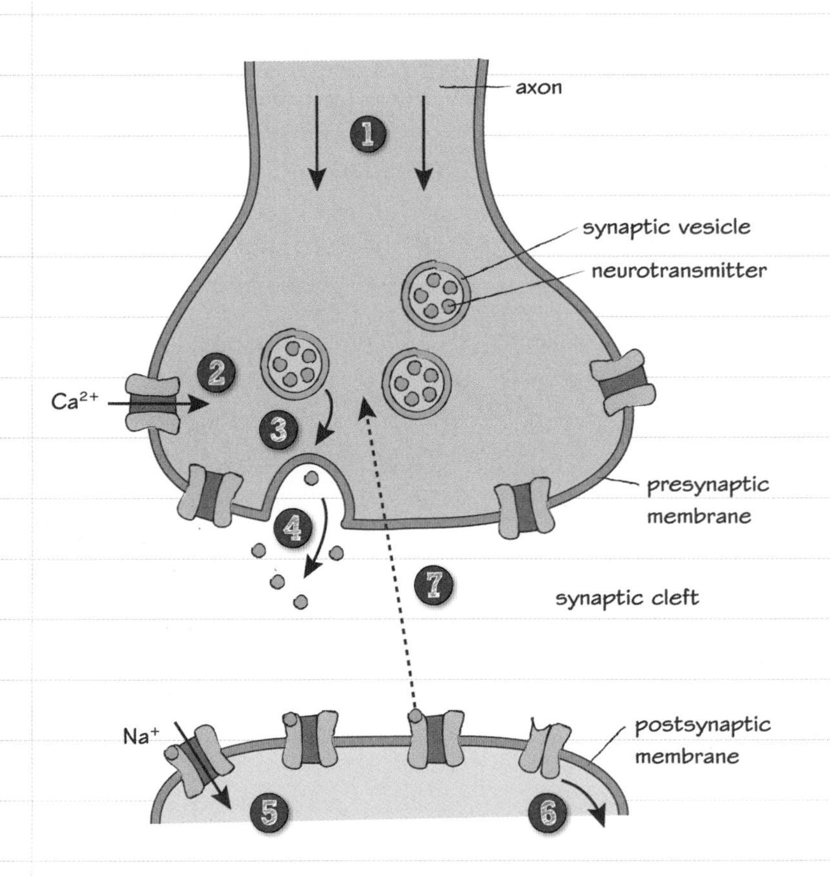

1. An action potential arrives.
2. The membrane depolarises. Calcium ion channels open. Calcium ions enter the neuron.
3. Calcium ions cause synaptic vesicles containing neurotransmitter to fuse with presynaptic membrane.
4. Neurotransmitter is released into the synaptic cleft.
5. Neurotransmitter binds with receptors on the postsynaptic membrane. Sodium channels open. Sodium ions flow through the channels.
6. The membrane depolarises and initiates an action potential.
7. When released, the neurotransmitter will be taken up across the presynaptic membrane (whole or after being broken down), or it can diffuse away and be broken down.

Acetylcholine

Acetylcholine was the first neurotransmitter to be discovered and has many functions, including the stimulation of muscles. It is largely made in the brain.

Acetylcholine is broken down in the synaptic cleft by acetylcholinesterase.

Now try this

Botulism toxin paralyses respiratory muscles, causing suffocation and death.

(a) State whether botulism toxin prevents or stimulates the release of acetylcholine.
(b) Explain the effect of botulism toxin.

Think about what acetylcholine does.

Had a look ☐ Nearly there ☐ Nailed it! ☐

Brain chemicals

Parkinson's disease and depression have been shown to be associated with imbalances of important brain chemicals. This knowledge is allowing the development of drugs for the treatment of these conditions.

Dopamine and Parkinson's disease

The symptoms of Parkinson's disease are muscle tremors (shakes); stiffness of muscles and slowness of movement; poor balance and walking problems; difficulties with speech and breathing; depression.

Parkinson's disease is associated with the death of a group of dopamine-secreting neurons in the brain. This results in the reduction of dopamine levels in the brain. Dopamine is a neurotransmitter that is active in neurones in the frontal cortex, brain stem and spinal cord. It is associated with the control of movement and emotional responses.

Treatment of Parkinson's disease

A variety of treatments are available for Parkinson's disease, most of which aim to increase the concentration of dopamine in the brain. Dopamine cannot move into the brain from the bloodstream, but the molecule that is used to make dopamine can. This molecule is called L-dopa and can be turned into dopamine to help control the symptoms.

Serotonin and depression

Serotonin is a neurotransmitter linked to feelings of reward and pleasure. A lack of serotonin is linked to clinical depression (prolonged feelings of sadness, anxiety, hopelessness, loss of interest, restlessness, insomnia). Ecstasy (MDMA) works by preventing the reuptake of serotonin. The effect is the maintenance of a high concentration of serotonin in the synapse, which brings about the mood changes in the users of the drug.

How synapses are affected by drugs

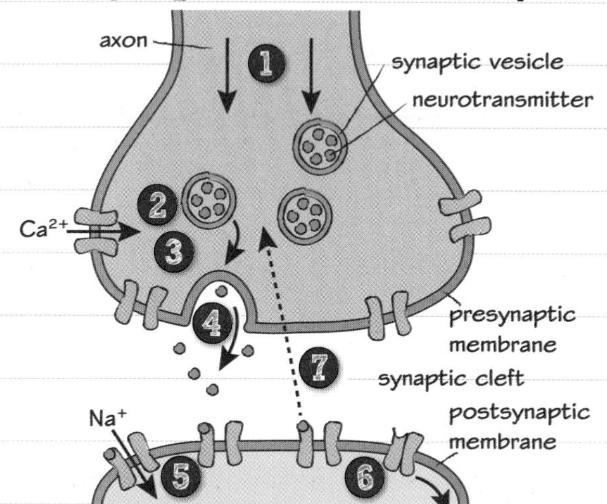

Many drugs affect the nervous system by interfering with the normal functioning of a synapse. Some drugs:

1. affect the synthesis or storage of the neurotransmitter.
2. may affect the release of the neurotransmitter from the presynaptic membrane.
3. may affect the interaction between the neurotransmitter and the receptors on the postsynaptic membrane.
4. may be stimulatory by binding to the receptors and opening the sodium ion channels.
5. may be inhibitory, blocking the receptors on the postsynaptic membranes and preventing the neurotransmitters binding.
6. prevent the reuptake of the neurotransmitter back into the presynaptic membrane.
7. may inhibit the enzymes involved in breaking down the neurotransmitter in the synaptic cleft.

Seretonin selective reabsorption inhibitor (SSRI)

Prozac is an example of an SSRI. It may be given to patients to reduce depression. Serotonin is not reabsorbed because the SSRI binds to reuptake proteins. This means there is a high level of serotonin. Consequently, the increased level of serotonin in the synapses continues to bind to receptors in postsynaptic membranes, increasing the feelings of reward and pleasure.

Now try this

Patients with Parkinson's disease have little of the neurotransmitter dopamine in their brains. Explain how 'dopamine agonists' that mimic dopamine might be a useful treatment for Parkinson's disease.

Dopamine agonists have a similar shape to dopamine.

Writing formulae and equations

Scientists use formulae and equations to summarise the changes that occur in reactions.

Formulae

There are some ions that it is helpful to remember to help you write chemical formulae.

Ion	Formula
nitrate	NO_3^-
carbonate	CO_3^{2-}
sulfate	SO_4^{2-}
hydroxide	OH^-
ammonium	NH_4^+
zinc	Zn^{2+}
silver	Ag^+

Learn these by covering, recalling them, uncovering to check until you are consistently getting them correct.

Working out formulae

You can deduce the formula of an ionic compound using the formulae and charges of the ions involved. For example, working out the formula of calcium nitrate:

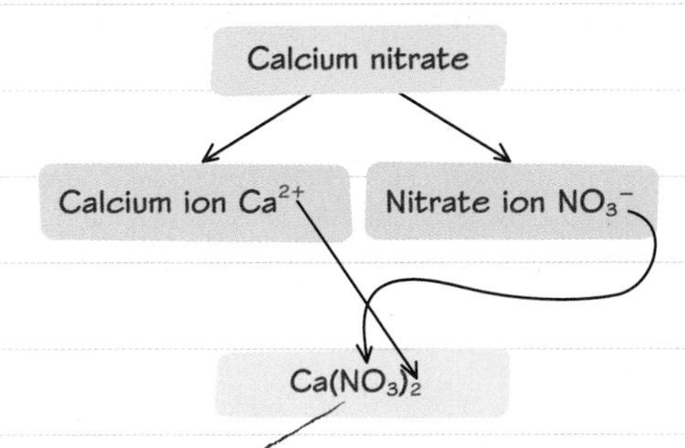

Brackets are used whenever you have more than one ion that consists of two or more types of atom, in this case, NO_3^-

If a molecule is covalent, a quick way to find out how many bonds each atom will form is to work out 8 minus its group number. So carbon dioxide contains C, which forms 8 − 4 = 4 bonds, and O which forms 8 − 6 = 2 bonds. Hence its formula must be CO_2.

Predicting charges on ions

You should be able to predict the charge on some ions based on their position in the periodic table by following some simple rules. For example:

metals in groups 1, 2 and 3 form 1+, 2+ and 3+ ions, respectively

non-metals in groups 5, 6 and 7 can form 3−, 2− and 1− ions, respectively.

Writing a balanced equation

You cannot simply create or destroy atoms in chemical reactions, so the numbers of each atom on both sides of an equation have to balance, even if you rearrange the atoms into new compounds. Look at the equation for the combustion of methane.

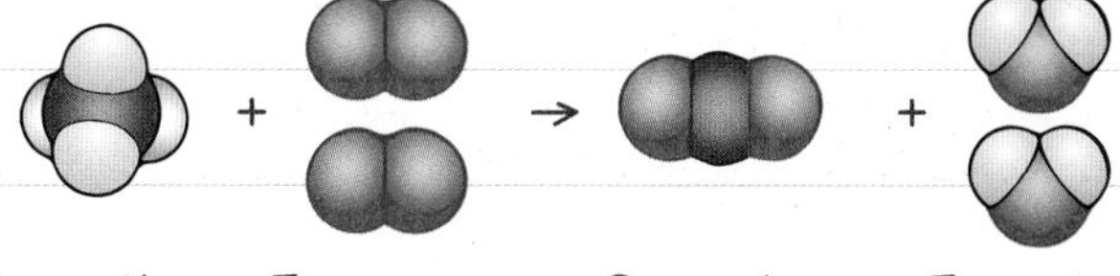

One methane molecule		Two oxygen molecules		One carbon dioxide molecule		Two water molecules
CH_4	+	$2O_2$	→	CO_2	+	$2H_2O$
1C 4H		4O		1C 2O		2O 4H

Remember the large numbers before a formula tell you how many of that molecule take part or form in the reaction, in this case two oxygen molecules and two water molecules. Once you have practised lots of examples, balancing most equations becomes quite easy, especially if you learn the charges on ions to help you write the correct formulae.

Now try this

The biofuel ethanol, C_2H_5OH, reacts with oxygen to form carbon dioxide and water.

Write an equation for the reaction.

Leave elemental reactants, for example, O_2, until last when balancing.

Practising writing equations

✓ Write the **correct formula** for all reactants and products, for example:

$$Na_2CO_3 + HCl \rightarrow NaCl + CO_2 + H_2O$$

✓ Check the numbers of each atom **balance**. If not, use numbers in front of reactants or products to balance atoms. In this equation, there are 2 Na atoms on the left but only one on the right, so '2' is placed before NaCl:

$$Na_2CO_3 + 2HCl \rightarrow 2NaCl + CO_2 + H_2O$$

✓ Add **state symbols** if required, (s) = solid, (l) = liquid, (g) = gas, (aq) = aqueous solution:

$$Na_2CO_3(s) + 2HCl(aq) \rightarrow 2NaCl(aq) + CO_2(g) + H_2O(l)$$

Had a look ☐ Nearly there ☐ Nailed it! ☐

Electronic structure of atoms

Bohr's model describes the shapes in space where electrons can be found in atoms.

Main energy levels or shells

Electrons orbit the nucleus in **shells**. The further the shell, or main energy level, is from the nucleus, the higher its energy. Each shell can hold a fixed number of electrons.

n	Shell	Number of electrons
1	1st shell	2
2	2nd shell	8
3	3rd shell	18
4	4th shell	32

You can find the total number of electrons in an atom by using the **atomic number**. This is defined as the **number of protons** in an atom, but for a neutral atom also equals the number of electrons.

Atomic orbitals

Each shell consists of **atomic orbitals**. These are regions in space where electrons may be found. The larger the main energy level, the more orbitals it is made up from. The first four types of are called s, p, d and f orbitals.

Shapes of orbitals

An s-orbital is spherical in shape and holds up to 2 electrons.

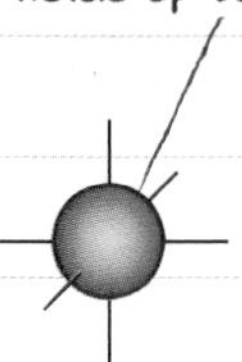

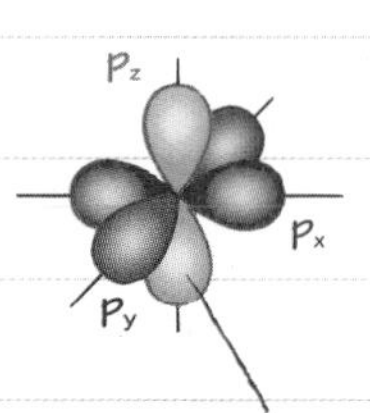

Each p-orbital is dumb-bell shaped and holds up to 2 electrons. Each p sub-shell has 3 p-orbitals, p_x, p_y and p_z.

Writing electron configurations

Electron configurations are the arrangement of the electrons in atoms or ions. Bohr's model recognises that the main shells are split into **sub-shells** and electron configurations should reflect that.

Element	Electron configuration
B	$1s^22s^22p^1$
C	$1s^22s^22p^2$
N	$1s^22s^22p^3$
O	$1s^22s^22p^4$

Electrons 'in boxes'

In these diagrams each 'box' represents an orbital. The arrows represent electrons in the orbitals.

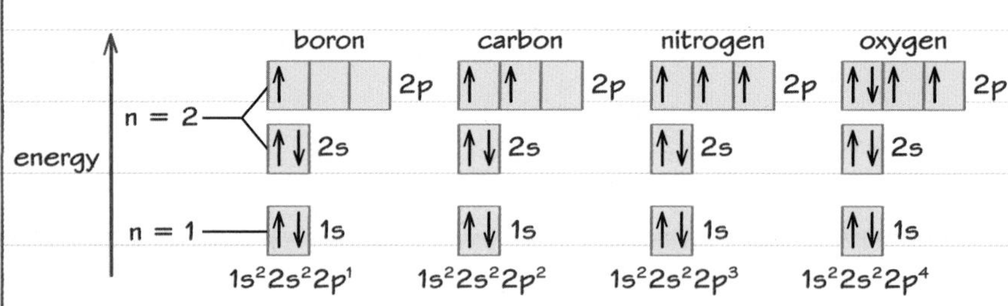

A third main energy level consists of one s-orbital, 3 p-orbitals and 5 d-orbitals.

The direction of the arrows shows the **spin of each electron**. A pair always spin in opposite directions to reduce repulsion.

Orbitals can be confused with sub-shells. Any orbital, regardless of the sub-shell it is in, can hold up to two electrons.

Rules for arranging electrons

1. Start at the lowest shell and add electrons one at a time to build up the configuration.
2. Fill each sub-level before starting on the next.
3. Fill each orbital singly in a sub-level before pairing electrons.
4. Paired electrons have opposite spins, so they are shown as arrows pointing in opposite directions.

The 4s orbital is lower in energy than the 3d when empty, but higher in energy when occupied.

The idea, shown by rules 1 and 2, that electrons fill lower shells before the higher shells, is called the **Aufbau principle**.

Now try this

Write the electron configuration of the following particles. The number of electrons is given in brackets.

(a) Na atom (11) (b) O^{2-} ion (10)
(c) Ti atom (22) (d) V^{3+} ion (20)

Ionic bonding

Ionic bonds are strong **electrostatic attractions** between positive and negative ions. The ions in ionic bonds can be shown using **electron configuration diagrams**.

Ionic compounds and giant ionic structures

Ionic compounds have giant structures.

In a giant ionic structure, the ions are arranged in a regular, three-dimensional pattern called a **lattice**. The electrostatic forces between the ions act in all directions and keep the structure together. The large number of these strong electrostatic attractions gives ionic compounds **high melting points**.

Sodium chloride lattice

In the sodium chloride lattice each sodium ion is surrounded by six chloride ions and each chloride ion is surrounded by six sodium ions. This repeating pattern continues for a vast number of ions.

Cl^- ion

Na^+ ion

Positive ions:

- are generally formed by **metal atoms** losing electrons
- have a **positive** charge equal to the group number if formed from a group 1, 2 or 3 element
- have different charges if formed from a transition metal (for example, Fe^{2+}, Fe^{3+})
- can be represented in an electron configuration diagram, for example a Na^+ ion
- are known as **cations**.

Outer shell now empty as sodium atom has lost one electron to become an ion

Must show charge on ion

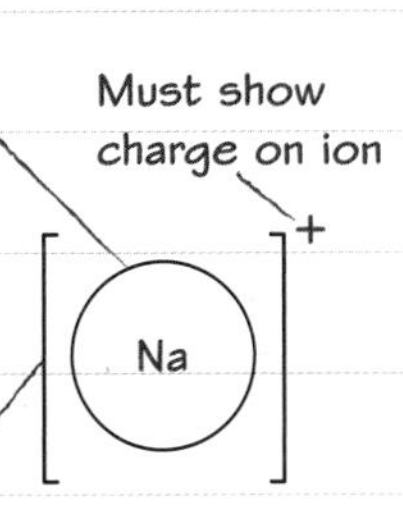

Square brackets to show charge is spread over whole ion

Negative ions:

- are generally formed by non-metal atoms gaining electrons from metal ions
- have a negative charge equal to 8 minus the group number of the element
- sometimes exist as polyatomic ions, such as CO_3^{2-}, SO_4^{2-}, NO_3^- and OH^-, whose charges should be learnt
- can be represented in an electron configuration diagram, for example an F^- ion.
- are known as **anions**.

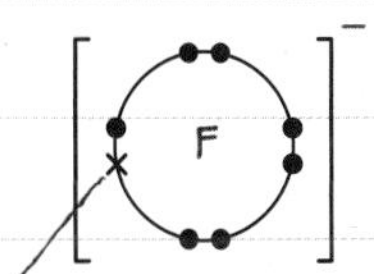

One of the electrons is shown as a cross to show that it has been gained by a fluorine atom.

Strength of ionic bonds

To compare the relative strength of ionic bonds, the **ionic charge** and **ionic radius** have to be considered, sometimes called the charge/size ratio. For instance, the ionic bonding in MgF_2 is much stronger than the ionic bonding in NaF. This is because the magnesium ion is smaller than the sodium ion and also has a greater charge. These factors increase the **electrostatic attraction** between the ions.

Now try this

Sodium chloride (NaCl) and potassium chloride (KCl) are both important in maintaining normal blood pressure.

Explain which compound has the strongest ionic bond.

Think about the size of each ion and their charges.

Had a look ☐ Nearly there ☐ Nailed it! ☐

Covalent bonds

A covalent bond is an **electrostatic attraction** between a shared pair of electrons and the nuclei of the bonded atoms.

How do atoms form covalent bonds?

A covalent bond forms when atoms **share** a pair of electrons. Generally, each atom in the bond contributes one electron to the pair, but a covalent bond consisting of an electron pair derived from one of the atoms is called a **dative covalent (coordinate)** bond.

Dot and cross diagrams

These are a way of showing the bonding between atoms, for instance in a chlorine molecule.

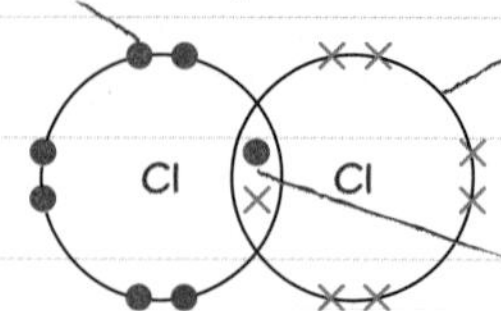

More dot and cross diagrams

Ammonia consists of 3 hydrogen atoms bonded to a nitrogen atom. Each hydrogen atom has only 1 unpaired electron ($1s^1$), so can form a single bond. The nitrogen atoms has 3 unpaired electrons in its outer shell, so can form 3 bonds.

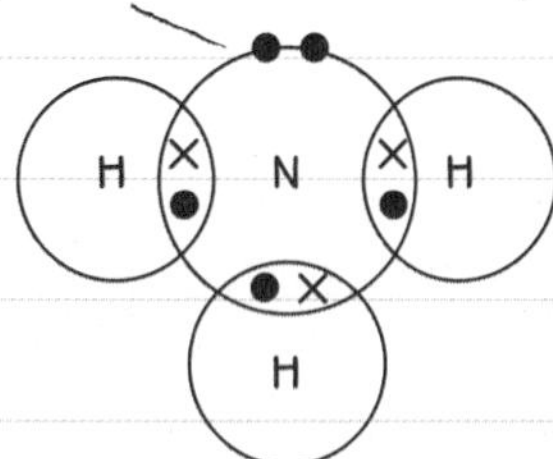

The **ammonium ion** forms from ammonia, NH_3 and a hydrogen ion, H^+.

The hydrogen ion has no electrons in its vacant orbital to form a bond, but the nitrogen atom in ammonia has a lone pair not involved in bonding. It uses this pair to bond to the hydrogen ion, forming a **dative covalent** (coordinate) bond.

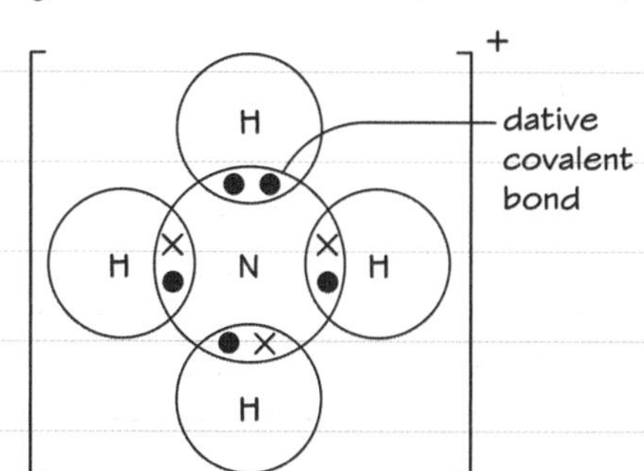

In **carbon dioxide**, each oxygen atom shares 2 pairs of electrons with the carbon atom.

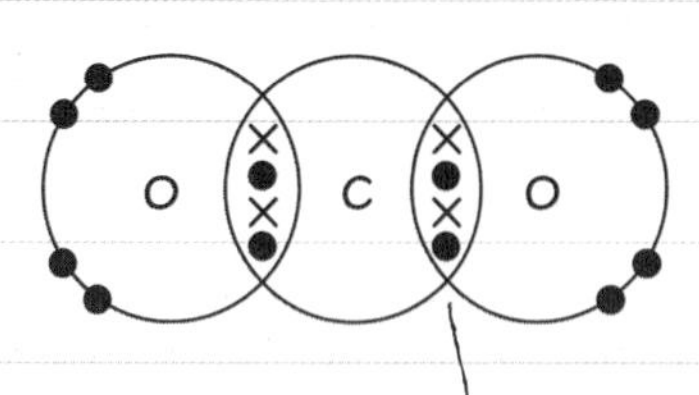

Methane consists of 1 carbon atom bonded to 4 hydrogen atoms. The orbitals containing the electron pairs repel as far away as possible, forming a **tetrahedral** shape, common around carbon atoms in organic chemistry.

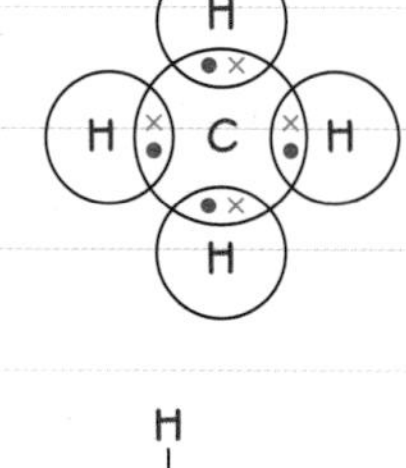

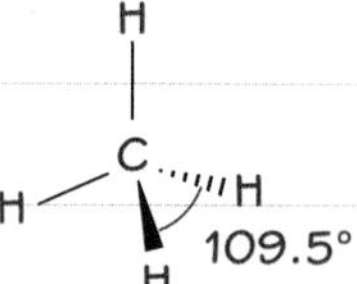

Strength of covalent bonds

Bond length and bond strength in covalent bonds are **inversely related**. This means that the shorter the covalent bond length, the greater the covalent bond strength.

Now try this

Boron trifluoride, BF_3, is used in the manufacture of semi-conductors. It reacts with ammonia, NH_3, to form the compound $BF_3.NH_3$.

Draw a dot and cross diagram of $BF_3.NH_3$, labelling any dative covalent bonds clearly.

Metallic bonding

You will need to be able to describe the model of metallic bonding and use it to explain properties such as **melting points**.

Metallic bonding

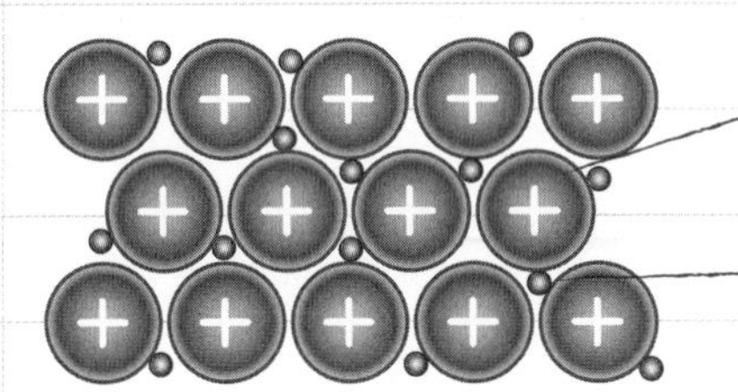

Delocalised electrons are the electrons from the outer shell of the metal atoms, but are not fixed to a particular atom, so can move freely throughout the structure.

The metallic bond is a strong **electrostatic** attraction between the positive metal ions and the delocalised electrons.

The positive ions are layered in three dimensions – a giant **lattice** structure.

Properties of metals

The key properties of metals due to the bonding are:

- **electrical and thermal conductivity** due to the delocalised electrons, which are free to move
- **high melting and boiling points** due to strong electrostatic attractions between positive ions and electrons
- **malleability** – can be shaped, as layers of positive ions slide over each other and the delocalised electrons move with the layers, so strong metallic bonds remain intact
- **ductility** – can be pulled into wires as positive ions roll over each other and the delocalised electrons move with the positive ions, so strong metallic bonds remain intact.

Trends in melting points

These tables show the melting points of some different group 1 and group 2 metals:

Group 1 metal	Melting point (°C)
lithium	181
sodium	98
potassium	63
rubidium	39
caesium	28

Group 2 metal	Melting point (°C)
beryllium	1278
magnesium	649
calcium	839
strontium	769
barium	729

The melting points decrease as the atoms get larger. Larger metals have more electrons and more electron shells. This means they have more shielding between the nucleus and delocalised electrons, so the electrostatic attraction force between them is weakened. This produces a **weaker metallic bond**, so less energy is required to break these bonds.

Each group 2 metal has a higher melting point than the group 1 metal **in the same period**. This is because the group 2 metal has **two** delocalised electrons per positive ion, rather than one. This gives it a greater electron density around the positive ions. This, alongside the +2 charge, produces a stronger electrostatic attraction between the nucleus and the delocalised electrons, and so a stronger metallic bond.

Now try this

Metals are often used in manufacturing because they are hard-wearing and can be shaped easily.

Use the metallic bonding model to explain why most metals can be hammered into thin sheets without easily splitting.

A diagram to show the model may help you explain this property.

Had a look ☐ Nearly there ☐ Nailed it! ☐

Intermolecular forces

Intermolecular forces are interactions between molecules caused by either **permanent** or **induced dipoles**.

London forces

Instead of 'London' forces, you may still see the older terminology, 'van der Waals' forces.

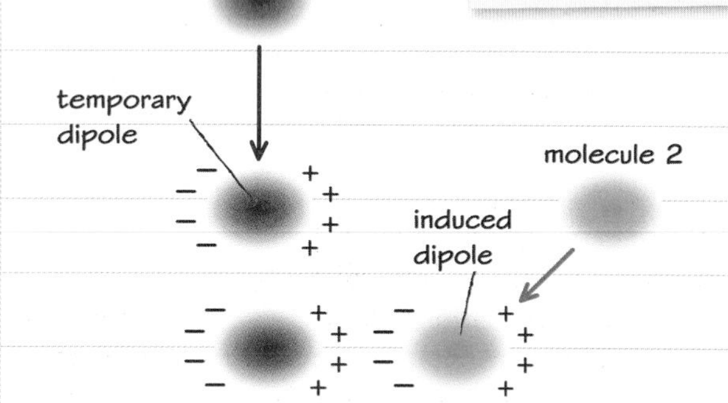

Electrons are moving randomly within the shells of a molecule or atom.

This can cause an uneven distribution in the molecule, resulting in an **instantaneous**, **temporary dipole**. This can **induce a temporary dipole** in a nearby molecule. This results in a weak attraction, called a **London force**. The more electrons a molecule has, the more likely this process occurs, so the stronger the London force.

Dipole–dipole interactions

Polar molecules such as HCl have permanent dipoles due to the much greater electronegativity of the chlorine atom and the fact that the molecule is not symmetrical. Hence the oppositely charged ends of two molecules are attracted to each other. This weak attractive force is called a **permanent dipole–dipole interaction**.

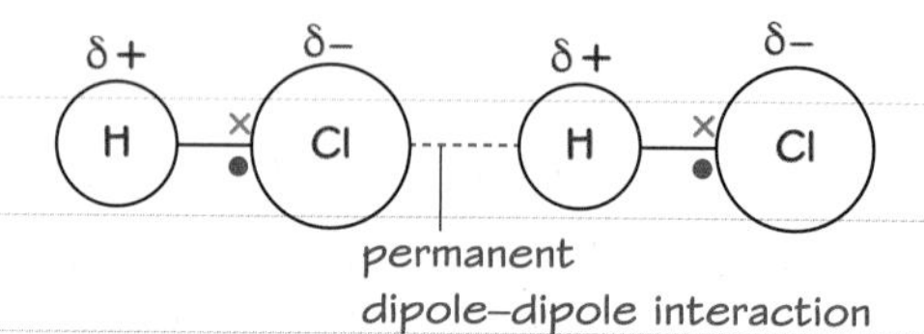

Hydrogen bonds

This type of intermolecular force is the attraction between an **electron-deficient hydrogen** atom (δ+) and a lone pair on oxygen, nitrogen or fluorine atoms. O, N and F are the only atoms that can form hydrogen bonds as they are small and highly electronegative, which means they pull pairs of electrons towards them.

Water molecules can form hydrogen bonds between each other.

Hydrogen bonds are an especially strong intermolecular force.

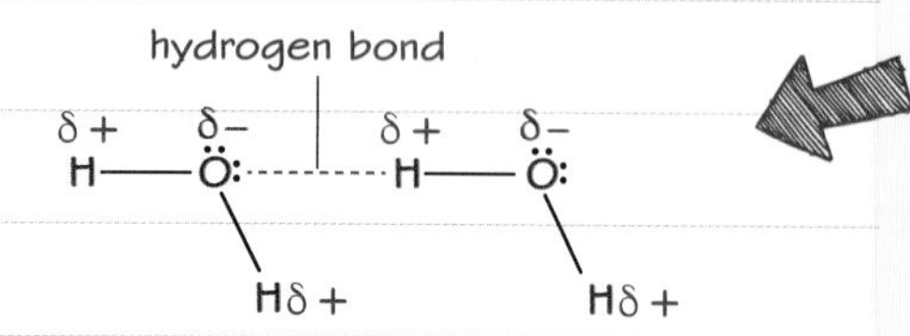

Only electron-deficient hydrogen atoms can form hydrogen bonds, so hydrogen atoms attached directly to a carbon cannot form hydrogen bonds, as the **electronegativities** of carbon and hydrogen are similar. This means hydrogen atoms attached to carbon atoms in hydrocarbons **do not** form hydrogen bonds.

For more information on electronegativity, see page 26.

Remember when drawing diagrams to show hydrogen bonds between molecules:

- Show the dipole charges on relevant atoms.
- Show the lone pairs of electrons on O, N or F.
- Indicate the hydrogen bond clearly, e.g. using a dashed line.
- The OHO bond angle should be 180°.

Now try this

The compound HF is used in the manufacture of polymers such as Teflon™. It is a liquid at 10°C but HCl, even though it has more electrons, is a gas.

Explain this observation in terms of intermolecular forces.

Relative masses

As atoms are so small, scientists use the idea of **relative masses** to compare the mass of atoms, elements and compounds.

Relative atomic mass

The atoms in a sample of an element may have slightly different masses, so **relative atomic mass** is often useful as a way of comparing the masses of different elements. It is defined as:

Mean mass of atoms of the element (in a sample) compared with $\frac{1}{12}$**th of the mass of a carbon-12 atom.**

These values are found on the periodic table in your exam paper, so can be looked up when required.

Relative molecular mass

This term is used when comparing the mass of molecules with simple covalent structures. It is the sum of the relative atomic masses of all the atoms present in the substance.

Relative formula mass

This term is used when referring to substances with giant structures. It is the sum of the relative atomic masses of all the atoms in the formula of the substance.

Using relative atomic masses

Relative atomic masses can be used to calculate the relative molecular masses of molecules, for example, ethanol, C_2H_5OH.

Ethanol contains 2 C atoms = 2 × 12.0 = 24.0
and 6 H atoms = 6 × 1.0 = 6.0
and 1 O atom = 1 × 16.0 = 16.0
so relative molecular mass = 46.0

You don't have to include all the stages of working out shown here, but you may find it helpful when practising such examples, especially with more complex compounds. Notice the values used from the periodic table are given to one decimal place.

They are also used to calculate the relative formula mass of more complex substances, for example, hydrated copper sulfate, $CuSO_4.5H_2O$

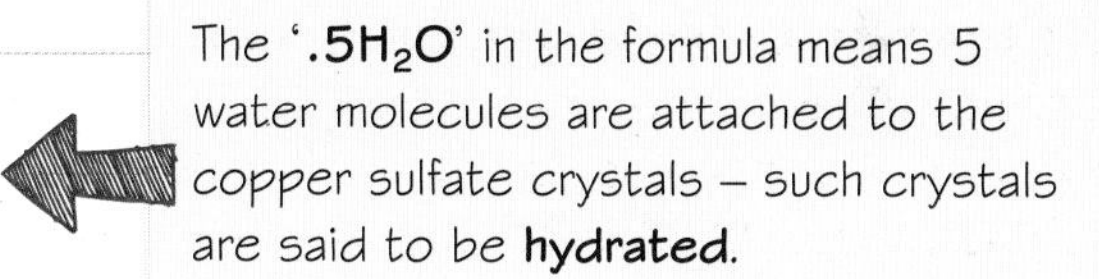

The '$.5H_2O$' in the formula means 5 water molecules are attached to the copper sulfate crystals – such crystals are said to be **hydrated**.

$CuSO_4.5H_2O$ contains 1 Cu atom = 1 × 63.5 = 63.5
and 1 S atom = 1 × 32.1 = 32.1
and 4 O atoms = 4 × 16.0 = 64.0
and 5 H_2O molecules = 5 × 18.0 = 90.0
so relative molecular mass = 249.6

Always use the more precise value from the periodic table, rather than any values you may have remembered. For instance, 32 is often recalled for sulfur, but its precise value is 32.1, as used here.

Units of relative masses

As relative masses are effectively a ratio of the mass of atoms in an element, compared to $\frac{1}{12}$ of the mass of an atom of carbon-12, they **do not have units.**

Now try this

Mohr's salt, $(NH_4)_2Fe(SO_4)_2{\cdot}6H_2O$, is used in experiments to measure an absorbed dose of gamma radiation.

Calculate the relative formula mass of Mohr's salt.

Had a look ☐ Nearly there ☐ Nailed it! ☐

Amount of substance: The mole

Scientists measure and calculate the amount of a substance using a quantity called the **mole**.

The amount of substance

The **amount** of a substance is measured in **moles** (mol). One mole of any substance contains the same number of particles as there are carbon atoms in 12.00 g of carbon-12.

The mass of 1 mole of an element is easy to work out. It is simply equal to its relative atomic mass in grams. So, for instance, 24.3 g of magnesium contains 1 mole of atoms and 48.6 g of magnesium contains 2 moles of atoms.

The mass of 1 mole of any substance is called its **molar mass** and has units of $g\,mol^{-1}$. It is equal to its relative formula (or molecular) mass.

$$\text{amount (mol)} = \frac{\text{mass, } m \text{ (g)}}{\text{molar mass, } M \text{ (g mol}^{-1})} = \frac{m}{M}$$

Calculating the mole

To know the amount of moles in 10 g of calcium, you calculate it using amount $= \frac{m}{M}$.

The molar mass of calcium is $40\,g\,mol^{-1}$ so amount $= \frac{10}{40} = 0.25$ mol.

This applies to molecules and ions as well as atoms. For instance, you might want to know the number of oxygen molecules in 16 g of oxygen gas. You would use the same equation but would need to realise the molar mass of oxygen gas, O_2, is $32\,g\,mol^{-1}$.

So the number of oxygen molecules $= \frac{16}{32}$ $= 0.5$ mole

Make sure you can remember and rearrange this equation. For instance, molar mass $= \frac{\text{mass}}{\text{amount}}$

To calculate the number in moles in 15.0 g of $CaCO_3$ you have to:

- Calculate the molar mass of $CaCO_3$
  $= 40.1 + 12.0 + (3 \times 16.0) = 100.1\,g\,mol^{-1}$
- Use the equation
  amount = m/M
  = 15.0/100.1
- Calculate the number of moles, with units
  = 0.150 mol

To calculate the mass of 1.40 mol of NaOH you have to:

- Calculate the molar mass of NaOH
  $= 23 + 16 + 1 = 40.0\,g\,mol^{-1}$
- Rearrange the equation
  $m = \text{amount} \times M$
- Calculate the mass, with units
  $= 1.40 \times 40.0 =$ **56.0 g**

Calculating molar mass using moles

If you know the mass and number of moles of a substance it can be used to calculate its molar mass. This can be helpful if you are trying to identify an unknown gas or volatile liquid.

For instance, the amount of 0.114 g of a substance, X, when vaporised into a gas, was found to be 0.00200 mol. To find its molar mass you have to:

Rearrange the equation

$$\text{molar mass} = \frac{\text{mass}}{\text{amount}}$$

Calculate the molar mass, with units

$$= \frac{0.114}{0.00200}$$

$= $ **$57.0\,g\,mol^{-1}$**

If you have other information about substance X, for instance, the results of any qualitative tests that might tell you the type of substance, you may be able to identify it.

Now try this

An organic compound, Q, is used as a solvent and in the synthesis of polymers. 0.62 mol of Q has a mass of 49.6 g.

Calculate the molar mass of Q.

Calculating reacting masses and gas volumes

Once the amount of substance in a reaction has been determined, it can be used alongside the balanced equation to calculate **quantities** of reactants or products.

Using chemical equations to determine amounts used in chemical reactions

A balanced equation shows the ratio, in moles, of reactants and products that take part in a chemical reaction. For instance, the equation for the formation of ammonia in the Haber process is:

$$N_2(g) + 3H_2(g) \rightarrow 2NH_3(g)$$

This shows that if **1** mole of nitrogen ($N_2(g)$) reacts with **3** moles of hydrogen (**3**$H_2(g)$) you would expect to form **2** moles of ammonia (**2**$NH_3(g)$). This ratio of reactants and products is called the **stoichiometry** of the reaction. However, if you tried to react 1 mole of nitrogen with 6 moles of hydrogen you would still only expect to form 2 moles of ammonia, as you would not have enough nitrogen to react with the extra hydrogen. In this case, the hydrogen is in **excess** and the nitrogen is the **limiting reactant**.

Calculating an amount in a reaction

Using ideas about moles, you can calculate the amount of a reactant or a product in a reaction. For instance, to calculate the amount, in mol, of oxygen gas formed when 3.50 g magnesium nitrate, $Mg(NO_3)_2$, is strongly heated you have to:

- ✓ Write or check you have been given the balanced equation for the reaction.

  $2Mg(NO_3)_2(s) \rightarrow 2MgO(s) + 4NO_2(g) + O_2(g)$
- ✓ Calculate the amount of the limiting reactant, that is, the one NOT in excess.

  Amount of $Mg(NO_3)_2$ = mass/molar mass = 3.50/148.3 = 0.0236 mol
- ✓ Use the ratio in the equation to deduce the amount of gaseous product formed.

  Ratio of $Mg(NO_3)_2 : O_2$ is 2:1, so amount of O_2 = 0.0236/2 = 0.0118 mol

Calculating a mass in a reaction

Using ideas about moles, you can calculate the mass of a reactant or a product in a reaction. For instance, to calculate the mass, in grams, of lithium oxide, Li_2O, formed when 1.24 g of lithium nitrate, $LiNO_3$, is heated strongly you have to:

- ✓ Write or check you have been given the balanced equation for the reaction

  $4LiNO_3(s) \rightarrow 2Li_2O(s) + 4NO_2(g) + O_2(g)$
- ✓ Calculate the amount of the limiting reactant, that is, the one NOT in excess.

  In this case there is only one reactant, $LiNO_3$, so amount of $LiNO_3$ = mass/molar mass
  = 1.24/(6.9+14+48)
  = 0.018 mol
- ✓ Use the ratio in the equation to deduce the amount of product formed.

  $4LiNO_3 : 2Li_2O$ so 0.018 mol of $LiNO_3$ would form 0.0090 mol of Li_2O.
- ✓ Convert the amount of product formed into a mass using mass = amount × molar mass,

  so mass of Li_2O = 0.0090 × 29.8 = 0.27 g

Now try this

Calcium nitrate is used in cement to accelerate the setting process.

Calculate the total mass, in g, of all gases formed when 4.20 g of calcium nitrate, $Ca(NO_3)_2$, is strongly heated.

$2Ca(NO_3)_2(s) \rightarrow 2CaO(s) + 4NO_2(g) + O_2(g)$

Had a look ☐ Nearly there ☐ Nailed it! ☐

Calculations in aqueous solution

Many chemical reactions take place in aqueous solution. Calculating chemical amounts in such reactions uses volumes and concentrations.

Concentration

Concentration is a measure of the amount of solute dissolved per unit of solvent. Concentration can be quantified using this equation:

$$\textbf{Concentration}, c, (\text{mol dm}^{-3}) = \frac{\textbf{amount}, n \text{ (mol)}}{\textbf{volume}, V (\text{dm}^3)}$$

Concentration is also given as **molarity** in mol dm^{-3}.

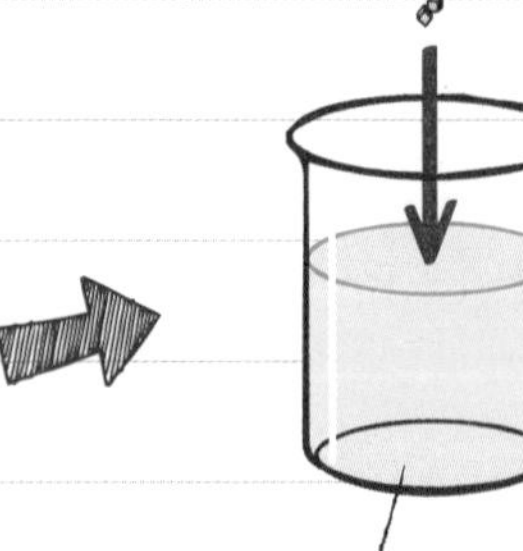

A small amount of copper sulfate is added – this solution is **dilute**.

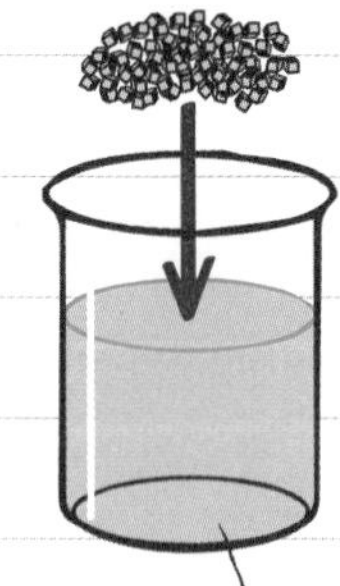

A larger amount of copper sulfate is added – this solution is more **concentrated**.

Calculating concentrations

To calculate the concentration of the solution formed when 0.20 mol of NaCl(s) is added to 400 cm^3 of water you need to:

- ✓ Recall the equation $c = n/V$.
- ✓ Make sure the volume used in the calculation is in dm^3.

  1000 cm^3 = 1 dm^3, so
  400 cm^3 = 400/1000 dm^3
- ✓ Use the data from the question to find c, using $c = n/V$:

  $c = 0.20/(400/1000) = 0.50$ mol dm^{-3}

To calculate the concentration of the solution formed when 16.0 g of NaOH is dissolved in 200 cm^3 of water you need to:

- ✓ Calculate the amount of NaOH using amount = mass/molar mass
  = 16.0 / 40 = 0.40 mol
- ✓ Make sure the volume used in the calculation is in dm^3.
- ✓ Use the data to find c, using $c = n/V$:

  $c = 0.40/(200/1000) = 2.0$ mol dm^{-3}

Percentage yield

Percentage yield calculations can also be carried out using the 'mass' technique.

Percentage yield can be found using the expression:

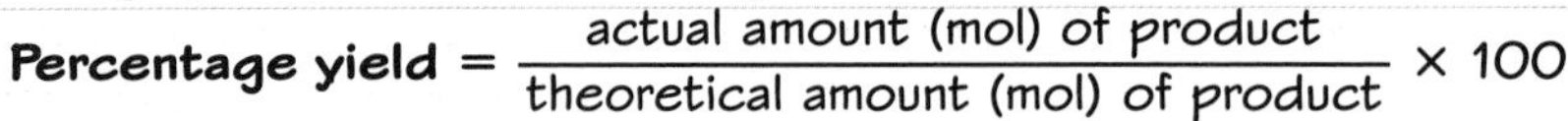

$$\textbf{Percentage yield} = \frac{\text{actual amount (mol) of product}}{\text{theoretical amount (mol) of product}} \times 100$$

For instance, a scientist trialled the manufacture of the fertiliser ammonium nitrate, using 100 cm^3 of 0.25 mol dm^{-3} nitric acid with excess ammonia. She formed 1.6 g of ammonium nitrate. To calculate the percentage yield you need to:

✓ Find the theoretical yield of ammonium nitrate

Amount of nitric acid used = $c \times V$ = 0.25 × (100/1000) = 0.025 mol of nitric acid.

As the equation for the reaction is $HNO_3 + NH_3 \rightarrow NH_4NO_3$, the theoretical yield of ammonium nitrate should also be 0.025 mol.

✓ Calculate the actual amount of ammonium nitrate

Using amount = mass/molar mass, actual amount of ammonium nitrate = 1.6/80 = 0.020 mol

✓ Calculate the percentage yield

= (0.020/0.025) × 100 = 80%

Now try this

A solution of sodium carbonate, Na_2CO_3, can be used as a mild disinfectant for animal cages.

Calculate the mass of Na_2CO_3 needed to make 500 cm^3 of a 0.016 mol dm^{-3} solution.

The periodic table

The periodic table shows all known elements in order of increasing atomic number.

Groups 1 to 0 indicate the number of electrons in their outer energy level (shell) of each element in the group.

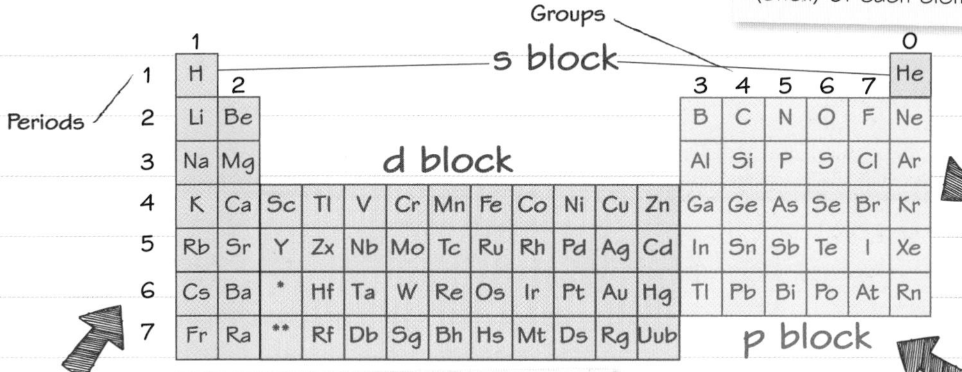

You are likely to be given a full periodic table in your Unit 1 exam. You should check the latest Sample Assessment Materials (SAMs) on the Pearson website to see exactly what you will have available to you.

Periods 1 to 7 indicate the number of main energy levels of each element in that period.

s, p and d blocks indicate the sub-level being filled with electrons as you go across the table.

First ionisation energy

This is the energy required to remove an electron from each atom in 1 mole of gaseous atoms. It is shown by the equation $M(g) \rightarrow M^+(g) + e^-$

Trend down a group

As a group is descended, the first ionisation energy decreases. This is because the outer electron being removed is further from the nucleus and there are more inner shells to **shield** the outer electron. This means the electrostatic attraction to the nucleus is less.

This can be seen in the data from group 1.

Element	First ionisation energy ($kJ\,mol^{-1}$)
lithium	520
sodium	495
potassium	420

Trend across periods 2 and 3

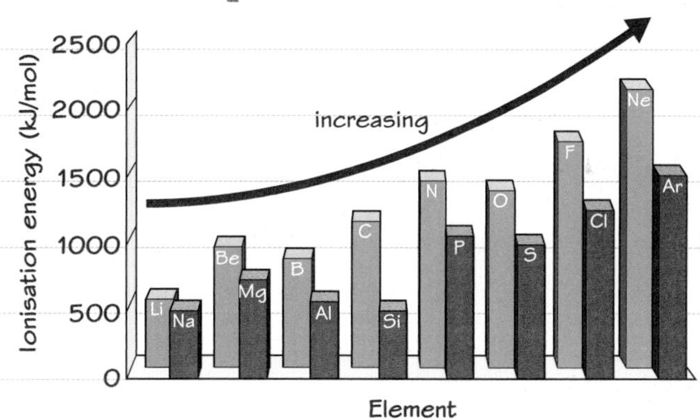

The general trend increases across the period because the nuclear charge increases, but the shielding due to the inner shells stays the same. This means the electrostatic attraction of the nucleus to the outer electrons increases. There are anomalies in group 3, as the electron is removed from a higher energy sub-shell and group 6, as electrons pair up and repel in an orbital.

First electron affinity

This is the energy change when an electron is added to each atom in 1 mole of gaseous atoms. It is shown by the equation $X(g) + e^- \rightarrow X^-(g)$.

The first electron affinity decreases down a group as the shell to which the electron is being added is further from the nucleus and has more shielding from inner shells. However, please note that F and O are exceptions to the trend in electron affinity.

Now try this

Some salts containing the hydride ion, H^-, are used in drying agents.

Write an equation for the first electron affinity of hydrogen, to form the hydride ion.

Trends in the periodic table

A number of repeating patterns in physical properties are observed in the periodic table.

Radii of atoms and ions

The **atomic radius** decreases across a period.

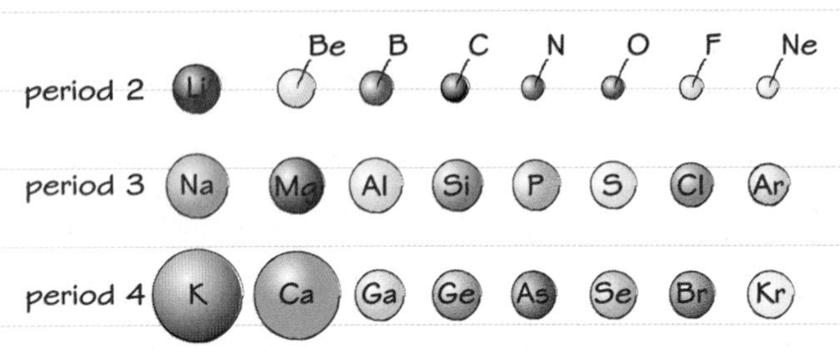

This is because the nuclear charge increases but the number of shells remains the same, so the outer electrons are more strongly attracted to the nucleus.

The **ionic radius** of positive ions decreases, as do the radii of negative ions. Positive ions are smaller than the atom they formed from. Negative ions are larger than the atom they formed from. In the same period, the negative ions that form from the non-metals are larger than the positive ions that form from metals, as they have an extra electron shell.

Electronegativity

Electronegativity measures the tendency of an atom to attract the electron pair in a covalent bond.

Electronegativity increases →

Period ↑																		
1	H 2.20																	He
2	Li 0.96	Be 1.57											B 2.04	C 2.55	N 3.04	O 3.44	F 3.98	Ne
3	Na 0.93	Mg 1.31											Al 1.61	Si 1.90	P 2.19	S 2.56	Cl 3.16	Ar
4	K 0.82	Ca 1.00	Sc 1.36	Tl 1.54	V 1.63	Cr 1.65	Mn 1.55	Fe 1.83	Co 1.86	Ni 1.91	Cu 1.90	Zn 1.65	Ga 1.81	Ge 2.01	As 2.18	Se 2.55	Br 2.96	Kr 3.00
5	Rb 0.82	Sr 0.95	Y 1.22	Zx 1.33	Nb 1.6	Mo 2.16	Tc 1.9	Ru 2.2	Rh 2.28	Pd 2.20	Ag 1.93	Cd 1.69	In 1.78	Sn 1.96	Sb 2.05	Te 2.1	I 2.66	Xe 2.6

Electronegativity increases up a group as the bonding electron pair in the covalent bond will be closer to the nucleus attracting it.

Electronegativity increases across a period as the nuclear charge attracting the bonding electron pair is greater, but the number of electron shells shielding the bonding pair from the nucleus is the same.

Melting points

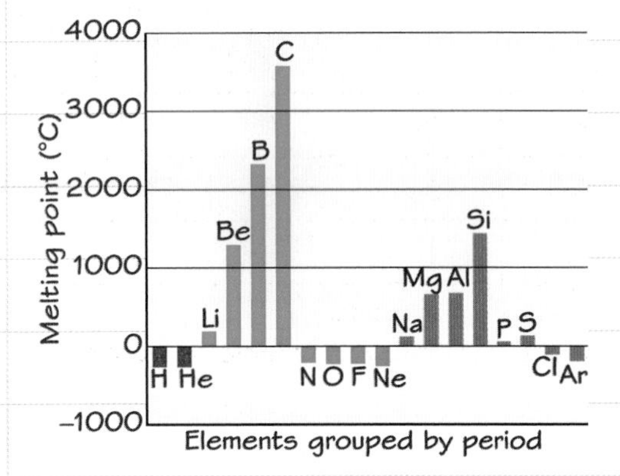

As you go across a period the melting points tend to increase across groups 1 to 4, then decrease across groups 5 to 0.

Boiling points

The boiling points across a period also tend to increase across groups 1 to 4, then decrease across groups 5 to 0.

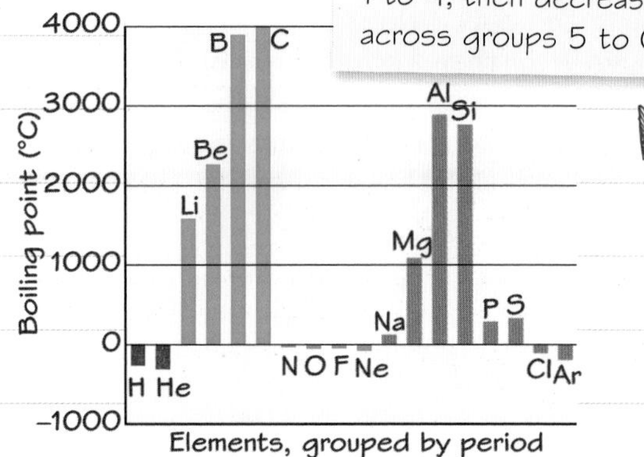

Links For more on London forces, see page 20.

Explaining the trends in melting and boiling points

The melting and boiling points of metals increase as the **metallic bond strength increases** (see page 19). This is because the positive metal ions get smaller across the period, their charge gets larger and there are more delocalised electrons. Hence, there is a stronger attraction between the positive ions and the delocalised electrons. The elements from group 4, carbon and silicon, both have very high melting and boiling points as both have **giant covalent**, or macromolecular, structures, with **many strong covalent bonds**. The elements from groups 5, 6, 7, and 0 are **simple molecules** or atoms, with only **weak London forces** between them.

Now try this

Materials made from carbon fibres are used in the brake discs of cars in Formula 1 racing. These discs reach high temperatures in use but do not melt.

Suggest reasons that explain the very high melting point of carbon fibres.

Reactions of periods 2 and 3 with oxygen

Elements in these periods react with oxygen to form oxides.

Period 2

Period 2 elements can form oxides, often by being burnt in pure oxygen. The properties of some of the oxides are summarised below.

Group	1	2	3	4	5	6	7	0
Formula	Li_2O	BeO	B_2O_3	CO_2	NO_2	†	F_2O	†
State	(s)	(s)	(s)	(g)	(g)		(g)	
Structure	GIL	*GIL	GCL	SM	SM		SM	
Acid or alkali		#						

Key: GIL = giant ionic lattice; GCL = giant covalent lattice; SM = simple molecular;
Purple = alkaline solution with water; Red = acidic solution with water; # = oxide insoluble in water;
* = with some covalent character; † = no oxide forms

Period 3

Period 3 elements can form oxides, often by being burnt in pure oxygen. The properties of some of the oxides are summarised below.

Group	1	2	3	4	5	6	7	0
Formula	Na_2O	MgO	Al_2O_3	SiO_2	P_2O_5	SO_2	Cl_2O	†
State	(s)	(s)	(s)	(s)	(g)	(g)	(g)	
Structure	GIL	GIL	GIL	GCL	SM	SM	SM	
Acid or alkali			#	#				

Key: GIL = giant ionic lattice; GCL = giant covalent lattice; SM = simple molecular;
Purple = alkaline solution with water; Red = acidic solution with water; # = oxide insoluble in water;
† = no oxide forms

Explaining the trends in the properties of the oxides

If the **metal oxides dissolve** in water they tend to form **alkaline solutions**, as they form **metal hydroxides**. For instance, sodium oxide reacts to form alkaline sodium hydroxide.

$$Na_2O(s) + H_2O(l) \rightarrow 2NaOH(aq)$$

If the **non-metal oxides** dissolve in water, they tend to form **acidic solutions**, as they form **acids**. For instance, sulfur dioxide reacts to form sulfurous acid.

$$SO_2(g) + H_2O(l) \rightarrow H_2SO_3(aq)$$

The metallic elements tend to form solid, high melting point oxides, as metal oxides have a giant ionic lattice structure. The non-metal elements tend to form gaseous, low melting point oxides, as non-metal oxides have simple molecular structures.

Now try this

Lime mortar is a breathable mortar, or cement, used in old building preservation. It contains calcium oxide.

Predict what safety precaution builders should take when handling lime mortar. Justify your answer by considering the chemical properties of calcium oxide.

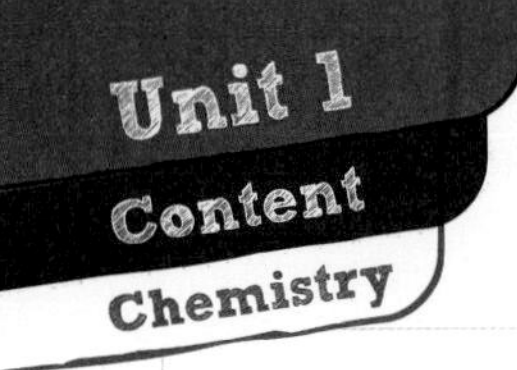

Had a look ☐ Nearly there ☐ Nailed it! ☐

Reactions of metals

Many metals will react with oxygen, water and dilute acids.

Reactions with oxygen

Most metals will react with oxygen to form an oxide. Reactive metals will burn vigorously. Less reactive metals, such as iron, will burn quickly when powdered.

If not ignited, many metals will still react with oxygen slowly, forming a layer of oxide on the metal surface. This is known as tarnishing, and even affects relatively unreactive metals such as silver.

Reactions with water

Some metals, for instance, those in groups 1 and 2, react directly with water to form hydroxides.

The metals in group 1 react immediately on contact with water, fizzing as hydrogen is produced, dissolving to form an alkaline solution, and sometimes catching fire or exploding.

The metals in group 2 react in a similar, but slower way. For instance, magnesium needs to be reacted with steam to form hydrogen.

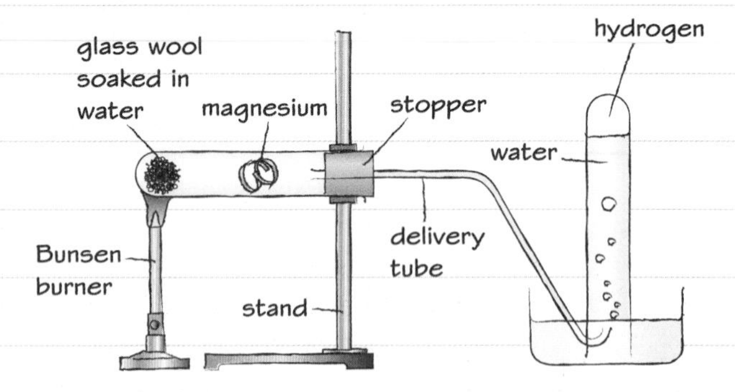

Less reactive metals do not tend to react directly with water, but may react with water **and** oxygen to corrode.

Reaction with dilute acids

Metals that are more reactive than hydrogen will react with dilute acids to form a salt and hydrogen gas. If collected, the gas can be confirmed as hydrogen as it will burn with a squeaky pop.

The salt formed depends on the dilute acid used. If you use:

- dilute sulfuric acid, $H_2SO_4(aq)$, you will form a salt called a sulfate
- dilute hydrochloric acid, $HCl(aq)$, you will form a salt called a chloride.

In each case the **salt forms** when a metal ionises and **metal ions replace hydrogen ions** in the acid. For instance, $Ca(s) + H_2SO_4(aq) \rightarrow CaSO_4(aq) + H_2(g)$

Metals that are less reactive than hydrogen, for example copper, do not react with acids.

Describing and explaining the trends in metal reactivity

As you go down a group, metals will become more reactive with oxygen, water and dilute acids. This is because the atoms become larger in size down a group. The distance between the nucleus and the outer electrons increases and more inner electron shells provide shielding from the nucleus. The electrostatic attraction from the nucleus is less, so the outer electron(s) can escape more readily.

As you go across a period, the metals become less reactive with oxygen, water and dilute acids. As you go across the period, more electrons have to be lost to form metal ions. Sodium has to lose only one electron to form an ion. However, magnesium, in the same period, has to lose two electrons. This requires more energy, so magnesium is less reactive.

Now try this

Aluminium powder is used as a fuel in reusable solid rocket motors for spacecraft. Safety recommendations for storing the powder say that 'Great care must be taken to prevent the contact of water with aluminium powders. Leaks in steam lines, water lines, radiators or roofs should be promptly repaired'.

Suggest reasons that explain these precautions, including equations for any reactions that take place.

Oxidation and reduction

Oxidation and reduction reactions involve the transfer of electrons.

A memory aid: OIL RIG

- **O**xidation **I**s **L**oss of electrons (OIL)
- **R**eduction **I**s **G**ain of electrons (RIG)

Displacement reactions of metals

A more reactive metal will displace a less reactive metal from a metal salt. For example, when pieces of metal are added to metal salt solutions.

	$MgSO_4$	$FeSO_4$	$CuSO_4$
Mg		brown solid	orange solid
Fe	no reaction		orange solid
Cu	no reaction	no reaction	

Displacement reactions of halogens

A more reactive halogen will displace a less reactive halogen from a halide salt. For example, when aqueous solutions of halogens are added to potassium halide solutions.

	Cl^- (aq)	Br^- (aq)	I^- (aq)
Cl_2 (aq)		orange solution	brown solution
Br_2 (aq)	no reaction		brown solution
I_2 (aq)	no reaction	no reaction	

Explaining the displacement reactions

In the displacement reactions of the metals, the more reactive **metal atoms lose electrons** to form ions, so are **oxidised**. Each **metal ion gains electrons** to form atoms, so is **reduced.** For instance, the equation for the reaction between magnesium and copper sulfate is:

$$Mg(s) + CuSO_4(aq) \rightarrow MgSO_4(aq) + Cu(s)$$

Mg atoms have lost electrons to form Mg^{2+} ions in $MgSO_4$. This change is shown using the **half equation**:

$$Mg(s) \rightarrow Mg^{2+}(aq) + 2e^-$$

Cu^{2+} ions, in $CuSO_4$, have gained electrons to form the orange solid Cu(s). This change is shown using the **half equation**:

$$Cu^{2+}(aq) + 2e^- \rightarrow Cu(s)$$

These equations can be combined to give the **ionic equation**:

$$Mg(s) + Cu^{2+}(aq) \rightarrow Mg^{2+}(aq) + Cu(s)$$

In the displacement reactions of the halogens, the more reactive **halogens gain one electron per atom** to form halide ions, so are **reduced**. Each **halide ion loses an electron** to form halogens, so is **oxidised.** For instance, the equation for the reaction between chlorine and potassium bromide is:

$$Cl_2(s) + 2KBr(aq) \rightarrow 2KCl(aq) + Br_2(aq)$$

Cl_2 has gained electrons to form Cl^- ions in KCl(aq). Br^- ions, in KBr, have each lost an electron to form the orange solution Br_2(aq). The equation can be simplified to:

$$Cl_2(s) + 2Br^-(aq) \rightarrow 2Cl^-(aq) + Br_2(aq)$$

In all displacement reactions, one species is oxidised, another is reduced. This is the reason such reactions are also known as **redox reactions**.

Variable oxidation states of transition metals

Transition metals can each form a range of different ions in redox reactions, depending on the extent to which they have been oxidised or reduced. The different ions are said to have **different oxidation states**. For instance, iron will react with hydrochloric acid to form iron(II) chloride. This contains the ion Fe^{2+} and has the oxidation state +2. If hydrogen peroxide is then added, the iron(II) chloride is oxidised to iron(III) chloride. This contains the ion Fe^{3+} and has the oxidation state +3.

Now try this

Iron(II) chloride is used to etch circuit boards. It is formed from the reaction of iron with hydrochloric acid.

Show, using OIL RIG, why this is defined as a redox reaction.

Had a look ☐ Nearly there ☐ Nailed it! ☐

Interpreting wave graphs

There is a lot going on in wave motion. A wave changes over time, but it also travels through space, so you need two different types of graph.

Displacement–time graph

Focuses your attention on oscillations at just one point in space.

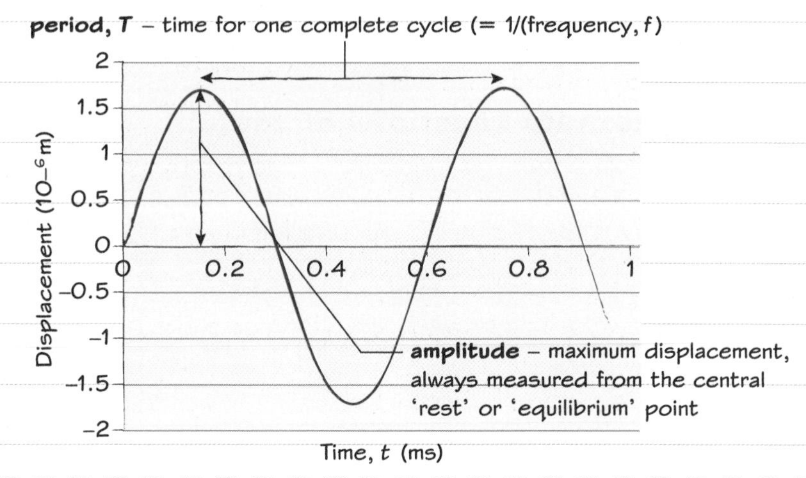

Displacement–distance graph

Takes a snapshot in time – shows you the wave shape over the whole space at that instant.

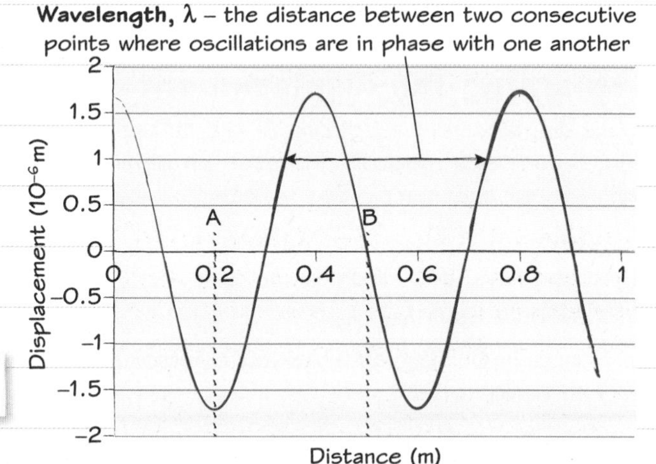

The same wave motion – different graph.

Oscillations out of phase

- Wave energy takes time to travel.
- Travelling one wavelength takes one cycle – that is, one periodic time.
- Two waves of the same frequency can be in phase or out of phase.
- Phase difference is measured as an angle representing a fraction of a cycle.

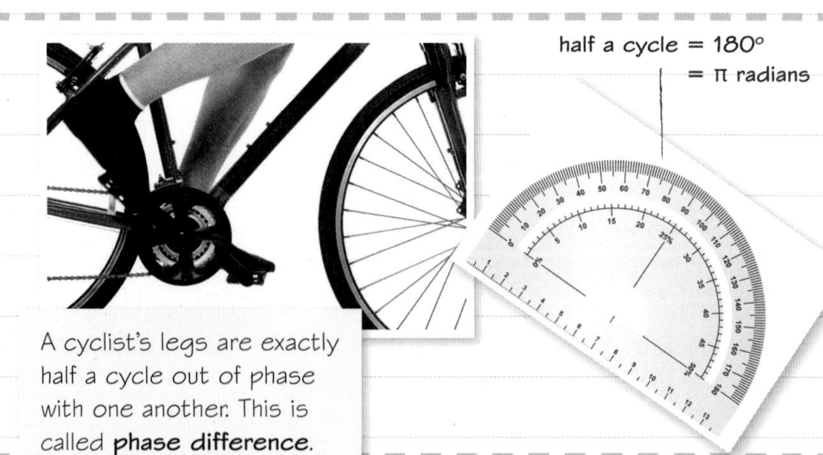

A cyclist's legs are exactly half a cycle out of phase with one another. This is called **phase difference**.

Now try this

1 Look at the displacement–time graph above. Give the periodic time of the wave.

You could now use the period to calculate the frequency and show it is 1.7 Hz.

2 Now look at the displacement–distance graph above.

(a) Give the amplitude of the wave.

Remember to measure from the displacement equals zero line.

(b) There is a phase difference between oscillations at the point A and those at point B. State what the phase difference is as a fraction of a whole cycle.

For more on frequency, see page 32.

Wave types

You need to be able to describe the difference between **longitudinal** waves and **transverse** waves, and to give examples of each.

What moves?

You need to know that in a wave:

- Energy is transferred from one place to another.
- There is no net transfer of matter.
- Something called 'the displacement' oscillates. The displacement could be physical movement of a particle, or it may be the change of another quantity, for example, of electric field for electromagnetic waves.
- Oscillations occur around a mean 'rest' value of the displacement.

Direction of propagation

The line along which wave energy travels and in which the wave oscillations propagate – that is, they reproduce the pattern of the wave.

Longitudinal wave

wave energy travel (propagation) direction

Oscillations back and forth along the direction of propagation

Transverse wave

wave energy travel (propagation) direction

Oscillations at right-angles to the direction of propagation – up and down or side to side

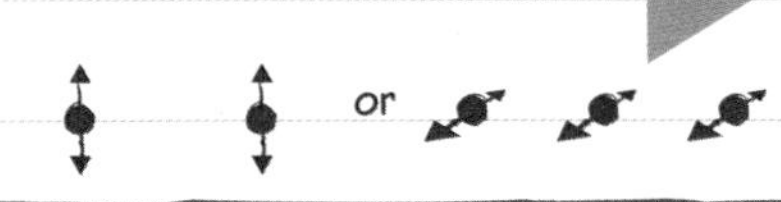

What can you see or measure?

In longitudinal waves the oscillations cause **compressions** (regions with particles closer together) and **rarefactions** (particles further apart) that move with the wave pattern.

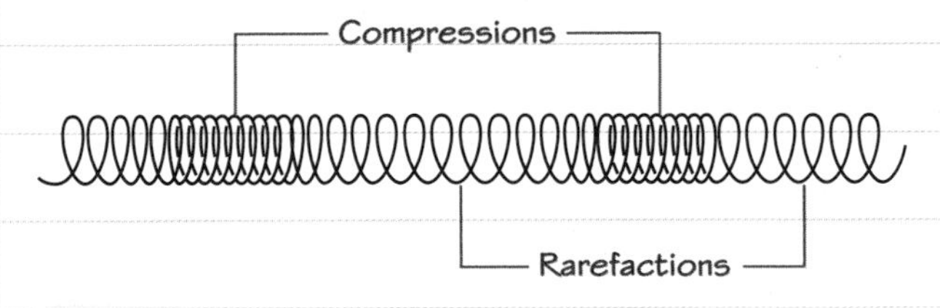

Longitudinal wave in a slinky spring

Microphone detects pressure variations caused by compressions and rarefactions of the air in a sound wave.

Transverse waves

Transverse waves in water or in stretched strings can be visible as travelling ripples or as standing waves.

Detectors for transverse waves must be aligned at right-angles to the direction of propagation, for example, TV and radio antennas.

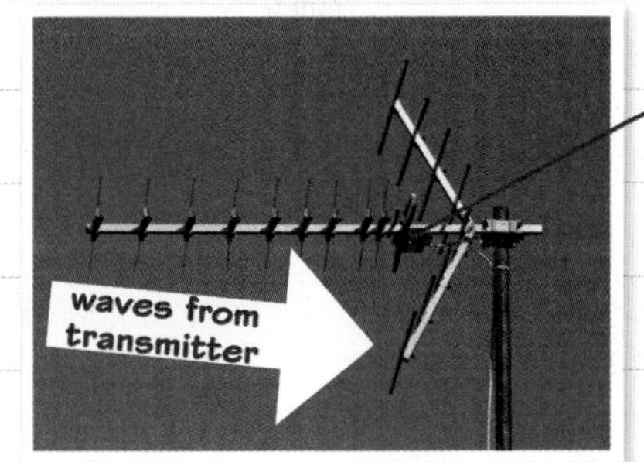

Now try this

1 Explain why sound is a longitudinal wave.
2 What evidence can you give for classifying radio and TV signals as transverse waves?

Make sure to describe what oscillates, how and in which direction the wave propagates, and how these two facts together fit the definition.

Wave speed

The speed at which a wave transfers energy from one place to another varies according to the medium through which it propagates (travels).

Wavelength, λ, and frequency, f

You need to know that:

- In a given medium, waves travel at a fixed speed or velocity, v.
- Travelling one wavelength, λ, takes one cycle – that is, one periodic time, T.
- Speed is distance over time taken, so: $v = \lambda/T = f\lambda$ (often called the 'wave equation').

Links For more on wavelength, see page 30.

Using the equation

You can use the fixed speed of waves, v, to calculate frequency when you know wavelength, or vice versa:

$f = v/\lambda$; $\lambda = v/f$

Some equations will be given to you in the exam. Check the latest SAMS on the Pearson website to see which equations appear on the formula sheet.

Musical instruments with strings

You can tune the notes without changing the length by making changes to the medium (the string) and so altering the speed of the waves, v:

- Thicker strings with more mass per unit length, μ, give a slower speed and hence a lower note.
- Tighter strings with more tension, T, give a faster speed and hence a higher note. Tension is a force measured in newtons (N).
- The equation for the wave speed is: $v = \sqrt{(T/\mu)}$

Take care not to confuse T used for tension force in this equation with T used elsewhere to mean the periodic time of a wave oscillation.

Air columns in pipes

The speed of sound in air is fixed for any given temperature and pressure, so you can only tune brass or woodwind instruments by altering the length of pipe being used.

The speed of light

- In free space (vacuum) light and other electromagnetic waves always travel at $2.998 \times 10^8\,\mathrm{m\,s^{-1}}$ (which is represented by the letter c).
- Nothing has ever been observed to travel faster than c.
- In solid or liquid media, light travels at a slower speed (represented by the letter v).
- The ratio $c/v = n$ is called the **refractive index** of the medium.

Refraction due to light moving more slowly in water

Now try this

1 Sound travels in air at $3.0 \times 10^2\,\mathrm{m\,s^{-1}}$. Calculate the wavelength in air for sound of frequency 250 Hz.

2 A guitar string of mass per unit length $0.001\,\mathrm{kg\,m^{-2}}$ is tuned to give a certain 'open' note – that is, without any fingers pressed onto frets to shorten it. In order to use a similar tensioning force, what weight of string should be used to produce an open note one octave lower (that is, half the frequency)?

How much slower will the waves be travelling?

3 Calculate the speed of light when it travels through glass of refractive index $n = 1.48$.

Wave interference

You first met the concept of **phase** when interpreting graphs of waves, but it is in explaining interference that it becomes really useful.

Links For more on interpreting wave graphs, see page 30.

Wave-fronts and phase

You need to know that **wave-fronts**:

- are lines (or planes in 3D) drawn to join points in a wave where all the oscillations are **in phase**
- are spaced **one wavelength apart**
- move forward in a direction **perpendicular** to the wave-front – that is, along a 'ray' line.

Between two successive wave-fronts, the wave oscillations go through one complete cycle – a **phase difference** of 360° or 2π rad.

The principle of superposition

Interference is a detectable pattern of different strengths (amplitudes) of wave oscillation, for example:

- large and small water ripples
- light and dark fringes (visible light)
- loud and soft (sound)
- good and poor reception (radio/ phone).

It is caused by waves from different sources crossing and adding together – that is, **superposition**.

Coherence and interference

Interference patterns only occur when the wave sources are **coherent** – that is, they have the same frequency and a fixed phase relationship.

Since no detector is fast enough to directly measure the frequency and phase of light, you have to use interference experiments to explore its coherence, and hence to understand the wave nature of light photons.

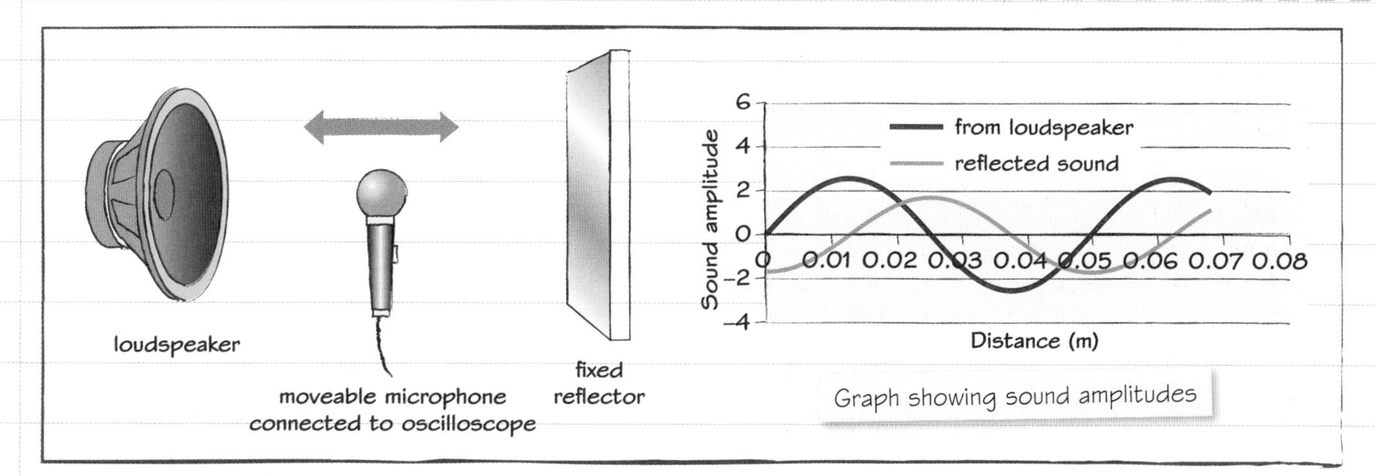

Graph showing sound amplitudes

Now try this

The graph in the box above shows how sound amplitudes might vary with distance in an experiment where a loudspeaker emits a pure note and it is bounced back by a vertical reflector.

1 Draw on the graph a curve showing the resultant sound amplitude that the microphone might detect.

2 Explain why, even at the points where this resultant amplitude is minimum (that is, nodes), the microphone will still pick up some sound.

Use the principle of superposition – that is, that the amplitudes add together mathematically.

Diffraction gratings

Gratings are flat arrays of regularly spaced lines, which are designed to break up a plane wave-front into a set of separate wave sources.

For more on interference fringes, see page 33.

Path difference and phase difference

To produce dark and bright interference fringes (by superposition):

- a path difference is created by dividing a light source so that separate rays of light travel different paths
- the difference in path causes a phase difference
- the path difference increases as the angle through which a grating scatters the light is increased
- when the path difference equals a whole number of wavelengths $n\lambda$ the light rays will be in phase and constructive interference – a bright fringe – occurs.

Grating line spacing, d, needs to be just a little larger than the wavelength, λ.

Spectra

When light has a mixture of wavelengths, the condition for constructive interference occurs at different angles for each wavelength. So a set of coloured line spectra is produced rather than just a set of bright fringes.

The straight-ahead (transmission) beam has no path difference, so is not split into a spectrum.

Coloured lines occur where $n\lambda = d \sin\theta$. The first order (i.e. where $n = 1$) is the brightest.

Emission spectra

The light emitted when the electrons in atoms are excited by heating or by an electric discharge is characteristic of the electronic structure of the element(s) present. Flame tests use this idea to identify elements.

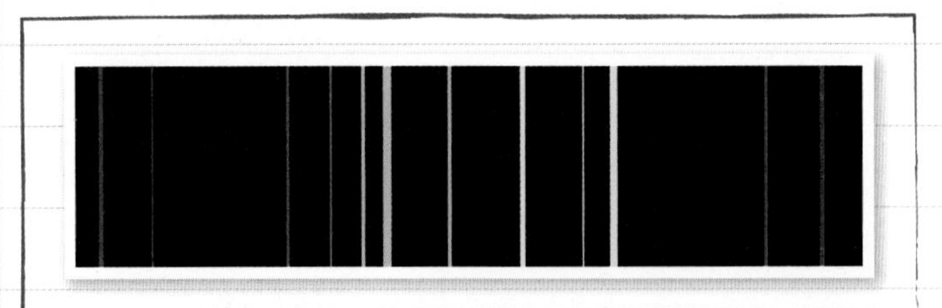

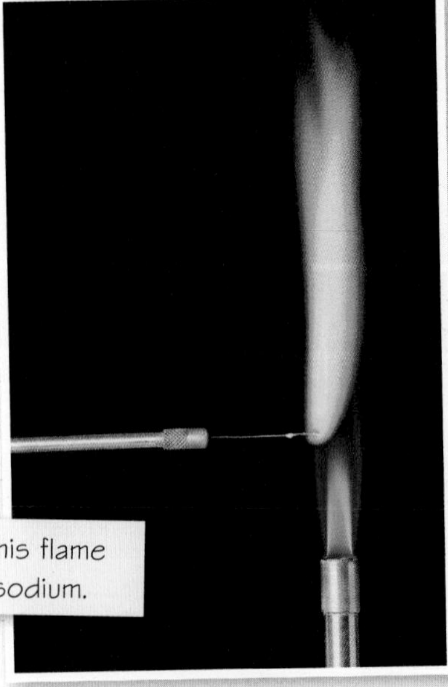

The colour of this flame test indicates sodium.

Now try this

1 The white light diffraction grating spectrum in the first photo on this page has a central white line and three coloured bands on either side. Use the concept of phase difference to explain this.

2 The box above shows emission spectra obtained from electric discharge through different gases and vapours. Explain what information you could deduce from measurements of the line frequencies in the emission spectrum of a gas sample of unknown composition.

Stationary waves

Stationary waves can be thought of as two progressive (travelling) waves moving in opposite directions and superimposed one on the other. The result is a fixed pattern of nodes and antinodes.

Links For more on wave types, see page 31.

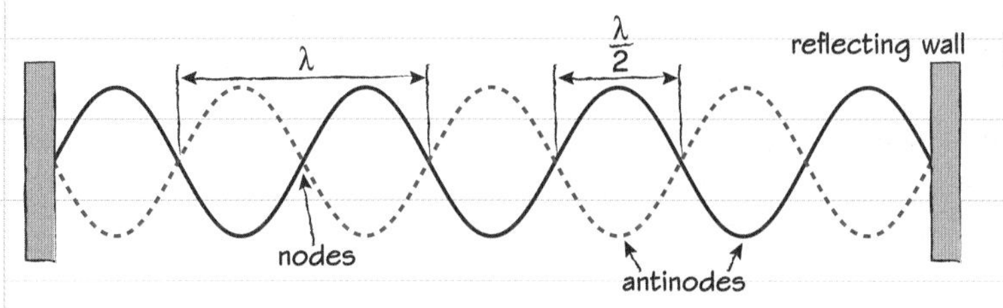

Nodes have minimum (ideally zero) oscillation and antinodes are points of maximum amplitude. The distance between a node and the adjacent antinode is a quarter wavelength.

Energy and motion

Progressive wave or travelling wave: **Energy is transferred** from one place to another

Stationary wave or standing wave: **Energy is stored** within a resonator

Resonators:

- store wave energy by reflecting the wave back on itself to form a **stationary wave** pattern
- only efficiently receive energy from an external source that has a frequency close to one of their own **natural frequencies**; this is called **resonance**
- have a **fundamental** (lowest) natural mode of oscillation, and higher **harmonics**.

Resonance can be damaging

Weights added for balancing

Vehicle wheels need to be carefully balanced. Otherwise at certain speeds they will resonate with the natural frequency of the steering or suspension and cause dangerous vibrations.

Uses of resonance

Reflection of the TV signal creates a node at the reflector. The dipole detector is placed at an antinode.

In a microwave oven, a resonant cavity is used to build up the microwave power. Reflections inside the oven cooking space cause hot spots (antinodes) and cold spots (nodes), so the food is usually rotated on a turntable to get even cooking.

Links Musical instruments use stationary wave resonators. For more information, see page 36.

Now try this

1 Give **two** differences that distinguish a stationary wave from a progressive wave.

Think about energy and about the wave motion.

2 (a) Explain how and why TV antennas make use of stationary waves.
(b) Sketch an antenna to show how its design relates to the TV signal wavelength.

Had a look ☐ Nearly there ☐ Nailed it! ☐

Musical instruments

Musical notes are produced in an instrument by a resonator that can be tuned to give the desired frequency as its fundamental mode.

Strings

- Stringed instruments use **transverse waves** in stretched strings to make a resonator.
- The fixed ends of the string are always nodes.
- The **fundamental mode** has one antinode, so $L = \lambda/2$.
- For higher **harmonics**, any number n of antinodes can fit in the length of the string, so $L = \lambda/2n$.

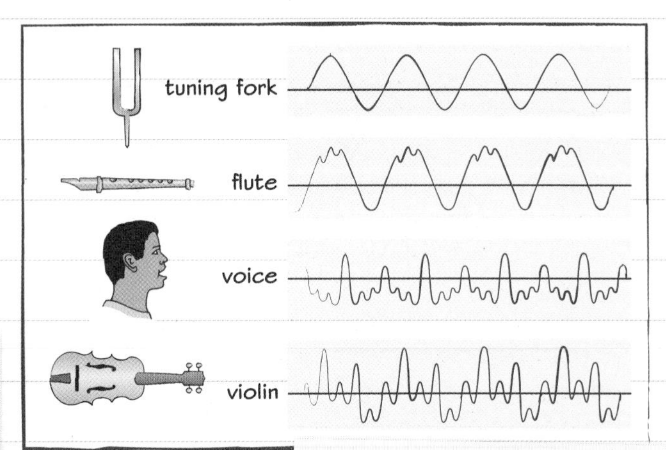

The tension, T, and the 'weight' – that is, mass per unit length, μ, of the string determine the wave speed and hence the resonant frequencies for the equations.

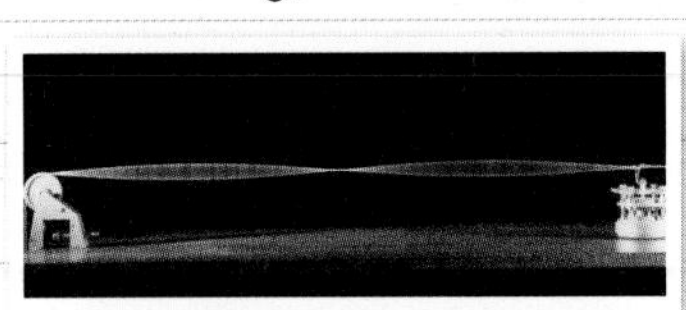

For more information on tuning string instruments, see page 32.

Smooth sounds like a violin or a human voice contain a wide mixture of harmonics giving a complex waveform.

Tubes and pipes

- Pipe organs, woodwind and brass instruments all use **longitudinal** sound waves in a tube or pipe to make a resonator.
- **Open ends of pipes are antinodes**: so a pipe open at both ends has $\lambda = L/2n$ – that is, a complete set of harmonics as in a stringed instrument.
- **Closing one end** of a pipe **creates a node** there: so the fundamental mode has $L = \lambda/4$, and only odd numbered harmonics will fit in the pipe for example, $L = 3\lambda/4$, $5\lambda/4$, etc.
- Missing out the even harmonics makes for a brighter/harsher sound.

Changing the note

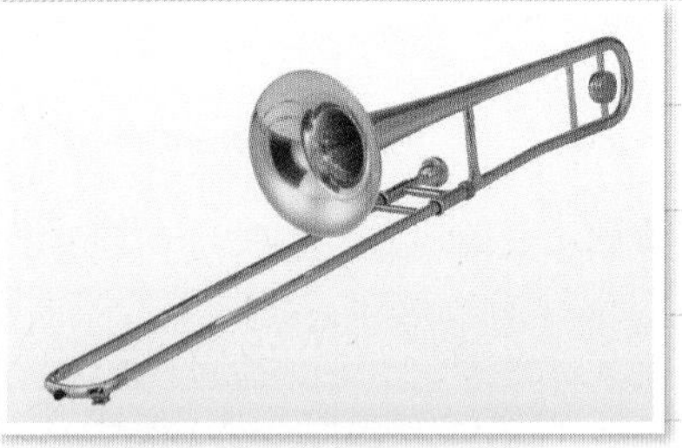

Brass instrument players change notes by (a) altering the length of the tube by means of valves or with a slide and (b) altering lip pressure and vibration to excite higher harmonics.

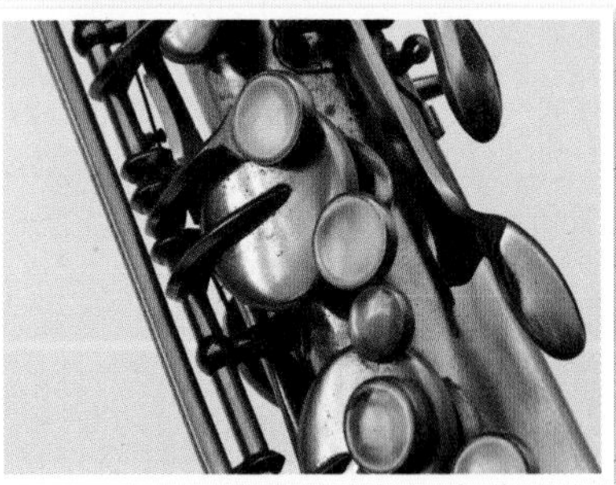

Woodwind players change notes by opening and closing holes carefully positioned along the length of the pipe.

The type of reed or mouthpiece determines whether an instrument behaves as if it is closed or open at that end.

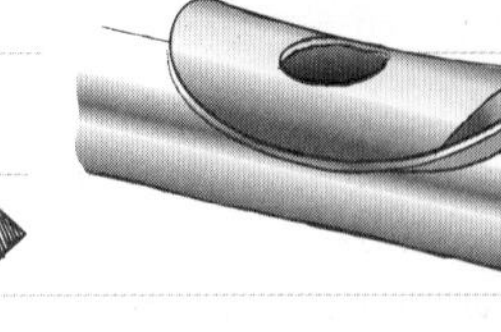

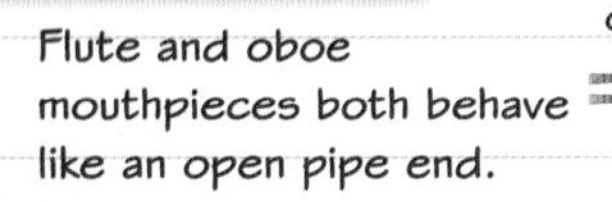

Flute and oboe mouthpieces both behave like an open pipe end.

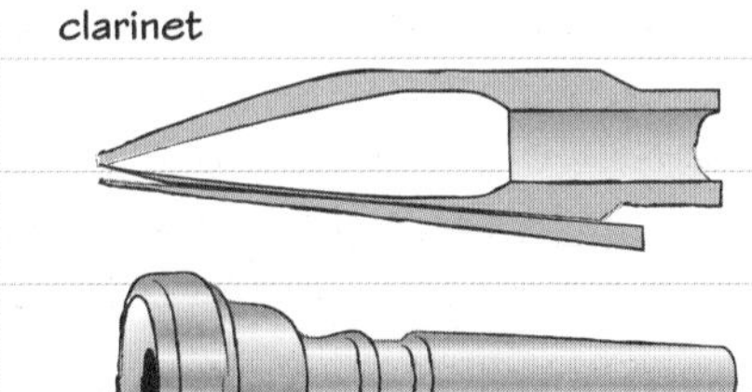

Clarinet and trumpet both behave like a closed pipe end.

Now try this

A guitar string has its effective length shortened by one third when the player's finger presses it against one of the frets.

(a) Explain why the wave speed remains unaltered.

(b) Calculate by what ratio the frequency will be increased.

Refer to the wave speed equations on page 32.

Optical fibres

Optical fibres made from high-density glass can carry light signals long distances without losing any light through their sides.

Critical angle

The critical angle, C, is the **least angle of incidence** at which total internal reflection occurs.

- Only applies when light tries to leave an optically dense (that is, high refractive index, n) medium at a boundary into a less dense (lower refractive index) medium.
- As angle of incidence increases, angle of refraction also increases and is always larger; some light is always reflected internally.
- For incidence at the critical angle, C, the refracted beam would be at 90° – that is, along the boundary surface – so it disappears and instead there is **total internal reflection**.

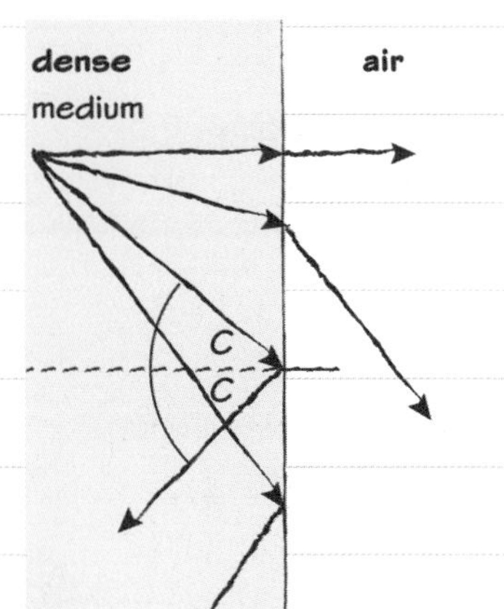

Glass/air boundary

Light entering glass:

$$n = \frac{c}{v} = \frac{\sin i}{\sin r}$$

Light leaving glass:

i and r are reversed.

At critical angle, $i = C$ and $\sin r = \sin 90° = 1$

so:

$$\sin C = 1/n$$

Some equations will be given to you in the exam. Check the latest SAMs on the Pearson website to see which equations appear on the formula sheet.

Light paths in a fibre

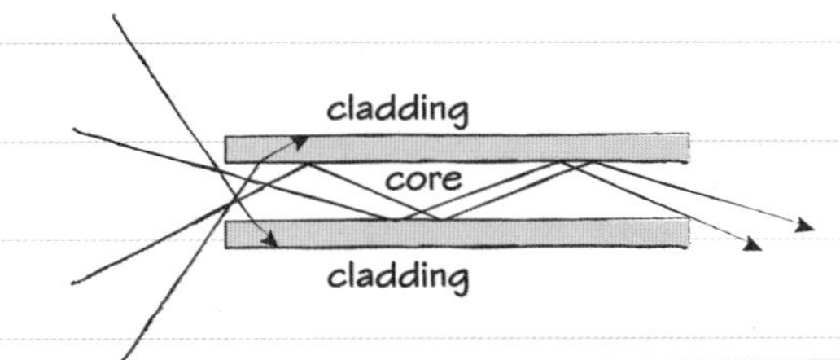

Light from a wide range of angles is refracted on entering the cut end of a fibre so that it hits its side surfaces at angles greater than the critical angle, C, and so is totally internally reflected.

Optical fibres versus copper wires

Optical fibres are better than copper wires because optical fibre signals:

- have **lower losses**, so travel further before needing amplification
- are **secure**, so cannot be tapped into
- are at higher frequencies, so provide **greater bandwidth**.

But optical fibres cost more and need specialist installation.

Broadband networks

- **Bandwidth** is a measure of the number of distinct signals at different frequencies that a network can carry.
- High-frequency carrier waves can accommodate many small 'bands' of frequencies.
- Fibre optic cables allow very high bandwidth indeed, and so networks using them are called 'broadband'. The long distance 'backbone' of the broadband network uses 'single mode' fibres in which the dense glass core is so narrow (~8 μm) that there is only one light path, straight down the centre of the fibre core.

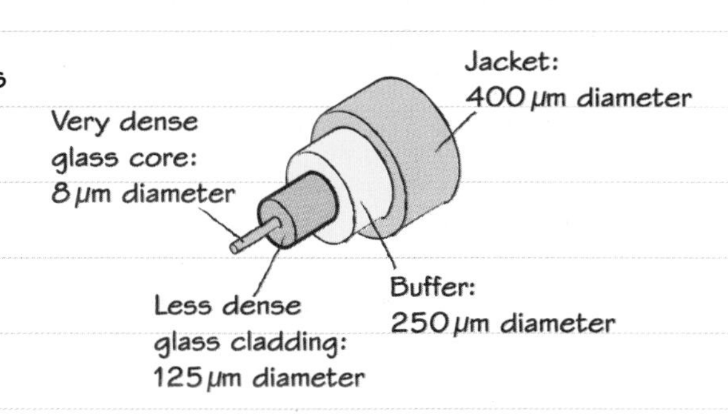

Make-up of a single-mode optical fibre

Now try this

1 Calculate the critical angle for light leaving glass of refractive index 1.40 to enter directly into air.

2 Give one example where you would choose to use copper cables to transmit signals, and another example where optical fibres would be preferred. Explain your choices.

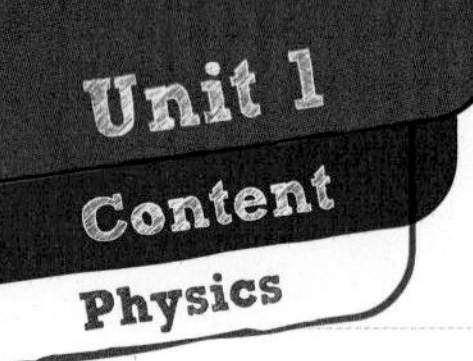

Endoscopy

Bundles of optical fibres make it possible to send light to and receive it back from places that would otherwise be inaccessible – for instance, inside the human body.

Illumination

A small fibre bundle is used for piping light from an external source down to the distal (remote) end, where it illuminates the area being investigated.

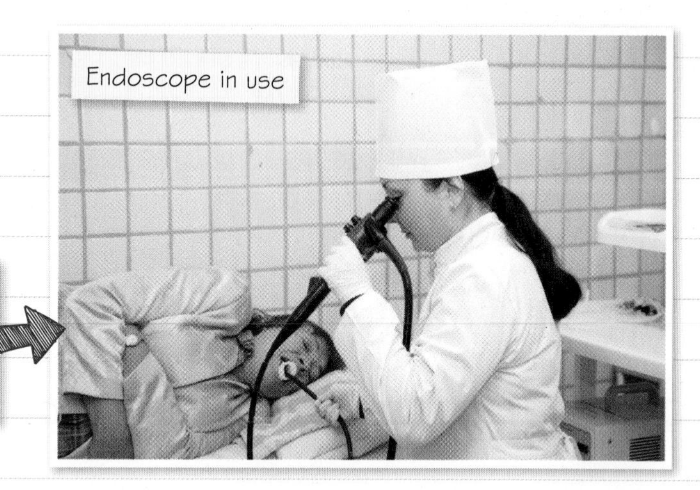

Endoscope in use

Along with the supply of light, water or air may be piped in and control wires for surgical instruments can be included too.

Forming an image

Lenses are used at both ends:

- A tiny **objective lens** at the distal end of an endoscope focuses an image onto cut ends of the optical fibres in the main bundle.
- Each fibre in the bundle collects and carries the light for one **pixel** (coloured dot) in the image.
- An **eyepiece lens** at the operator end takes light exiting from the fibre bundle and refocuses it to be viewed as an image directly by eye or with a camera.

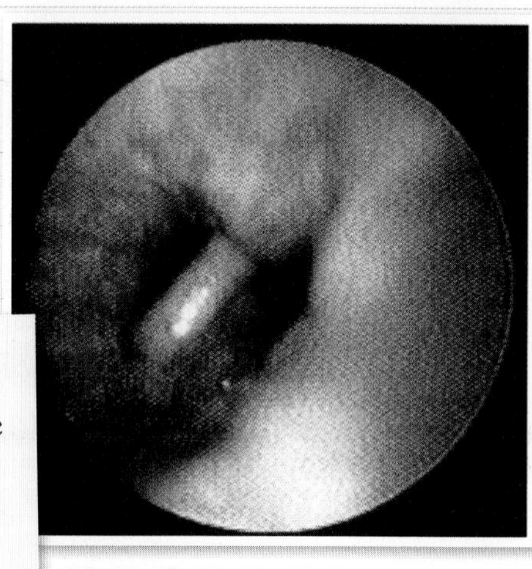

An image obtained from a fibrescope. The pixelation of the image can be seen – it is made up of many small coloured dots.

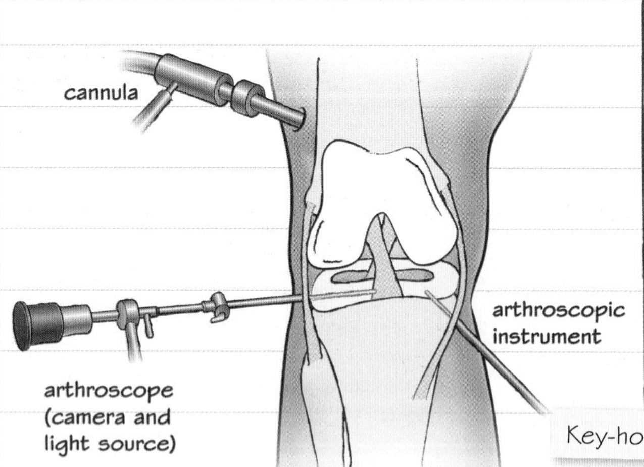

Key-hole surgery on a knee joint

Analogue images

Even though divided into pixels, these images are **analogue** – that is, the brightness and colour of each pixel vary with time in direct proportion to the light collected by the objective lens.

Links For more on how analogue and digital images differ, see page 39.

Now try this

The 'arthroscope' shown in the box above is a small fibre optic camera endoscope designed for use inside joints. Compare this endoscope with a microscope, highlighting the roles played by optical fibres.

Analogue or digital?

A signal may either mimic the variation in size and quality of the physical quantity it represents – an **analogue signal** – or it can encode that information as a series of numbers – a **digital signal**.

Meter displays

In a moving coil analogue meter, the needle moves in proportion to the current flow in it that the voltage causes.

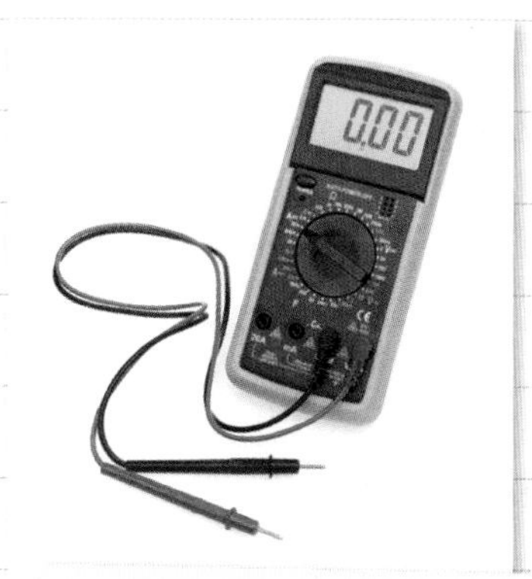

A digital voltmeter samples the voltage at regular intervals and displays the result each time as a number.

Digital accuracy

- **Sampling rate** is the number of times per second that the quantity is measured.
- **Sampling sensitivity** is the smallest increment in the quantity that is measured and recorded.

These set a limit on the accuracy of digitised information.

Analogue to digital converter (ADC)

To set up:

1. Connect a sensor (transducer) to give an analogue signal into the ADC input.
2. Select sampling rate and sensitivity.
3. Send the digital output signal to a display or a data storage device.

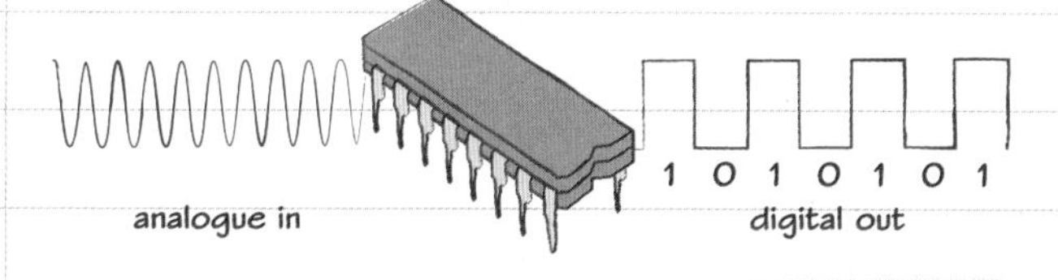

ADCs are standard electronic components.

Images

All images, whether printed, detected by the rods and cones in the retina of your eye, or displayed on a TV screen, are made up of coloured dots, grains, pixels, etc. However, each dot starts out as an analogue signal.

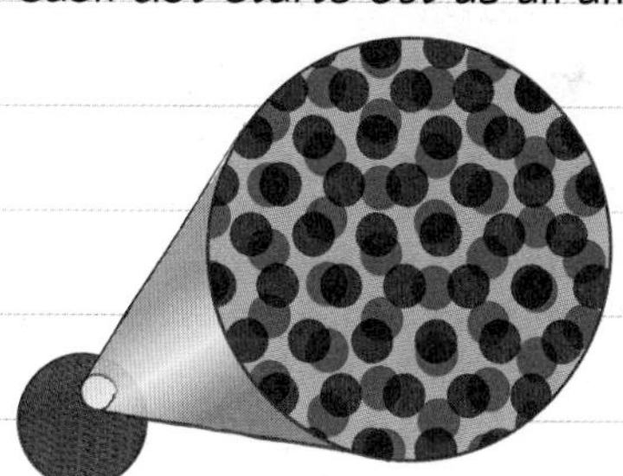

One of the dot techniques used in printing

Digitised video converts each pixel into a set of numbers representing its colour and brightness.

Why transmit in digital?

Most data is being switched to digital transmission, for example, radio, TV and all internet traffic. Analogue sounds and pictures are recreated in the receiver.

Advantages of digital

- More data fitted in the same bandwidth.
- Most interference can be eliminated giving higher quality.
- Can be stored and processed by a computer.

Drawbacks

- Time delay due to signal processing.
- Data accuracy limited by sampling settings.

Now try this

An experiment requires temperature to be monitored over a 24-hour period. A thermistor makes a useful small temperature sensor. When connected in a suitable electronic circuit, it will produce a voltage signal that varies directly with temperature and can be displayed on a meter.

1. Describe, using a diagram, how you could adapt this set-up to provide digital temperature data.
2. Explain what advantages this could have over the simple analogue output to a voltmeter dial reading.

Had a look ☐ Nearly there ☐ Nailed it! ☐

Electromagnetic waves

All electromagnetic waves have the same speed, c, but because they have widely differing wavelengths, frequencies and photon energies you detect and use them differently.

Intensity – the inverse square law

- Waves that radiate in all directions from an antenna have spherical wave-fronts.
- So they lose intensity, I, with radius from the source, r, according to the equation:
  $I = k/r^2$
  where k is the intensity 1 m from the centre of the source.

Double the distance ⇒ a quarter the intensity

Some equations will be given to you in the exam. Check the latest SAMS on the Pearson website to see which equations appear on the formula sheet.

Dish antennas

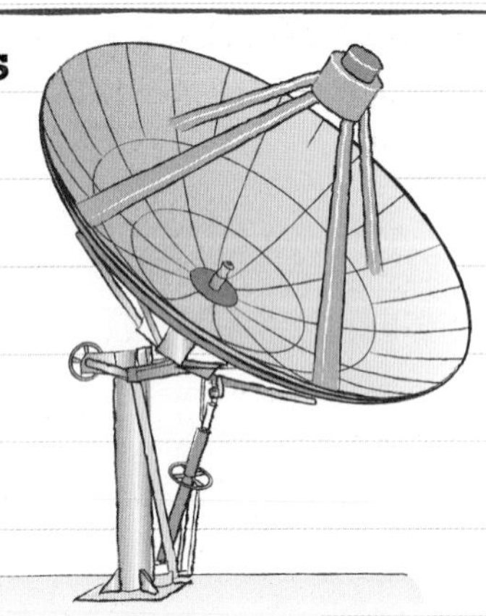

Satellite communications use dish antennas that concentrate the waves into a directional beam with flat plane wave-fronts. So those waves lose intensity more slowly.

The electromagnetic spectrum

Visible light – that is, the colours from deep red (λ = 740 nm) to indigo blue (λ = 370 nm) – is the tiny part of the electromagnetic spectrum that your eyes can detect.

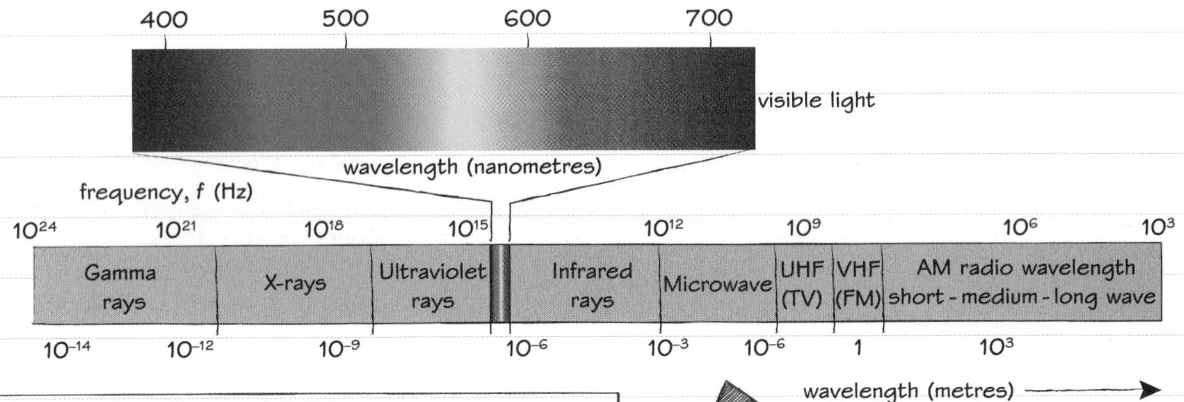

What to memorise

You only need to memorise *either* the wavelengths *or* the frequency range for each part of the spectrum.

Then use $c = f\lambda$ to calculate the other ($c = 3 \times 10^8\,\text{m s}^{-1}$).

Infrared and longer wavelengths are used for various communications purposes. Long waves penetrate better. High frequencies/shorter waves carry more data.

Now try this

The chart shows which sections of the electromagnetic spectrum are blocked by the atmosphere.

1. Identify the dangerous ionising radiations that are strongly absorbed, making life on Earth possible.
2. Explain which frequencies are used for satellite communications.
3. Wavelengths above about 30 m are reflected by a layer of charged particles in the upper atmosphere. Comment on the impact of this for broadcasting.

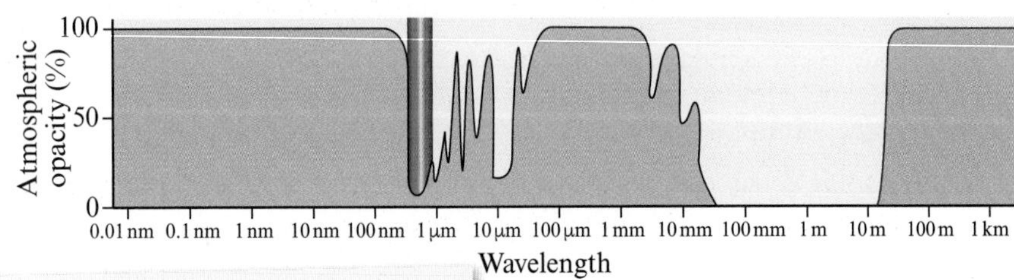

Chart showing atmospheric opacity

Waves in communications

You need to know the similarities and differences among satellite communications, mobile phones, Wi-Fi, Bluetooth® and infrared data transfer.

Geographical range

- Beaming signals up and down between **satellites** and **ground stations** allows them to be sent right round the world.
- Your **mobile phone** talks with the transceiver tower in a **cell** of a few km in radius.
- Each **Wi-Fi hub** has a range of 10 to 100 m.
- **Bluetooth®** is limited to about 10 m.
- **Infrared** signals reach only a few metres.

Changing frequency

To avoid **interference**:

- Satellites have two dish antennas and use quite different frequency bands for uplink and down link. A **transponder** unit on board receives, filters, amplifies and retransmits on the new frequency.
- Mobile phone **transceivers** (transmitter-receivers) also use separate frequencies for uplink and downlink, but these are nearby channels.
  In addition, each cell also uses a different frequency from all its adjacent cells.
- Bluetooth® devices frequency-hop across a range of channels many times a second to limit their interference with Wi-Fi, which operates in the same frequency band.

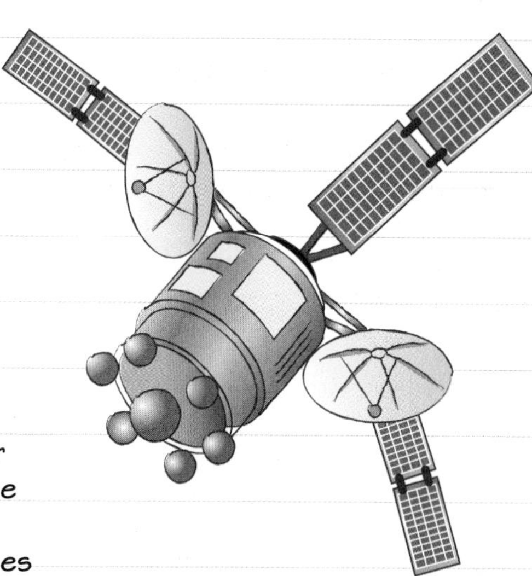

Networking

- **Wi-Fi hubs** connect to a wired local network and together can cover a whole building.
- Mobile phone cells are **networked** together to cover the country and also joined to the wider telephone network.
- Similarly, satellite ground stations are linked into **communications networks**.
- Networks use a combination of cable and wireless (for example, beamed microwave) links.
- Bluetooth® and infrared are **not** networked – just for device-to-device links.

Handshaking

- A series of short messages sent backwards and forwards allows one device to recognise another and to set up communication parameters. This is called handshaking.
- Handshaking also stops information getting lost or jumbled by acknowledging when a message has been received.
- Mobile phones swap from one cell tower to another as a user travels.

Now try this

1 Bluetooth® and Wi-Fi both operate in the unlicensed 2.4 to 2.4835 GHz Industrial, Scientific, Medical (ISM) band. Compare their operation and use for data communications, and explain how they co-exist without interference.

2 Mobile phones (800 MHz to 2.6 GHz) and satellite communications (1 to 40 Ghz) both operate in the same two regions of the spectrum.

(a) Name those regions.

(b) Compare how each of these technologies makes use of the frequencies allocated by the licensing authorities, and describe what equipment is involved.

Had a look ☐ Nearly there ☐ Nailed it! ☐

Your Unit 1 exam

Your Unit 1 exam will be set by Pearson and could cover any of the essential content in the unit. You can revise the unit content in this Revision Guide. In this Revision Guide, pages covering Structure and functions of cells and tissues have the heading "Biology", those covering Periodicity and properties of elements are headed "Chemistry" and those covering Waves in communication are headed "Physics". The section uses selected content and outcomes to provide examples of ways of applying your skills.

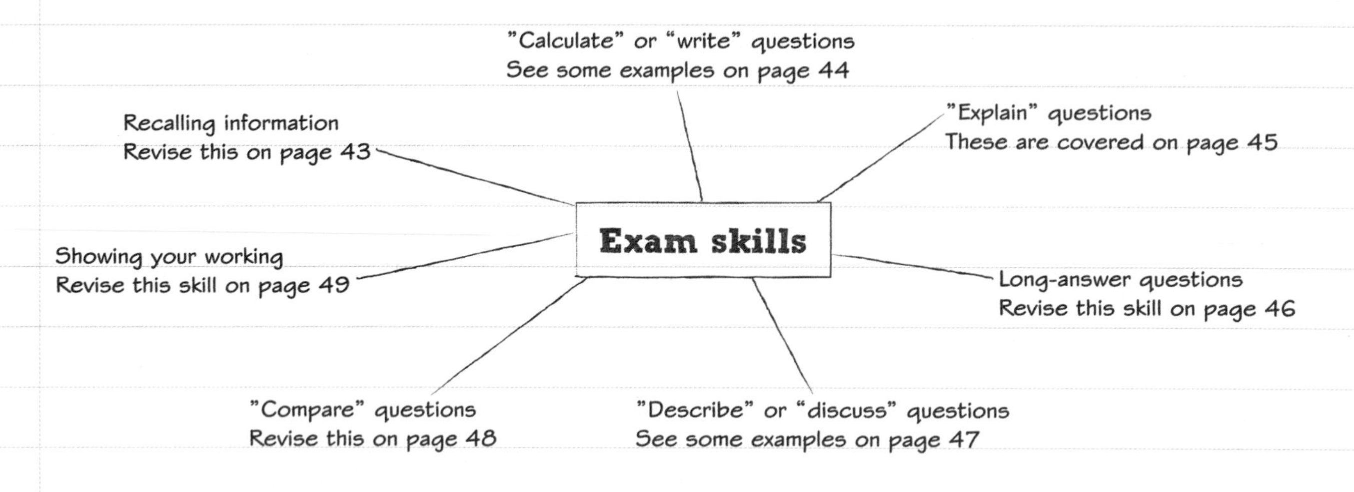

Exam checklist

For your exam, make sure you:

- ✓ Get a good night's sleep
- ✓ Have a black pen you like, a pencil and at least one spare of each
- ✓ Have double-checked the time and date of your exam
- ✓ Always show your working
- ✓ Remember to use the formula sheet and any other useful information you are given at the end of the paper
- ✓ Use any extra time to check your answers

Check the Pearson website

The questions and sample response extracts in this section are provided to help you to revise content and skills. Ask your tutor or check the Pearson website for the most up-to-date **Sample Assessment Material** and **Mark Scheme** to get an indication of the structure of your actual paper and what this requires of you. The details of the actual exam may change so always make sure you are up to date.

Now try this

Visit the Pearson website and find the page containing the course materials for BTEC National Applied Science. Look at the latest Unit 1 Sample Assessment Material (SAM) to get an indication of:

- the number of papers you have to take
- whether a paper is in parts
- how much time is allowed and how many marks are allocated
- what types of questions appear on the paper
- what additional information, such as a periodic table or formula sheet, may be given to you in the exam.

Your tutor or instructor may already have provided you with a copy of the Sample Assessment Material. You can use these as a 'mock' exam to practise before taking your actual exam.

'Give', 'state' and 'name' questions

'Give', **'state'** and **'name'** questions require you to recall one or more pieces of information. **'Label'** questions are similar, except you need to add information to a graph or diagram.

Worked example

Links Revise optical fibres on page 37.

State **two** advantages of sending signals by optical fibre cable rather than using copper wires. **2 marks**

This question asks for two advantages, so just two brief statements are needed.

Sample response extract

1 There is less attenuation so the signal travels further.

2 It is more secure as fibres cannot be hacked into.

The answer clearly identifies advantages by using the words 'less and 'more'.

Worked example

Links Revise how microphones pick up sound on page 31.

The graph below shows the signal produced when a microphone picks up a pure note. Give the amplitude of this electrical wave. **1 mark**

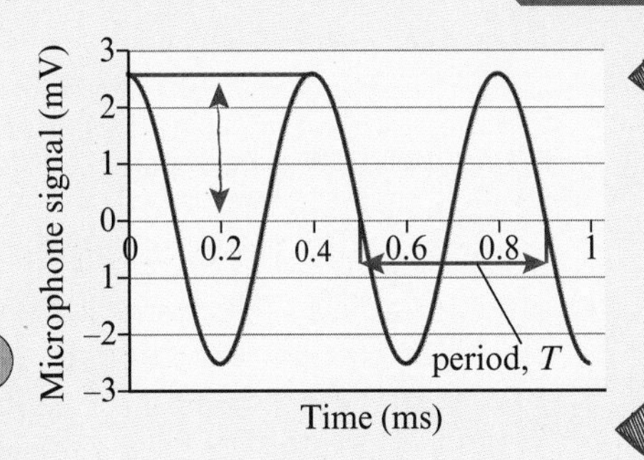

Mark on the graph to show where you are taking your measurement – in this case, from the central rest value to the peak.

Sample response extract

Amplitude = 2.6 mV

The answer is stated with an appropriate degree of accuracy and the unit is included.

Worked example

This question is asking you to list features you would find in a plant cell and in bacterial cells.

A soup was identified as causing food poisoning. Microscopic examination showed that some cells had cell walls and others did not. The cells with cell walls seemed to be two types: one from plant material and one was bacteria. State features that would distinguish the plant cells from the bacterial cells. **4 marks**

There are 4 marks for this question, so you need two distinguishing features for bacterial cells and two for plant cells.

Sample response extract

Bacteria	Plants
smaller than plant cells	larger than bacteria
just an area of DNA	membrane-bound nucleus
70S ribosomes	80S ribosomes
presence of plasmids	no plasmids
peptidoglycan cell wall	cellulose cell wall
presence of capsule outside cell wall	no capsule outside cell wall

A table is a good way to display this information.

For plant cells, you could also mention vacuole, chloroplasts and lack of capsule; for bacteria, pili and or flagella.

Links Revise the difference between bacteria and plant cells on pages 3 and 4.

Now try this

State what is meant by the term 'covalent bond'.

'Calculate' and 'write' questions

'Calculate' questions require you to obtain a numerical answer, showing relevant working. You must include the unit, if the answer needs one. 'Write' questions often ask for an equation.

Worked example

A sound check is done with a microphone just 20 cm away from a guitar soundboard. The musician then changes position so that the distance becomes 60 cm. Calculate by how much this will change the intensity of the sound picked up by the microphone. **2 marks**

Sample response extract

$\frac{I_2}{I_1} = (\frac{r_1}{r_2})^2 = (20/60)^2 = (1/3)^2 = 1/9$

Intensity will be 9 times smaller at a distance of 60 cm.

In 'calculate' questions you need to select the right formula and show you can apply it.

The inverse square law formula is correctly applied, the steps of working are shown, and the answer is then stated clearly.

Links Revise sound amplitudes on page 33.

Worked example

For the microphone signal shown on page 43, calculate the frequency of the sound wave that caused it. **3 marks**

Sample response extract

$f = \frac{1}{T} = \frac{1}{(0.4\,\text{ms})} = 2.5\,\text{Hz}$

Mark on the graph on page 43 to show how you measure the period correctly.

This question has three steps: recalling the formula for frequency, measuring the period from the graph, and then calculating the frequency from it, including the correct unit.

Links Revise wave frequency on page 32.

Worked example

A researcher has a photomicrograph of a blood smear, but unfortunately did not record the magnification of the image. However, they do know that they had previously calculated the diameter of the red blood cell as 8 micrometres. Calculate the magnification of the image. **3 marks**

You need to measure the diameter of the red blood cell in the image.

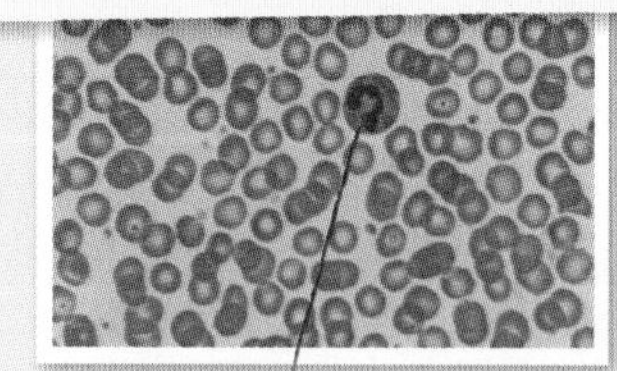

red blood cell

Sample response extract

$\text{Magnification} = \frac{\text{image size}}{\text{real object size}}$

$\text{Magnification} = \frac{4 \text{ millimetres}}{8 \text{ micrometres}}$

$\text{Magnification} = \frac{4000}{8} = \times 500$

Remember that you need to make sure the units are the same. There are 1000 micrometres in a millimetre. Magnification is a ratio so it does not have units.

Now try this

Nitric acid is used to manufacture fertilisers. A sample of 20.0 cm^3 of nitric acid required 17.8 cm^3 of 0.250 mol dm^{-3} potassium hydroxide, KOH, to exactly neutralise it. The equation for the neutralisation reaction is:
$HNO_3 + KOH \rightarrow KNO_3 + H_2O$

Calculate the concentration of the nitric acid, in mol dm^{-3}.

Links Revise concentration calculations on page 24.

'Explain' questions

'Explain' is probably the most common type of question. They require you to justify a point, using a clear process of reasoning. For maths questions, you must show your working. Some questions ask just for two things to be linked while others look for a longer series of connections and reasons.

Worked example

Explain how a closed-ended tube can act as a resonator, what determines the fundamental frequency, and which harmonics can also be produced. **4 marks**

This question has 4 marks and three sections, so be sure to tackle all the parts and to make at least four separate points.

Although the question asks about frequency, the answer needs to talk about wavelengths.

The closed end of the tube forces a node because no displacement can take place there. An antinode forms at the open end because the free air allows big displacements. The fundamental **mode** has a quarter wavelength in the pipe, and other harmonics with an odd number of quarter wavelengths in the pipes are also possible.

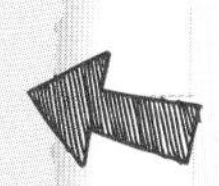

It is not sufficient just to say that a stationary wave forms or to mention nodes and antinodes. You need to explain **why** or **how** they are created.

Revise stationary waves on page 36.

Worked example

The ionic compound magnesium fluoride, MgF_2, is used in lens manufacture for telescopes.

(a) Draw an electronic configuration diagram of the ions in magnesium fluoride **2 marks**

'Draw' and 'Explain' may be parts of the same question. You might find that the drawing helps you with the 'Explain' part of the question.

Sample response extract

[F]$^-$ [F]$^-$

(b) Explain, using ideas about structure and bonding, why it has a high melting point. **4 marks**

Sample response extract

The structure consists of a giant ionic lattice, so contains many strong ionic bonds. These bonds are electrostatic attractions between oppositely charged ions. In order to melt the compound, these attractions have to be broken. This requires a lot of heat energy, hence the high melting point.

The answer not only discusses the structure and bonding in the compound but links these ideas to the energy needed to melt it to make sure the high melting point is explained.

Now try this

Digital television and radio transmission is allowing many more channels to be broadcast and received within the available frequency bands.

Explain how this is achieved, and comment on the compromises involved in analogue to digital signal conversion.

Links For more on the difference between analogue and digital signals, see page 39.

You should clearly and logically outline the differences between analogue and digital broadcasts.

Answering longer questions

Some questions build on a topic, and combine several command words and assessment outcomes, such as in the example below.

Worked example

The graph shows the first ionisation energies of elements in period 3.

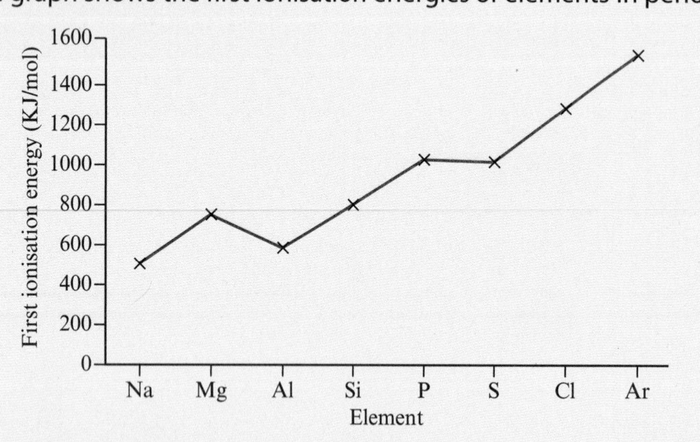

You may have to interpret data from graphs or tables and use it to support explanations of key chemical concepts.

(a) Write an equation to show the first ionisation of sulfur. **2 marks**

Write the equation clearly. If you make a mistake, cross it out and write the equation again. Remember to include state symbols.

Sample response extract

$S(g) \rightarrow S^{+}(g) + e^{-}$

(b) Explain the trend in first ionisation energy across the period. **4 marks**

If you are asked to explain a trend, outline the trend first.

Sample response extract

Across the period the ionisation energy increases. This is because the number of protons in each successive element increases, so the nuclear charge increases. In each element the electron removed comes from the same shell, with similar shielding from the nuclear charge by the inner shells. Hence the electrostatic attraction of the nucleus to the outer electrons increases across the period.

(c) Explain the slight anomaly in the trend, where sulfur has a lower first ionisation energy than phosphorus. **2 marks**

You may find it helpful in questions like this to work out the electronic configuration of each element, using 'electrons in boxes'.

Sample response extract

The electron removed in phosphorus comes from a singly occupied p-orbital, $3p^3$. The electron removed from sulfur comes from a p-orbital containing a pair of electrons, $3p^4$. The pair of electrons repel each other so one electron is now easier to remove, resulting in a slight drop in first ionisation energy.

Now try this

Explain why the second ionisation energy of an element is always greater than the first ionisation energy.

Take care with precision of language in an answer to this type of question. The attraction the nucleus has on the electrons increases, but not the nuclear charge.

'Describe' and 'discuss' questions

'Describe' questions require you to give an account of something but you do not need to include a justification or reason. **'Discuss'** questions ask you to explore all aspects of an issue or problem or argument by showing clear reasoning or argument.

Worked example

Scientists found that some bacteria were resistant to antibiotics. They also found they were gram-negative.

Describe how the scientists discovered this. **2 marks**

Sample response extract

The bacterial cells were stained with crystal violet, then washed with acetone and absolute alcohol. This removed the stain.

This question only asks what you do to see if a bacterial cell is gram-positive or negative. It does not ask why. That would be an explain question.

We know that gram-negative bacteria do not retain the gram stain because their cell wall has an outer layer and this is the reason they are more resistant to antibiotics. If you were asked to explain the link that would be the answer, but you are not!

Links For more on bacteria, see page 3.

Worked example

Discuss how the position of an element in the periodic table influences the properties of its oxides. **4 marks**

Sample response extract

Metallic elements tend to form oxides with giant ionic structures, and so have high melting and boiling points. If dissolved, these oxides tend to form alkaline solutions.

Non-metallic elements tend to form oxides with simple molecular structures, so have low melting points and boiling points. If dissolved, these oxides are likely to form acidic solutions. There are exceptions, though, and these tend to involve elements close to where metals and non-metals meet.

It is helpful to include examples to support your points in this type of question. For instance, you could use calcium oxide as an example of a typical metal oxide and sulfur dioxide as a typical non-metal oxide. Use the marks allocated to help determine the number of points you need to make.

Links For more information on oxides, see page 27.

For instance, silicon dioxide is not a simple molecule, even though it is a non-metal oxide. It has a giant covalent lattice structure.

Now try this

Discuss how path difference and phase difference contribute to the production of spectra by a diffraction grating.

Explanations are needed for both kinds of difference, and each must link to grating spectra. First explain how path difference arises, and then how it relates to phase difference – wavelength being the key. Explain conditions for the bright spectral lines using 'if … then …', then explain how destructive interference leads to darkness between them.

Had a look ☐ Nearly there ☐ Nailed it! ☐

'Compare' questions

'Compare' questions require you to look for at least one similarity and one difference between two or more things. You do not necessarily need to draw a conclusion.

See page 40 for more information about electromagnetic waves.

Worked example

Compare the uses of electromagnetic waves made by switched landline networks, cellular mobile phone operators and satellite communications in providing a telephone service.

Your answer should refer to range, flexibility and reliability, and to the frequencies used. **6 marks**

Note the key focus of the question: 'uses of electromagnetic waves'.

To help plan your answer, you could first note down what facts you know about each method of sending telephone messages.

Sample response extract

Be aware that in longer answer questions you get credited specifically for the clarity of your arguments and your ability to make links. So plan the order first, then write in clear sentences.

Mobile phone operators relay digitised sound data by means of UHF radio frequency signals broadcast from a network of cell antennas. Handsets transmit back on a slightly different frequency, and adjacent cell masts use different frequencies to avoid interference.

As well as recalling the frequency range of the waves used by each type of telephone network, the answer explains how different frequency bands are used and why.

Each mast transmits radially and has a range of just a few km, which defines the cell. Mobile phone handsets transmit to, and receive back from, their nearest cell mast and can switch cells when travelling. This makes them more flexible than landlines, but signal strength variations make them less reliable.

The answer compares both the flexibility and the reliability of mobiles and landlines, each with a reason for the difference. Later, it comments on the reliability of satellite links.

Landline networks use infrared frequency signals channelled down optical fibre cable. With amplification every few km, they can reliably cover the whole country. Satellites and their ground stations use dish antennae to transmit directional beams of UHF or microwave frequency signals and also to receive them. This gives them a much longer range of many hundreds of km, enabling them to span continents. Atmospheric conditions can limit their reliability.

Range for each type of wave signal has been linked with the way it is used to connect users to one another.

Leave out other things you may know about telephones that the question does not ask for, for example, details of how A to D conversion is done.

Now try this

Compare a red blood cell with a white blood cell.

You could give one difference and two similarities, or two differences and one similarity. In the exam, you should check on the number of marks and use that as a guide to how many points you should make.

See page 6 for more information about specilaised cells.

Showing your working

Whether you are answering questions in the biology, chemistry or physics sections, you must clearly show your working, especially if there is a maths element.

Worked example

A student weighs out 0.140 g of a metal hydroxide, MOH, and dissolves it in exactly 100 cm^3 of distilled water. 25.0 cm^3 of this solution is neutralised exactly using hydrochloric acid, HCl, of concentration 0.0500 mol dm^{-3}.

17.2 cm^3 of the hydrochloric acid is needed to neutralise the MOH solution.

Use this information to identify the metal, M. **5 marks**

Links You may have carried out experiments like this in Unit 2. They are called titrations.

Sample response extract

MOH + HCl → MCl + H_2O

Amount of HCl

= (17.2 ÷ 1000) × 0.0500

= 8.60×10^{-4} mol

So we can deduce that the **amount of MOH (in 25.0 cm^3) =**

8.60×10^{-4} mol as the ratio in the equation is 1:1

Hence the **amount of MOH in 100 cm^3** is 4 times greater than the amount in 25 cm^3

= $8.60 \times 10^{-4} \times 4 = 3.44 \times 10^{-3}$ mol

So the **relative formula mass of MOH**

= mass ÷ amount

= $0.140 \div 3.44 \times 10^{-3} = 40.7$

So subtract 17 (for relative mass of O and H atoms) and we are left with 23.7

Hence **M must be Na** (relative atomic mass = 23.0)

Carefully show your working, so that if you make a slip the examiner can follow and award part-marks. This is especially important in calculations, such as this example, without a clear structure to guide you through, so try to label each step to help them see what you are doing.

Notice the final answer in the calculation is quoted to three significant figures. This is to keep it consistent with the data in the question.

Now try this

Look at the sample assessment materials (SAMs) on the Pearson website, or ask your teacher to provide them to you. Analyse the mark schemes shown at the end. Some questions specifically ask you to show your working, which may include equations and calculations. Practise all the questions you can find that ask you to show your working, and check your answers against the sample answers.

These types of questions are most likely to come up in the chemistry and physics sections, as they are more likely to include calculations. However, you also need to ensure that you include enough detail when answering biology questions.

Had a look ☐ Nearly there ☐ Nailed it! ☐

Developing a hypothesis

A **hypothesis** is an idea that you can test using practical experiments.

Two types of hypothesis

Some people call a hypothesis an 'educated guess'. In fact, in science it is more than that. To develop a hypothesis you will already have some facts or made some observations. You will then use these to develop a hypothesis that you can test. Most hypothesis statements are **if/then statements**.

If more water is added to the soil **then** the plants will grow more.

Another way of writing a hypothesis is to suggest that one variable is dependent on a second variable.

The growth of a crop is dependent on the type of the fertiliser in the soil.

In this hypothesis, you are not talking about more or less fertiliser, but about **different types** of fertiliser. Remember, a null hypothesis always states that there will be no relationship between variables and is useful when performing statistical tests on your results.

Examples of hypotheses

Here are four examples of hypotheses:

If the temperature of the reactants is increased then the rate of reaction will increase.

If the mass attached to the spring increases then the extension of the spring will increase.

The bigger the molecules in the solution, the fewer molecules will pass through a membrane.

Crops will grow better with fertiliser containing potassium and sodium rather than fertilisers that just contain one of those elements.

Hypotheses from observations

You can use your observations of the plants in the picture to make a hypothesis.

One possible hypothesis is:

If a plant gets more sunlight then it will grow taller.

It would be important to test this as a lot of other factors may make a difference, for example, type of plant or minerals in the soil. Another hypothesis could be:

The height of a plant is dependent on the type of plant.

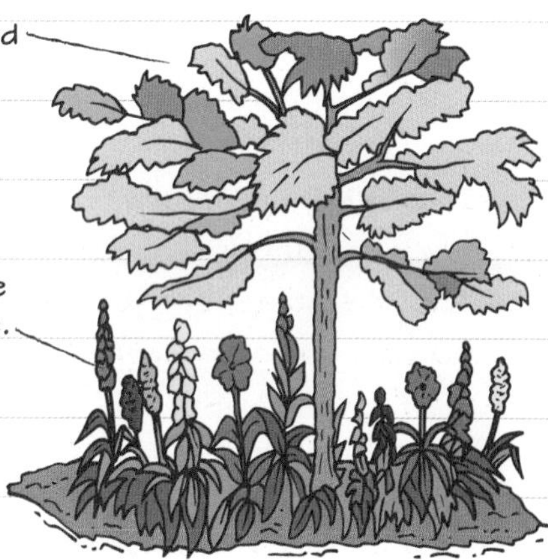

Key features of a good hypothesis

A good hypothesis must:

- ✓ be able to be tested scientifically
- ✓ relate one dependent variable to one independent variable.

Null hypothesis

A null hypothesis states that there is no relationship or causality. An example of a null hypothesis would be that there is no relation between the temperature of the reactants and the rate of reaction. Any differences between the rate of reactions at different temperatures is due to chance.

Now try this

Write a hypothesis for each of the following observations.

(a) When a 1 mol dm^{-3} solution of acid was added to 1 g of magnesium, the magnesium dissolved in less time than when a 0.5 mol dm^{-3} solution was added to another 1 g of magnesium.

(b) A lamp in a circuit was dim when there was one cell in the circuit. When there were three cells in the circuit the lamp was very bright.

Write an if/then statement relating the dependent variable to the independent variable.

Planning an investigation

When planning an investigation, you will need to select appropriate equipment, techniques and standard procedures.

Choosing equipment

- What is it for?
- How many do you need?
- How precise does it have to be?
- What size does it have to be?

You should always choose the smallest measuring equipment available for the volume you wish to measure. In this case measuring 10 cm³ with a 10 mL measuring cylinder is more accurate than using a 100 mL measuring cylinder.

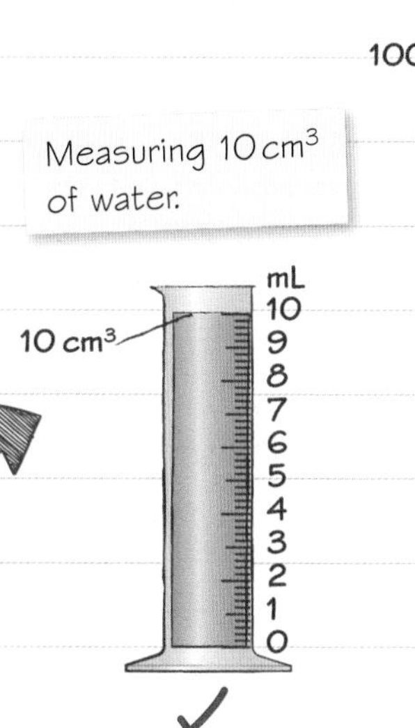

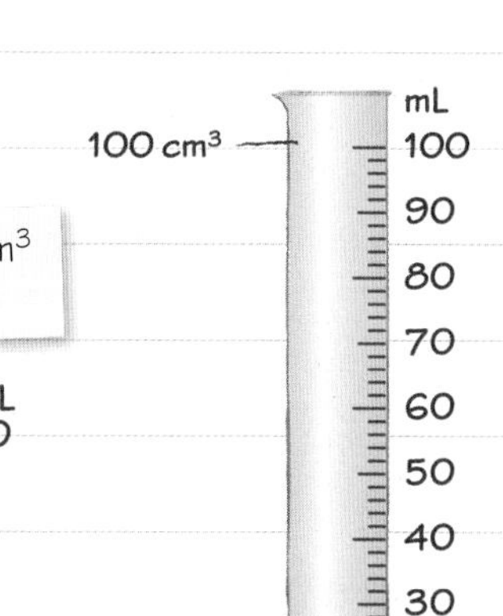

Some pieces of equipment have a smaller in-built error that others, here a volumetric pipette is more accurate than a measuring cylinder as it has a smaller in-built error.

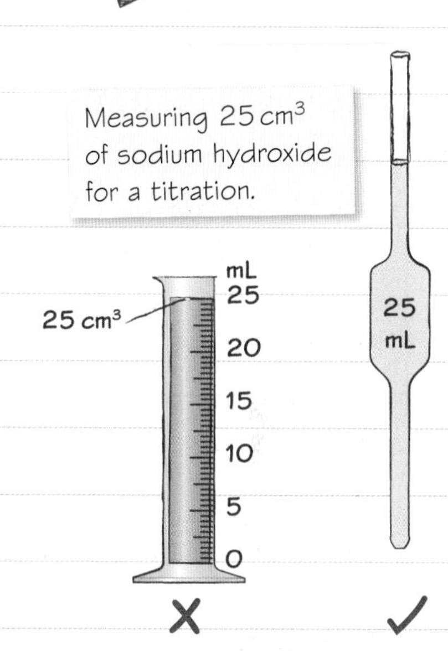

Do not forget other resources you will need, such as biological samples or chemical substances.

Choosing appropriate techniques

- What are you investigating?
- What method(s) are you using?
- Do you have the equipment available to carry out the technique?
- Do you have enough time to carry out the technique?
- Do you have the skills/knowledge to carry out the technique?

Qualitative and quantitative

Some investigations will generate **qualitative** information, while some investigations will generate **quantitative** data. Qualitative refers to the appearance or value of something, including examples such as colour observed, or whether bubbles are produced. Quantitative data refers to numerical measurements made, including examples such as volumes and times measured.

Standard operating procedure (SOP)

This is an established procedure or method used to carry out a routine activity.

Using an SOP ensures consistent results every time the activity is carried out.

Examples of SOPs:

- titration
- handling of chemicals
- disposal of waste
- qualitative analytical tests
- preparing solutions.

Now try this

You are going to investigate how temperature will affect the rate of decomposition of hydrogen peroxide by catalase (an enzyme in potato).

(a) List the equipment and resources you will need and justify why each piece is required.

(b) Name/describe any specific techniques you will use.

(c) Explain any standard operating procedures you will have to follow.

Had a look ☐ Nearly there ☐ Nailed it! ☐

Risk assessments

A risk assessment identifies hazards, evaluates the risks associated with those hazards, and determines ways to eliminate or control those risks.

It is important not to confuse **hazards** and **risks**.

Hazard – the equipment or substance that can cause harm, for example hot plate or hydrochloric acid.

Risk – The harm that the hazard could cause and the chances of it happening, for example the risk from using a hot plate is getting burnt. The risk from using hydrochloric acid is skin corrosion or chemical burns.

Risk assessment template

You might be provided with a risk assessment template to fill in. If not, you can produce your own. The template below shows you the headings that may be present on the template.

Hazard	Harm that could be caused	Severity of harm	Likelihood of harm	Control measures	Procedures if harm occurs
catalase enzyme					
hydrogen peroxide					
oxygen gas					
glassware					

Resources to help you fill in a risk assessment form:

- http://www.hse.gov.uk/
- COSHH regulations
- CLEAPSS Hazcards

When you complete a risk assessment for your practical investigations you may wish to consider the common risks (harm that could be caused), control measures and emergency procedures:

Common risks

- Chemical burns
- Eye damage or irritation
- Burns or fire
- Hearing damage
- Poisoning
- Inhalation of toxic fumes
- Cuts to skin from broken glassware
- Injury from dropping heavy equipment
- Slipping or tripping over

Common control measures

- Wearing googles or face shields
- Wearing a laboratory coat
- Wearing gloves
- Carrying out the experiment in a fume cupboard
- Using correct manual handling techniques
- Cleaning up broken glass or spillages as soon as they occur

Emergency procedures

- First aid
- Washing chemical contact with skin or eyes under running cold water
- Running burns under running cold water
- Removal of affected person(s) to fresh air
- Call emergency services for assistance

Now try this

Use the Hazcards in your school or college to complete the risk assessment above.

Variables in an investigation

There are three types of variables in an investigation: **independent**, **dependent** and **control** variables. You need to know what they are and how to measure them.

Independent variables – variables you change during an investigation

In the experiment shown, the independent variables can be:

- height ball is dropped
- type of ball
- size of ball
- surface ball will be dropped on.

Remember, you can only change one variable to test a hypothesis with validity.

Dependent variable – the variable you are investigating and measure during an investigation

In this experiment, the dependent variable would be how high the ball will bounce.

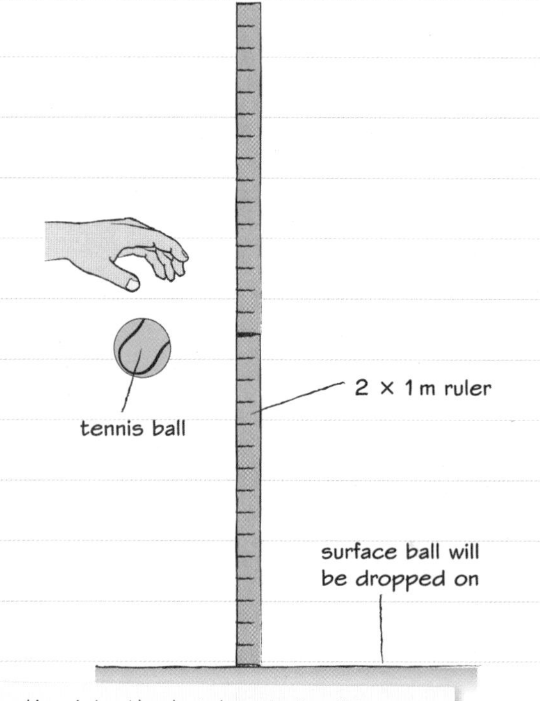

Investigation into the height a ball will bounce

Control variables – the variables you keep constant during an investigation

These are the same as the independent variables:

- height ball is dropped
- type of ball
- size of ball
- surface ball will be dropped on.

Decide which is your independent variable (the variable you are testing) and the rest become controls. These you need to keep the same.

You will need to state how you will control these variables.

For example, you could control the height the ball is dropped by using a metre rule to ensure you always drop it from the same height.

Now try this

You are investigating how different surfaces affect the height that the ball will bounce.

(a) List the possible independent variables.
(b) Name the dependent variable.
(c) Write a hypothesis relating your independent variables to the height of the bounce.
(d) State how you will control the other variables.

Links You can revise writing a hypothesis on page 50.

Had a look ☐ Nearly there ☐ Nailed it! ☐

Producing a method

The method you produce should be repeatable. Someone else should be able to follow it exactly as you performed it and get the same results.

A method should:

- be a step-by-step set of instructions
- be logically ordered
- allow you to take sensible measurements
- allow you to test your hypothesis.

You need to think about:

1. what you are testing
2. what you want to measure
3. how you are going to measure
4. what techniques you will use
5. what variables you must control
6. how many repeated readings you will take
7. what you are going to do with the data.

Validity of data

Age (years)	Height (cm)
20	164
25	158

The above data is meaningless.

✗ There are only 2 results.

✗ There are no repeats.

✗ No pattern can be identified.

✗ A meaningful graph/bar chart cannot be produced.

✗ A meaningful conclusion cannot be made.

One way to present this data is in a bar chart.

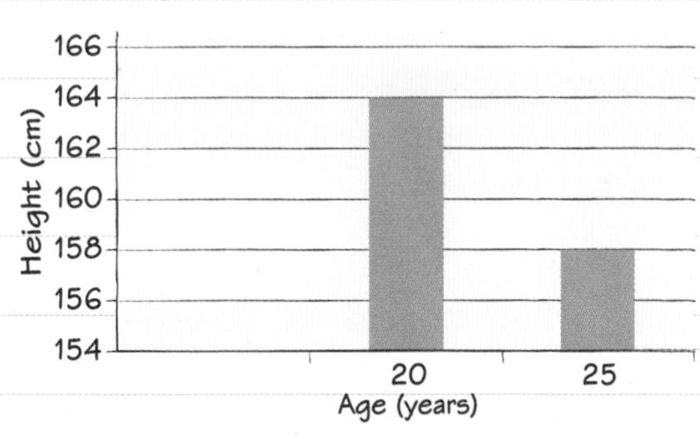

The only conclusion from this bar chart is that older people are shorter than younger people.

You know this is not always true. The data is meaningless because:

✗ There is not a big enough range of data. The method used to produce these results may not have been sufficient.

✗ There are no repeats of data.

✗ The conclusion from the data is not true.

Good planning

When planning your method make sure it will produce:

- a range of results
- reliable results
- precise results
- accurate results.

You should be able to present and analyse your data in such a way that you can be confident that any conclusions you make are valid.

Remember

✓ You should only change one variable – this is the factor you are testing.

✓ You must control all other variables.

✓ You must be able to measure/monitor your variables.

Now try this

Write a method to investigate the following hypothesis:

'In a field with trees, a smaller variety of plant species grow in the area under the trees than the area not under the trees. (There is a smaller species richness in shaded areas.)'

Recording data

You will need to collect data accurately and reliably. You will then have to record the data in an appropriate format.

Recording data in a table

Most data is recorded in a table.

The example shows the results from an investigation into how light intensity affects photosynthesis.

Distance between plant and light source (m)	Number of bubbles of oxygen produced in 1 minute			Mean number of bubbles of oxygen produced in 1 minute
	plant 1	plant 2	plant 3	
0.10	98	102	103	101
0.25	100	104	101	101.7
0.50	28	9	29	28.5
0.75	13	13	14	13.3
1.00	7	8	6	7

Clear headings with units

Good range

anomalous result

repeated results

incorrect precision (different number of decimal places)

average of result from plant 1 and 3 only, as the result from plant 2 is clearly anomalous

Precision refers to the number of decimal places or significant figures that the data has been recorded to.

Recording data

Results of calculations should be to the same **precision** as the results – in this case, only whole numbers are measured so the averages should also be in whole numbers. **Anomalous** results should be repeated where possible and not used in calculations. A **range** of at least three results allows for a valid graph to be produced from the mean of these repeats.

Qualitative data

Some investigations give qualitative data, for example, a flame test.

You must be able to make inference from **observations** as well as from measured results.

This can include: colours, smells, solids produced/dissolving, gases produced.

The colour of the flame can tell us the cations present in a compound.

Now try this

Think about the results for the investigation into how light intensity affects photosynthesis.

(a) How could you change the method to get more precise results?

(b) Explain when you would repeat results.

(c) Explain the minimum number of results you would need for each distance.

Had a look ☐ Nearly there ☐ Nailed it! ☐

Processing data

You record results and observations when carrying out an investigation. You usually have to process this data using calculations. You must also display the data in graphs, charts or tables.

Types of data

Data can be continuous, such as age of plant, or discrete such as number of plants.

Some data is a mixture of numbers and names, for example, number of different types of plants found in a garden.

For more on data tables, see page 55.

The first stage of data processing

The data in the table on page 55 has been processed as the means for each row have been calculated. This is often the first stage in processing data.

Transposition of formulae

When you are using formulae, you may sometimes need to re-arrange the formula to change the subject (the part being calculated). The best way to do this is to draw a formula triangle and then use this to re-write the equation with a different subject.

Ways to process data

You will need to decide what is the appropriate method to process the data.

For example:

- mean and standard deviation
- error bars
- statistical tests
- use of formulae
- conversion of units
- standard form
- percentage error in measuring equipment.

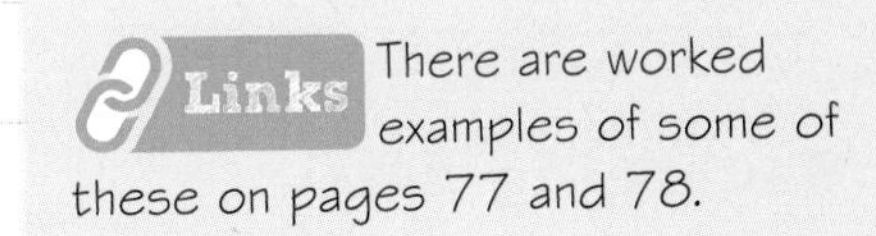

There are worked examples of some of these on pages 77 and 78.

Line graphs

Continuous data is normally displayed on a line graph.

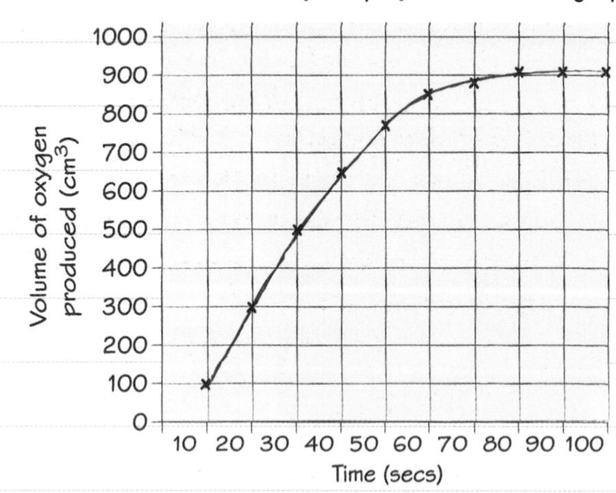

This is the data collected from a chemical reaction that produces oxygen.

- It is a smooth curve of best fit.
- The curve goes through or near most points.
- The line fills most of the graph paper.

Bar charts

Bar charts are used for discrete data, but can compare different sets of data.

The chart below compares hair colour to eye colour.

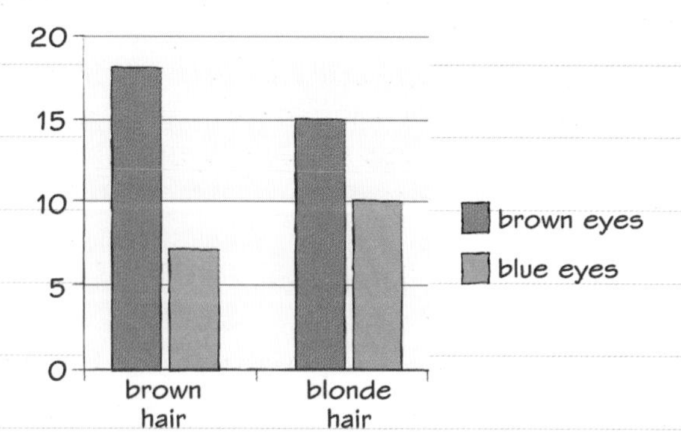

Things to consider

When displaying data you need to think about:

- What type of data is it?
- What type of chart or graph is appropriate?

Make sure you:

- use appropriate scales
- plot the data correctly
- draw a smooth curve or single line
- use all of the graph paper
- label both axes correctly and with units.

Now try this

Use a graph to display the mean data in the table on page 55.

Which graph will you choose to display the data?

Interpretation and analysis of data

Once you have drawn your tables and graphs, you will need to use the data to identify patterns/trends and draw conclusions.

Patterns – the shape of the graph.

Trends – the relationship between factors on the graph or in the table.

Conclusion – how well your results support your hypothesis.

Patterns

In the graph on page 56, the **pattern** is a line that goes up as time goes by, but that eventually levels off.

There is a simple trend – 'the volume of oxygen produced increases over time'.

However, it is more complex than that because the line starts to level off.

Trends

The full **trend** is that the volume increases as time increases but then levels off and after a certain amount of time no more oxygen is produced.

Identifying a pattern or trend

When identifying a pattern or a trend you must:

- describe the whole graph/chart
- ignore any anomalous plots
- recognise that the pattern/trend may change across the graph and these changes must also be described.

For an example of how to see a pattern on a line graph, see page 56.

Sample size

The number of people sampled for the hair and eye colour investigation on page 56 is very small.

To be confident in your conclusion you need a large enough sample to give strong evidence.

Primary data

This is data you gathered for yourself through experiments and investigations. If you carried out the investigation into hair and eye colour, that would be primary data.

Secondary data

You could get more data by researching in books, scientific papers or on the internet.

The data you get is called secondary data.

You should compare your primary and secondary data to draw a conclusion.

If they do not show the same patterns/trends, then you should question why this is?

- Did you use the same method?
- If not, is one method better than another?
- Was the secondary data a larger data set?
- Is the hypothesis correct?

Hypothesis

The hypothesis for the investigation that produced the bar chart on page 56 was:

'People with brown hair are more likely to have brown eyes than people with blonde hair.'

50 people were observed, 25 who had brown hair and 25 who had blonde hair.'

The pattern does show that more people with brown hair also had brown eyes compared to people with blonde hair having brown eyes.

The conclusion you would draw from this would support the hypothesis.

Now try this

(a) Describe the pattern(s) and trend(s) on the graph you produced of the average data from the table on page 55.

(b) Write a conclusion based on the pattern(s) and trend(s).

Had a look ☐ Nearly there ☐ Nailed it! ☐

Evaluating an investigation

You will need to evaluate your methods and your data. You will need to look for areas that can be improved and explain how they can be improved.

Improvements to the investigations

Consider your:

- equipment
- reactants
- methods
- techniques.

You should be able to explain how your investigation was valid and explain any changes you could make to improve your investigation further.

You should be aware that human error may occur. Where you make a human error, you should repeat the procedure to correct it.

Explain anomalous data

How do you know it is anomalous?

✓ It does not fit the pattern of other results.

✓ When plotted on a graph it does not lie close to the line/curve of best fit.

What caused the anomalous data?

✓ not following method correctly

✓ inaccurate measuring/recording

✓ impurities.

Sources of error

Systematic error – this is an error that is built in to the method or equipment you are using. For example, if a balance is not calibrated correctly you will get an identical error every time you use it. This can be corrected by calibrating the balance. These errors are **quantitative**.

Random error – these are due to not following the method correctly and can be minimised by repeating results and taking averages.

Misreading of observations can cause qualitative errors, for example, observing an incorrect colour in a flame test.

Accuracy, reliability and precision

When you carry out an evaluation you must look at the accuracy, reliability and precision of the data collected.

Accuracy – is the data close to the true value? **Error due to the equipment** could make the results less accurate.

Reliability – how trustworthy the data is. Usually ensured by repeating readings to identify any anomalous results

Precision – how close repeat readings are to each other. Can be improved by using measuring equipment that measures to more decimal places.

Identify strengths and weaknesses

How did your method/equipment/standard procedure/technique ensure you got reliable results?

How might the method/equipment/standard procedure/technique you used cause errors?

Explaining strengths and weaknesses

When writing an evaluation, it is important that you explain the strengths and weaknesses and not just describe them.

You can then explain your suggested improvements based on the strengths and weaknesses you have considered.

Now try this

Consider a practical you have carried out. List all the possible errors that might have occurred or been due to the equipment used and explain how you would prevent them in future.

Enzymes: Protein structure

An **enzyme** is a **protein molecule** that acts as a **catalyst** in a biochemical reaction.

Amino acids

Proteins are made of **amino acids**.

In an amino acid, there is a central (alpha) carbon with four chemical groups attached to it:

 a hydrogen atom

 an amino group

 a carboxyl group

4 a variable R group (side chain).

There are 20 different R groups and so there are 20 different amino acids.

alpha carbon with H atom attached

amino group

carboxyl group

R group

Structure of an amino acid molecule

Peptide bonds

Amino acids are linked together by **peptide bonds**.

- A **condensation** reaction (water is produced) occurs between two amino acids.
- A **dipeptide** is formed by two amino acids linked by a peptide bond.
- Three amino acids form a **tripeptide**.
- Many amino acids join to form a **polypeptide**.
- A polypeptide with more than 50 amino acids is called a **protein**.

glycine + alanine

H_2O removed via condensation

peptide bond

Condensation reaction forming a dipeptide

Protein chain

NH_2—CH—C—N—CH—C—N—CH—C—N—CH·········C—N—CH—C—N—CH—COOH

N terminal

C terminal

Now try this

Research the structure of the R groups of the amino acids cysteine and threonine.

Draw structural diagrams of the amino acids and then produce a drawing to show the condensation reaction between these two amino acids. Include the product(s) made and label the peptide bond present.

Had a look ☐ Nearly there ☐ Nailed it! ☐

Enzymes: Active sites

The **active site** is the part of the enzyme where the biochemical reaction takes place.

Secondary structure

- Polypeptide chains form orderly shapes (**secondary structure**).
- The secondary structures are held in shape by **hydrogen bonds** between the -NH of one peptide link and the -C=O of another amino acid.
- The most common shapes of secondary structures are coiled **α-helix** and folded **β-pleated sheet**.

Tertiary structure

- The secondary structure folds into **three dimensional** shapes to make the **tertiary structure**.
- It is the tertiary structure that gives the enzyme the shape of its active site.

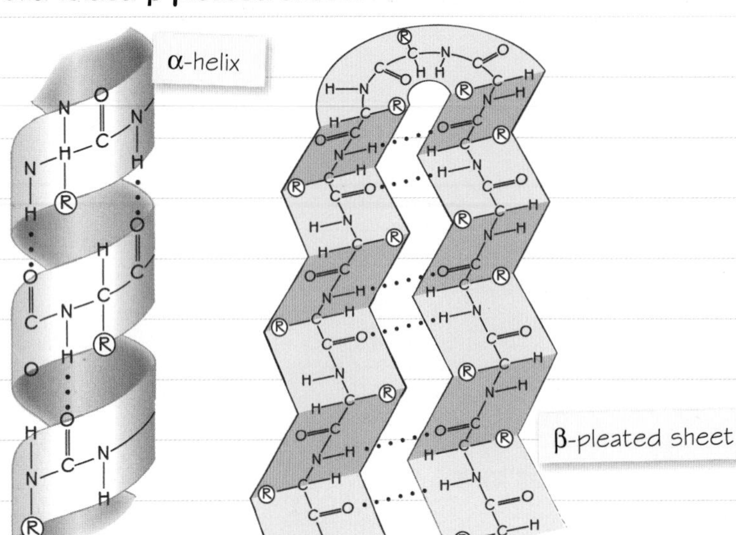

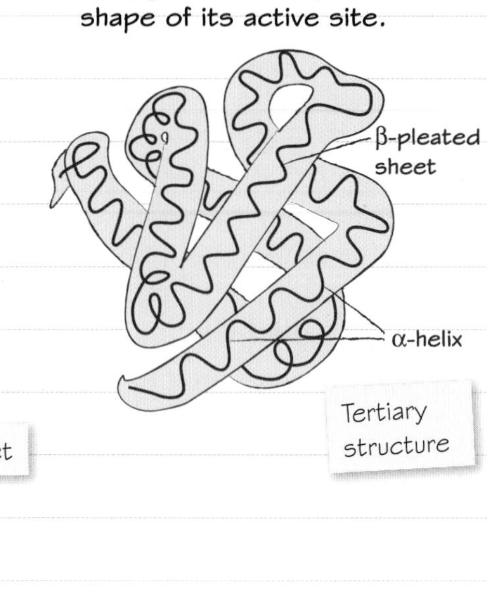

Tertiary structure

Active site

When an enzyme catalyses a reaction, the reaction takes place on products are the active site of the enzyme. The active site consists of a small number of amino acids, which form a specific shape. When an enzyme catalyses a reaction, the reaction takes place on the active site of the enzyme. The active site consists of a small number of amino acids, which form a specific shape because of the protein's secondary and tertiary structure.

Enzymes work **specifically** to catalyse one type of biochemical reaction. For example, the enzyme amylase specifically breaks down starch but does not break down protein.

In a biochemical reaction, **substrates** are converted into **products**.

The substrate binds to the active site. Only specific substrates will bind to a particular enzyme's active site. The substrate has a **complementary shape** to the shape of the active site of the enzyme.

After the reaction, the product is released and the active site is available to catalyse further reactions.

enzyme E + substrate S ⇌

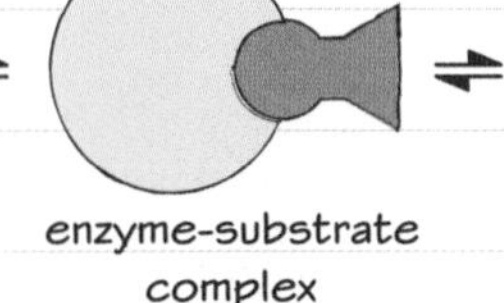

⇌

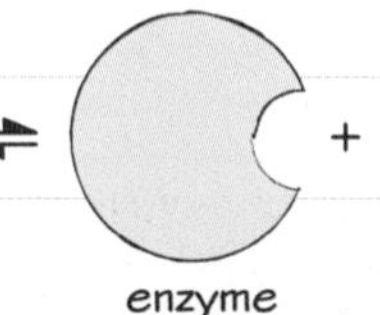

\+ products P

Lock and key mechanism

The active site is a specific shape that only the substrate will fit.

- If the temperature goes too high, for example, above 40°C, the hydrogen bonds will break.
- This means the enzyme changes its shape.
- So the active site loses its shape.
- The enzyme is **denatured** and can no longer act as a catalyst.

Now try this

In your own words, describe the lock and key mechanism of enzyme action.

Enzymes: Biological catalysts

A **catalyst** is a substance that speeds up a reaction. It can take part in the reaction but is left **unchanged** at the end of the reaction. Enzymes are biological catalysts.

Collision theory

For a reaction to occur:

- particles must **collide**
- particles must have enough energy to **react**.

The energy of the collision depends on the speed of the particle and the angle they collide.

The minimum energy needed for particle to react is the **activation energy**.

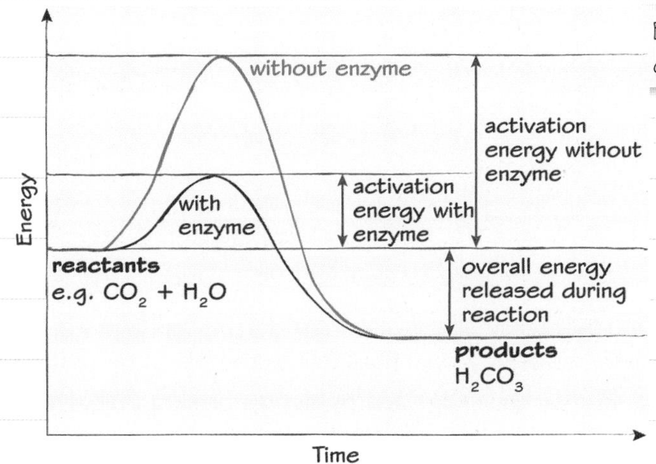

Energy profile for an enzyme-catalysed reaction

The enzyme lowers the activation energy needed:

→ more particles have the required energy to react

→ there are more succesful collisions

→ there is a faster reaction.

Enzymes

Catalytic activity – increase in the rate of a reaction caused by the inclusion of an enzyme.

Substrate – the molecule that is affected by the action of the enzyme.

Active site – the area of an enzyme where a substrate binds.

Denatured – when the tertiary structure of the enzyme is changed.

Measuring rate of reaction

Enzyme reactions can be measured by:

- measuring decrease in substrate
- measuring increase in product.

Measuring increase in product is easier because you know the starting measurement for the product will be 0.

The initial rate of reaction must be recorded because the rate will decrease as the substrate is used up and its concentration decreases.

Now try this

1 What is necessary for an enzyme catalysed reaction to occur?

2 How does an enzyme speed up a reaction?

3 Why is it important to record the initial rate of an enzyme-catalysed reaction?

Enzymes: Factors affecting activity

Enzymes need specific conditions to work effectively. If these conditions are not present, then the enzyme-catalysed reactions may slow down or not work at all.

Temperature

Each enzyme has an **optimum temperature** at which it works fastest.

For enzymes in humans, this is about **40°C**.

As the temperature increases up to the optimum temperature, the reaction rate also increases.

Below optimum temperature, there is less energy → fewer collisions → slower reaction.

Above optimum temperature → hydrogen bonds in enzyme break → enzyme and active site loses its shape (denatured) → no catalysis.

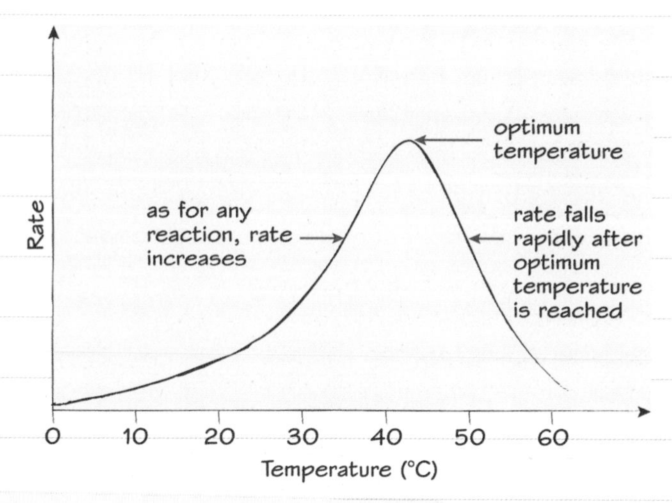

Graph showing effect of temperature on enzyme-catalysed reaction

pH

Enzymes have an **optimum pH**.

- For many enzymes in the body, it is between 7 and 8 as this is the pH of most body cells and the blood.
- Some enzymes have a more extreme optimum pH, for example, pepsin, a stomach enzyme, has an optimum pH at about 1.
- pH affects the charge on the active site. If pH changes then the charges on the active site change and the substrate may no longer be able to bind to it.
- For example, the carboxyl group of an amino acid (COOH) will be uncharged at low pH. But at a high pH it will donate H^+ and become negatively charged (COO^-). The effect of this is that the active site may only attract a substrate with a positive charge if the pH of the environment is high.

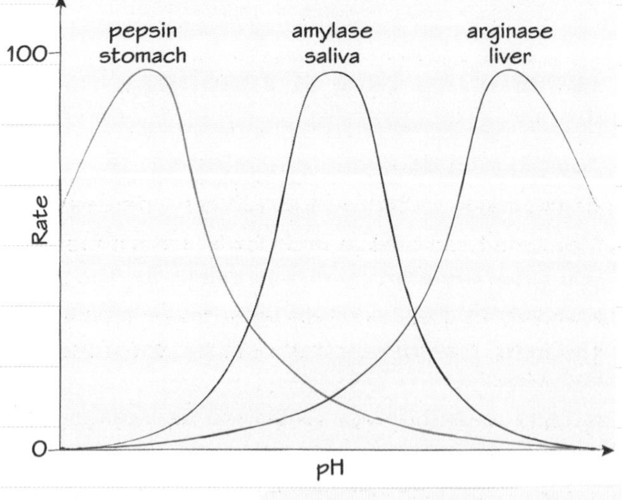

Graph showing effect of pH on enzyme catalysed reactions

Concentration

The concentrations of the enzyme and of the substrate affect the rate of reaction.

Increase in substrate concentration → increase in rate of collisions → increase in reaction rate.

Increase in enzyme concentration → increase in rate of collisions → increase in reaction rate.

If the concentration of the substrate is too high, then there will be no more active sites for them to bind to. At this point, increasing the concentration of the substrate will no longer affect the rate of reaction.

If the concentration of enzyme is too high, there are too many free available active sites compared to substrate molecules – and so the rate of reaction will no longer increase.

Now try this

Describe how temperature, pH and concentration can affect the rate of reaction of the enzyme protease. Protease is an enzyme found in the stomach, where the pH is around 1–2.

Think about the optimum conditions for this enzyme and how changing these conditions will affect its activity. Make sure you consider temperature, pH and concentration.

Diffusion of molecules

Diffusion is the movement of substances from a region of **high to low concentration**.

Diffusion

In the experiment shown on the right, when the dye molecules are added, they move from an area of high concentration to areas of low concentration until they are evenly spread. The molecules are then described as being in dynamic equilibrium.

The more dye molecules that are added, the faster the rate of spread because there is a bigger concentration gradient. This is diffusion – the movement of molecules down a **concentration gradient**. Diffusion is **passive** – it does not require an input of energy.

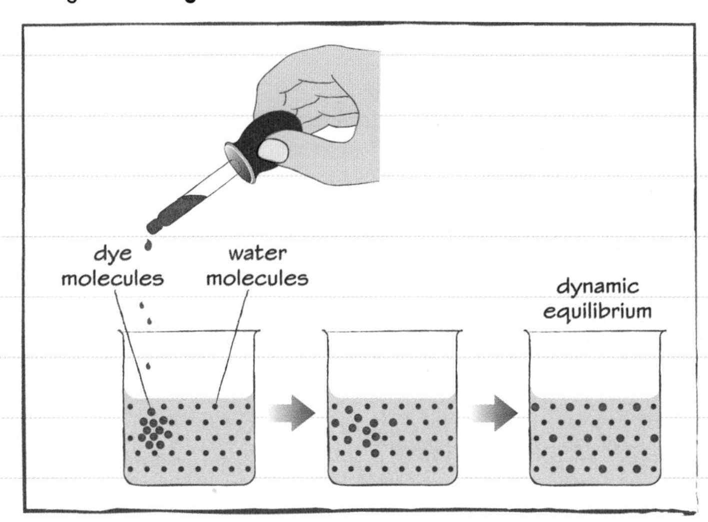

Diffusion can occur across a membrane

As long as the molecules are small enough to pass through the membrane then diffusion will occur.

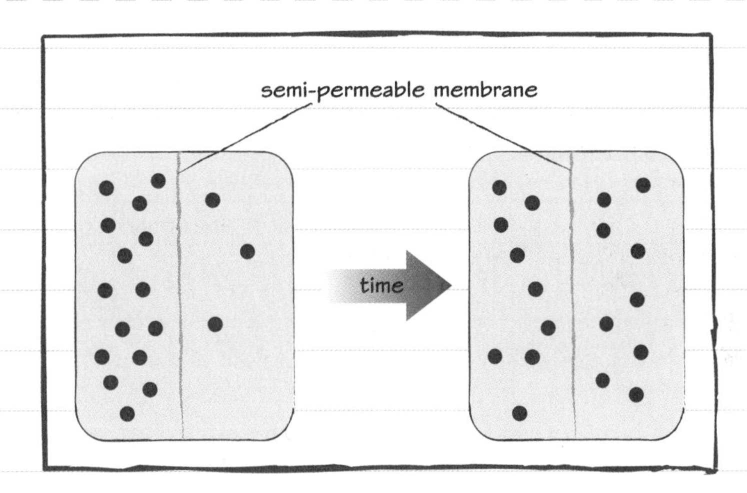

Factors affecting rate of diffusion

- **Size** – Larger molecules move more slowly than smaller ones and so the rate of diffusion decreases.
- **Temperature** – A higher temperature increases the energy molecules have and so they move faster, increasing the rate of diffusion.
- **Distance through a substance** – The greater the distance that the molecules must travel through a substance, the slower the rate of diffusion. For example, leaves are thin, which reduces the distance that gases diffuse through and so increases the rate of diffusion.
- **Surface area** – As surface area increases, the area of which molecules can spread is increased and so the rate of diffusion increases.
- **Shape of the surface of which the molecule is diffusing** – the shape of the surface of which the molecule is diffusing can affect the rate of diffusion. For example, the shape of a cell can affect its surface area, therefore affecting the rate of diffusion of molecules into or out of the cell.

Now try this

1 Explain why you can smell a cake cooking in all rooms in a house and not just the kitchen.

2 Where would you be able to smell the cake (a) first? (b) last?

Had a look ☐ Nearly there ☐ Nailed it! ☐

Kinetic theory and diffusion

Kinetic theory explains the properties of different states of matter. It also explains diffusion.

States of matter

	Solid	Liquid	Gas
Arrangement of particles	close together regular pattern	close together random	far apart random
Movement of particles	vibrate about a fixed position	move around each other	move quickly in any direction

Diagram of particles in the three states of matter

Knowing the behaviour of the particles in solids, liquids and gases helps us to understand why diffusion does or does not occur.

Properties of states of matter

The properties of the three states of matter are dependent on how their particles behave.

Property of solids	Property of liquids	Property of gases
fixed shape	take shape of container	completely fill container
cannot flow	they flow and move randomly	they flow and move randomly
cannot be compressed or squashed	cannot be compressed or squashed	can be compressed or squashed

Gases and liquids can flow. They move randomly. This allows diffusion in gases and liquids but not in solids.

Gases move more quickly than liquids so diffusion occurs faster in gases than in liquids.

Diffusion can happen within:

- a mixture of liquids
- a mixture of gases
- a mixture of liquids and gases.

Now try this

1 How does the arrangement of particles in a gas allow diffusion to occur?

2 Explain why diffusion does not happen between two solids.

Your answer should include reference to the amount of energy the gas particles have, as well as the way they are arranged.

Plant growth and distribution

You need to understand and be able to investigate the factors that affect plant growth.

Human effects on plant growth

Humans put a lot of effort into improving the growth of crops by ensuring:

- correct soil pH
- optimum watering
- correct nutrients present.

Humans can also have negative effects:

- **trampling**
- picking wildflowers
- removing hedges/trees (habitat destruction).

Humans have an effect on the distribution of plants:

- Sowing of crops means fields with only one type of plant.
- Trampling can cause unusual or more widespread distribution patterns due to seeds being carried on the bottom of shoes.

Soil pH and aeration

Most plants grow best in a soil with pH between 4.5 and 7.5.

Acidic soil can mean there is a deficiency in necessary minerals, such as iron and manganese.

Soil aeration is necessary to allow sufficient oxygen to get to the plant and for carbon dioxide to be removed.

Toxins also form where there is little **aeration**, such as hydrogen sulfide gas.

Soil of pH between 5–6.5

Soil with low pH. The acidic soil causes yellow leaves and stunted growth.

Temperature

Different plants prefer different temperature conditions to grow, for example, rye grass and white clover

Most plants have an **optimum temperature** they will grow at.

Temperature is important in **seed germination**. Seeds need the soil to be within a specific temperature range in order for them to germinate.

Relative rate of growth

Air temperature (°C): 9, 11, 13, 15, 17, 19, 21, 23, 25, 27, 29, 31, 33

Rye grass
White clover

Make sure your plan includes detailed enough instructions for someone to be able to follow, and that you consider your independent, dependent and control variables.

Now try this

Plan a simple experiment to find out the optimum temperature for germination of cress seeds.

Links You can revise variables in an investigation on page 53, and how to produce a method on page 54.

Improving plant growth

Light, water and minerals are important for plant growth.

Light intensity

The intensity of a light source can affect plant growth.

- Plants need light for **photosynthesis**.
- Plants in high light intensity are likely to have better growth than those in shade.
- Plants used to low light intensity can get sun scorched in the sun.
- Different plants prefer different light intensity and many prefer to be in the shade at some point of the day.
- Leaves in the shade may grow larger in order to absorb more light. Other differences can be noticed also, for example, difference in colour.

Soil moisture

If there is not enough soil moisture:

- The rate of photosynthesis decreases.
- Plants cannot take up necessary minerals and nutrients.

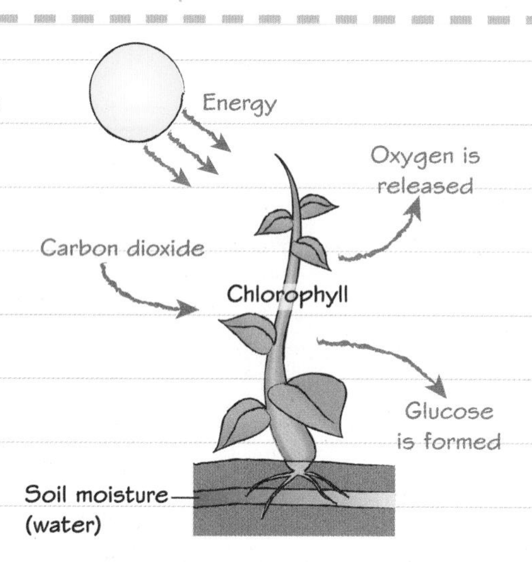

Rainfall

Most water enters the soil through rainfall. Plants then take up water using their roots. A plant can be in soil that has too much or too little rainfall.

Effects of too little rainfall	Effects of too much rainfall
yellow leaves	yellow leaves
dry leaves	rotted roots
	minerals/nutrients washed away
ground cracks	soil erosion

Mineral ions

Plants take mineral ions in through their roots. The diagram shows the effect of deficiency of some minerals.

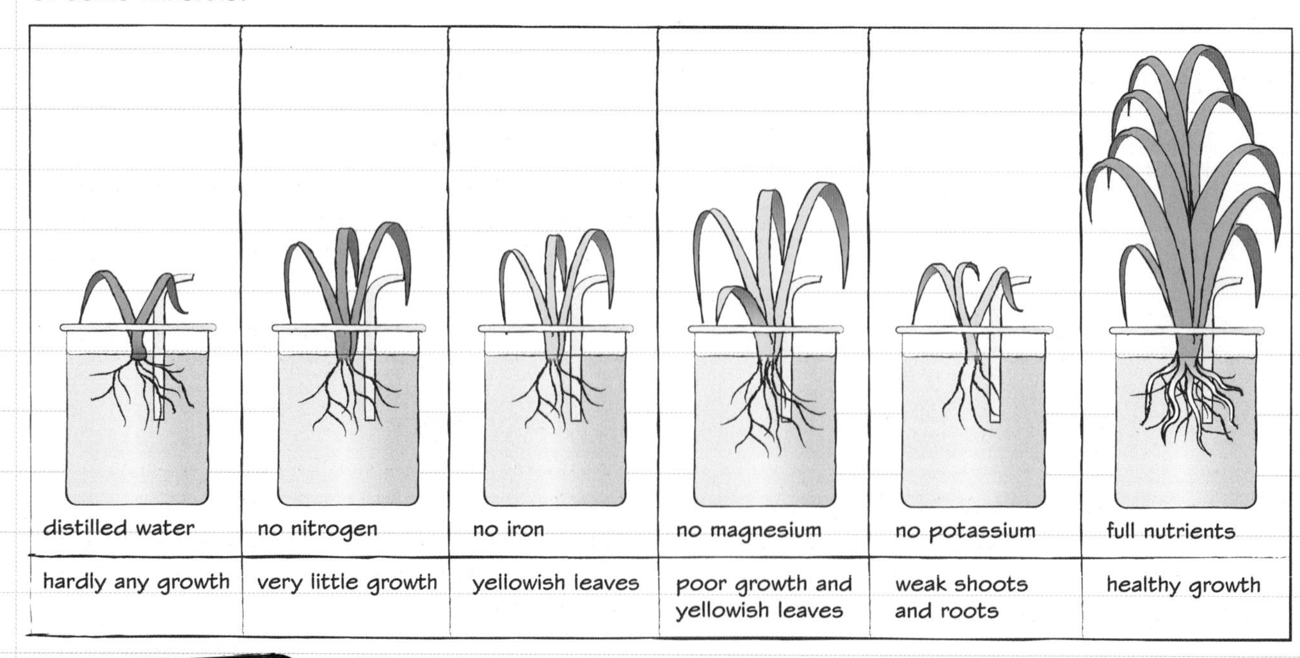

Now try this

1 What mineral or minerals may be missing from the soil if a plant has weak shoots and roots and yellow leaves?

2 What are the effects of too little water in soil?

Sampling techniques

Random sampling

This means that each potential sample has the same probability of being selected. Used when:

- Area under study is large.
- Area under study is fairly uniform.
- There is limited time.

Pilot studies of the area can help you to work out your sampling size. The minimum sample size is the number of samples at which the number of different species found starts to level off. Ecologists often double this sample size to be sure they are doing enough.

Abiotic factors

Examples of abiotic factors that can affect plant population are:

- soil type/moisture/pH
- topography, for example, steepness of slopes
- weather/sunlight
- temperature.

It is important that the appropriate techniques are used to investigate the effects of these factors.

Quadrats

A quadrat is a small area of habitat (normally 1 m^2), which sometimes has a grid within it. They can be different sizes and shapes, as long as the same quadrat is used for all sampling. The frame is placed on the ground, and plants and animals inside the frame are counted. The percentage cover of different plant species can also be estimated.

Placing a quadrat

Quadrats must be placed randomly to get a representative sample

1. Draw a map of area.
2. Overlay a grid on the map.
3. Use a random number generator to decide where the quadrat is placed.

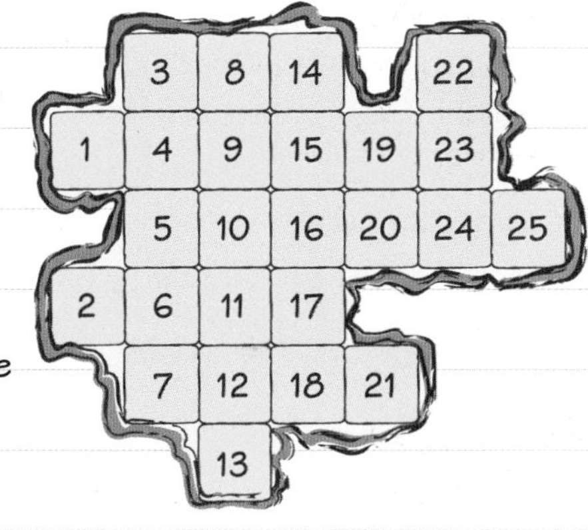

Point frames

A point frame contains a number of pins. When the frame is placed on the ground the plant or soil type at each point is measured.

There are a range of point frames you can use.

A line of pins in a frame.

Wherever the pin or pointer touch the ground, the ground cover is recorded.

Transects

Line transect

Sampling lines are set up across the area to be sampled. Species can be recorded along the whole length of the line – this is continuous sampling.

Or, species can be recorded at set intervals along the line – this is systematic sampling.

Belt transect

This is similar to a line transect, but gives information on the abundance of species.

Quadrats are placed at set intervals along the line and numbers of each species are counted within them.

Now try this

Explain why random sampling is important and describe one way it can be carried out.

Had a look ☐ Nearly there ☐ Nailed it! ☐

Investigating fuels

Different fuels release different amounts of energy. This will depend on the length of the carbon chain in the molecule.

Points to consider

You will need to investigate fuels:

- How much energy they produce.
- What pollutants they produce.
- How a fuel is chosen for specific purposes, for example, fuelling cars, heating rooms, cooking.

Fuels you could investigate:

- petrol
- food
- methanol
- propan-1-ol
- pentan-1-ol
- paraffin
- cooking oil
- ethanol
- butan-1-ol
- wax

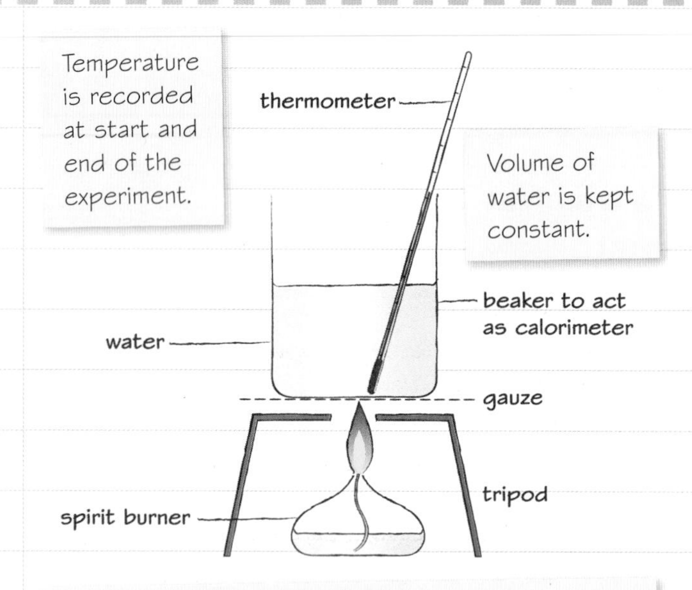

The mass of the fuel and spirit burner is recorded at the start and the end of the experiment.

The change in temperature of the water and the change in the mass of the spirit burner can be used to calculate the heat of combustion of the fuels.

Combustion of fuels

The reaction for combustion of fuels is **exothermic**. Therefore energy is released as heat.

fuel + oxygen → carbon dioxide + water (+ heat energy)

For example,

methanol + oxygen → carbon dioxide + water

$2CH_3OH + 3O_2 \rightarrow 2CO_2 + 4H_2O$

The longer the carbon chain the more energy is released during combustion.

Other factors are considered when choosing a fuel:

- how easy it is to ignite
- how easy it is to store
- how easy it is to transport
- how much/type of pollution it produces.

For more information on using temperature, see page 70.

Now try this

(a) Write a plan for the experiment in the diagram, based on the following hypothesis:
'The longer the carbon chain in a fuel, the more energy it will produce.'

(b) Make a list of the equipment you would select and justify why you have chosen each piece.

You can revise planning an investigation on page 51.

Remember to include a range of different fuels to test which have different length carbon chains (you could test 5 different alcohols). Be specific about the size and number of each piece of equipment you would use. Read your answer back to yourself and make sure that someone else would be able to follow your plan without needing to ask any further questions about what they need to do.

Risks of investigating fuels

Care has to be taken when working with fuels. Remember you must always produce and follow a **risk assessment** for all investigations you carry out.

Flammability

One of the most useful properties of fuels is that they are **flammable** (catch fire easily).

This is also a hazard and should be treated as such. There is a risk of fire and burns when working with fuels.

Toxicity

Although many fuels themselves are not **toxic**, often they contain **impurities** that are.

For example, fossil fuels often contain benzene, which is a toxic substance.

Products of combustion of impurities, such as sulfur impurities being released as sulfur dioxide, can also be toxic.

In car exhausts, toxic nitrogen oxides are produced.

Risk of explosion

With some fuels there is a risk of explosion.

Sulfur dioxide

Sulfur dioxide reacts with moisture in the air to produce **acid rain**.

Acid rain:

- erodes limestone buildings and statues
- acidifies lakes and rivers
- kills aquatic plants and animals.

Carbon monoxide

Links: On page 68 you saw the chemical equation for combustion of fuels.

As covered on page 68, the equation for combustion of fuels is:

fuel + oxygen → carbon dioxide + water (+ heat energy)

That reaction occurs when there is plenty of oxygen present.

If there is limited oxygen, then **incomplete combustion** occurs which can produce **carbon monoxide** and **soot**:

fuel + oxygen → carbon monoxide + water (+ heat energy)

For example:

methane + oxygen → carbon monoxide +water (+ heat energy)

$2CH_4 + 3O_2 \rightarrow 2CO + 4H_2O$

Carbon monoxide is a poisonous gas and can cause headaches, dizziness, nausea and death.

Soot can also be produced:

$CH_4 + O_2 \rightarrow C + 2H_2O$

Soot powder can cause respiratory problems.

Now try this

Explain the risks of burning a fuel with a limited oxygen supply.

Had a look ☐ Nearly there ☐ Nailed it! ☐

Units of energy

The **energy content** of substances such as food and fuels can be investigated. The results of the practical investigation can then be used to calculate the energy released.

Energy units

Energy can be measured in different **units**.

In many cases, you will see energy measured in **joules**. It takes 4.2 joules of energy to raise the temperature of 1 g of water by 1°C. This is known as the **specific heat capacity** of water.

Energy can also be measured in **calories**. 1 calorie is needed to raise the temperature of 1 g of water by 1°C.

Converting units

Remember, putting kilo in front of a unit increases the size by 1000.

For example:

1 kilojoule (kJ) = 1000 joules (J)

1 kilometre (km) = 1000 metres (m)

1 kilogram (kg) = 1000 grams (g)

1 kilocalorie (kcal) = 1000 calories (cal)

Food labels

The energy provided by food can be measured in kilojoules, but is often measured in kilocalories.

Each slice of bread (40 g) contains:

Energy 397 kJ 94 kcal	Fat 0.9 g	Saturates 0.2 g	Sugars 1.4 g	Salt 0.4 g
5%	1%	1%	2%	7%

of an adult's reference intake.
Typical values (as sold) per 100 g: energy 993 kJ/235 kcal

You will usually see both units on food labels.

You can investigate the energy content of foods practically, then compare your results to the package labels.

Calorimetry

You can investigate the energy in food or fuel by burning the food or fuel under a certain volume of water using a **calorimeter**.

You would measure the temperature change of the water and the mass of the food or fuel used. Temperature can be measured in °C or K (Kelvin). 1°C = 274 K. For the calculation it does not matter which units are used, as only the change in temperature is needed.

The heat energy that was supplied to the water can be calculated using the equation:

Heat energy (J) = mass of water (g) × specific heat capacity of water ($J\,g^{-1}\,°C^{-1}$) × temperature rise of water

Remember the specific heat capacity of water is $4.2\,J\,g^{-1}\,°C^{-1}$. $1\,cm^3$ of water has the mass of 1 g.

Energy released from a fuel

If you calculate the energy supplied to water by a fuel (such as ethanol), you can then calculate the energy released by a mole of the fuel.

$$\text{heat of combustion (kJ mol}^{-1}) = \frac{\text{heat energy supplied to water (kJ)} \times \text{relative molecular mass of fuel}}{\text{mass of fuel burnt (g)}}$$

Remember, you will need to convert the heat energy supplied to the water to kJ.

The molar mass of a fuel is the atomic mass of all the atoms in the fuel added together.

Now try this

1 Calculate the heat energy supplied to $50\,cm^3$ of water when the temperature is raised by 44 °C by burning ethanol.

2 Calculate the molar mass of ethanol.

3 Use this information to calculate the (heat of combustion) energy released in $kJ\,mol^{-1}$ when 3 g of ethanol is burnt. Express your answer to 2 significant figures.

Remember to include a minus sign in front of your answer, as the reaction is exothermic (gives out heat).

Symbols in electrical circuits

When drawing circuit diagrams, scientists use standard symbols for the components so that the diagrams can be interpreted by anyone.

Battery

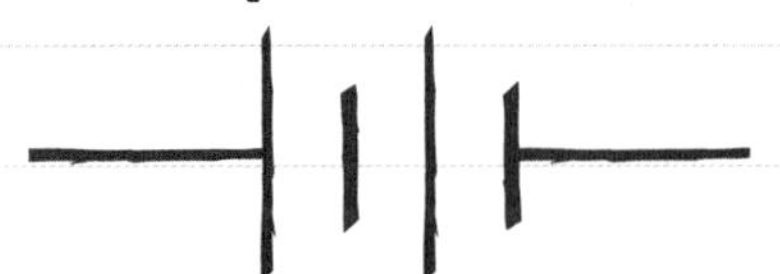

A battery is made up of more than one cell.

(A cell is what you would call a battery that you use in a remote control, for example.)

Ammeter

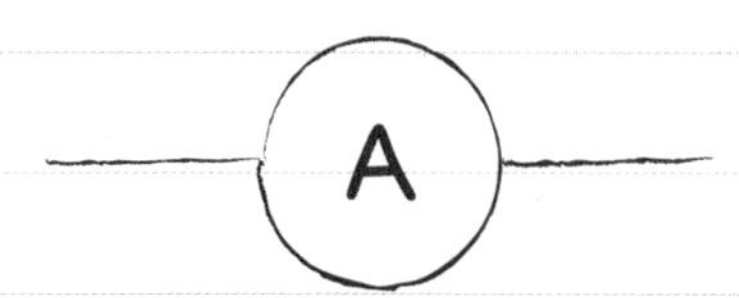

An ammeter measures current that flows through a component in a circuit. It is placed in series with the component.

Voltmeter

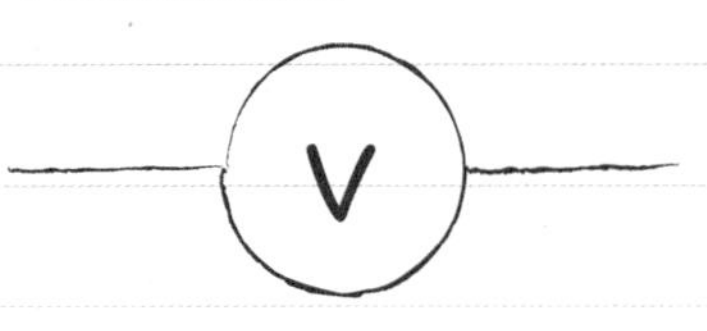

A voltmeter measures voltage across a component in a circuit. It is placed in parallel with the component.

Bulb

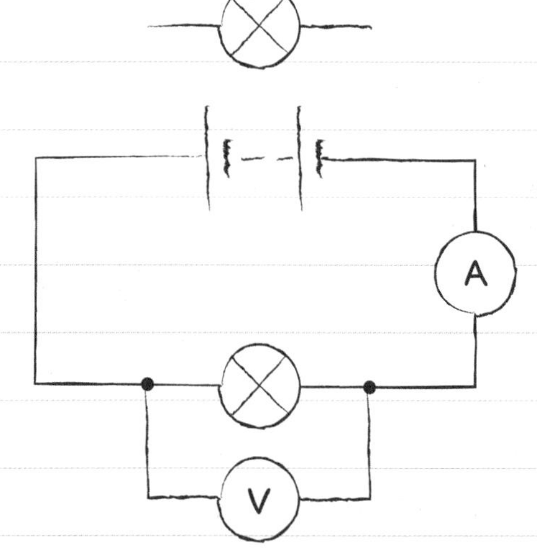

The circuit shows the ammeter in series (with the current in one single path) with the bulb and the voltmeter in parallel (with the current split into more than one path) to the bulb.

Resistor

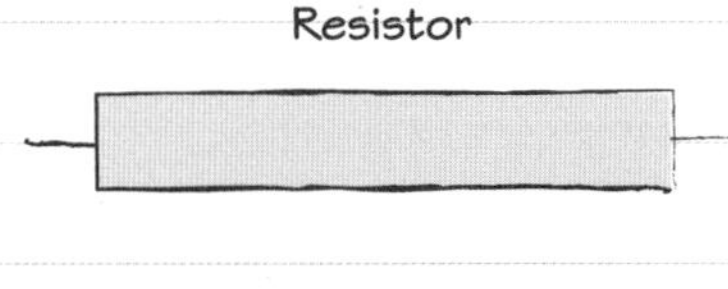

Variable resistor

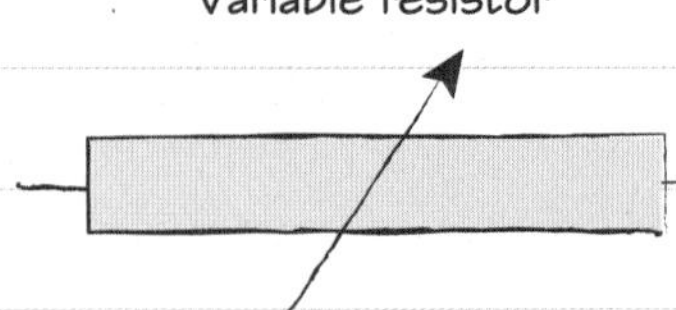

A **resistor** controls the current in a circuit.

A **variable resistor** is used to change the current in a circuit. It goes in series in the circuit.

Diodes

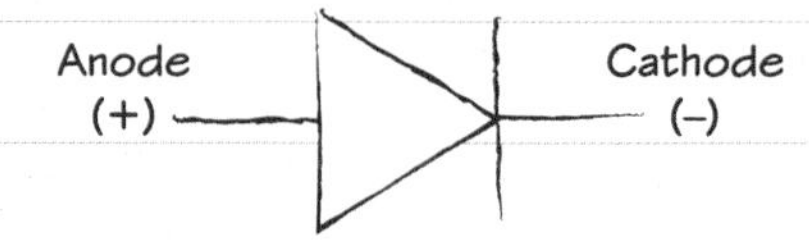

A **diode** only allows current to flow one way through a circuit.

It is in series in the circuit.

Now try this

Draw a circuit diagram containing all of the components on this page.

Only use one type of resistor in the circuit. Remember which components go in parallel and which go in series.

Had a look ☐ Nearly there ☐ Nailed it! ☐

Power equations

You need to understand concepts mathematically. To do this you use equations.

Power

Power is the rate of doing work. It is the **amount of energy consumed per unit of time**. It is another measure of energy, but here you calculate how quickly the energy is transferred.

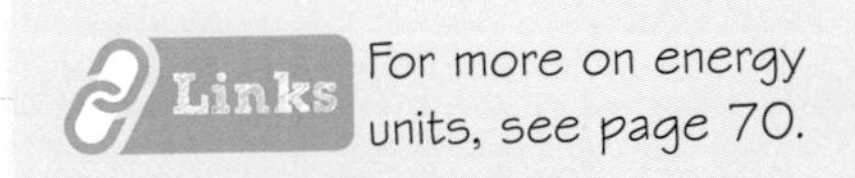

For more on energy units, see page 70.

Power can refer to **physical work**, for example, a car moving, or **electrical power** (the rate of energy consumption in a circuit).

Work done = the amount of energy supplied or transformed.

Electrical power

power = VI (voltage × current)

For example, the current flowing through a lamp is 2 A and the voltage is 240 V.

Then the power is 2 × 240 = 480 W

Power is measured in watts (W).

Physical power

$$\text{power} = \frac{\text{work done}}{\text{time}}$$

An engine does 1200 J of work in 60 seconds. What is the power of the engine?

$$\text{power} = \frac{1200}{60}$$

power = 20 W

Using a power triangle

You can use a triangle to help you with both the power calculations.

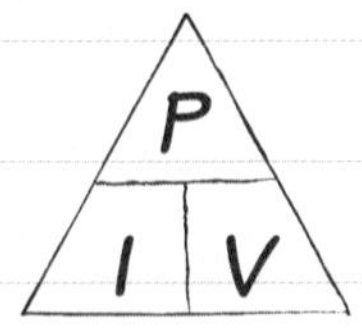

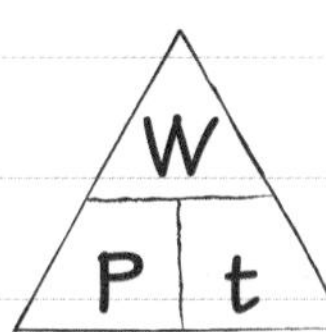

When you put your finger over the value you are calculating, the triangle shows you how to calculate it. So, putting your finger over power in the first triangle shows you that you have to multiply voltage and current to work it out.

Putting your finger over power in the second diagram shows you that you have to divide work done by time to get the value.

You should always show your working. This includes the equations you use and how you substitute and rearrange the numbers in the equations. The triangle is there to help you, but does not count as showing your working.

Now try this

1 Calculate the current flowing through a lamp with a power of 60 W when the voltage in the circuit is 120 V.
2 Calculate the work done by an engine that is using a power of 800 W in 60 seconds.

Start by writing out the formula triangle you would need, then write out the correct equation for the quantity you are trying to calculate. Substitute any numbers you have into the equation and use a scientific calculator to find the answer. Think about how many decimal places you will give in your answer, and make sure you double check each calculation to confirm your answer.

Energy usage

Different domestic appliances use different amounts of energy.

Washing machine

A washing machine uses 220 volts and 10 amps for a specific cycle.

This means its energy consumption is 2200 W.

Laptop

A laptop uses about 200 volts and 0.5 amps.

This means its energy consumption is 100 W.

Kettle

A laptop uses about 230 volts and 13 amps.

This means its energy consumption is 3000 W.

Toaster

A toaster uses about 220 volts and 9 amps.

This means its energy consumption is 2000 W.

Fuse

A **fuse** is a component of a circuit that will 'blow' if there is a fault in the circuit.

The wire in the fuse melts and breaks the circuit, stopping the flow of electricity.

Fuses are **rated** as different currents, for example, 3 A, 5 A, 10 A and 13 A.

The plug on an appliance should have a fuse that is rated the same or just above the current needed by the appliance.

Appliance	Current used	Suitable fuse rating
washing machine	10 A	10 A
laptop	0.5 A	3 A
kettle	13 A	13 A
toaster	9 A	10 A

Now try this

Complete the table.

Appliance	Current used (A)	Power (W)	Voltage used (V)	Suitable fuse rating (A)
printer	0.5	50		
hairdryer		2200	220	
tumble dryer	11	2500		

Had a look ☐ Nearly there ☐ Nailed it! ☐

Your Unit 3 set task

Unit 3 will be assessed through a task, which will be set by Pearson. In this assessed task you will need to complete an experiment and record your results, as well as processing these results and planning a separate investigation.

Set task skills

Your assessed task could cover any of the essential content in the unit. You can revise the unit content in this Revision Guide. This skills section is designed to revise skills that might be needed in your assessed task. The section uses selected content and outcomes to provide examples of ways of applying your skills.

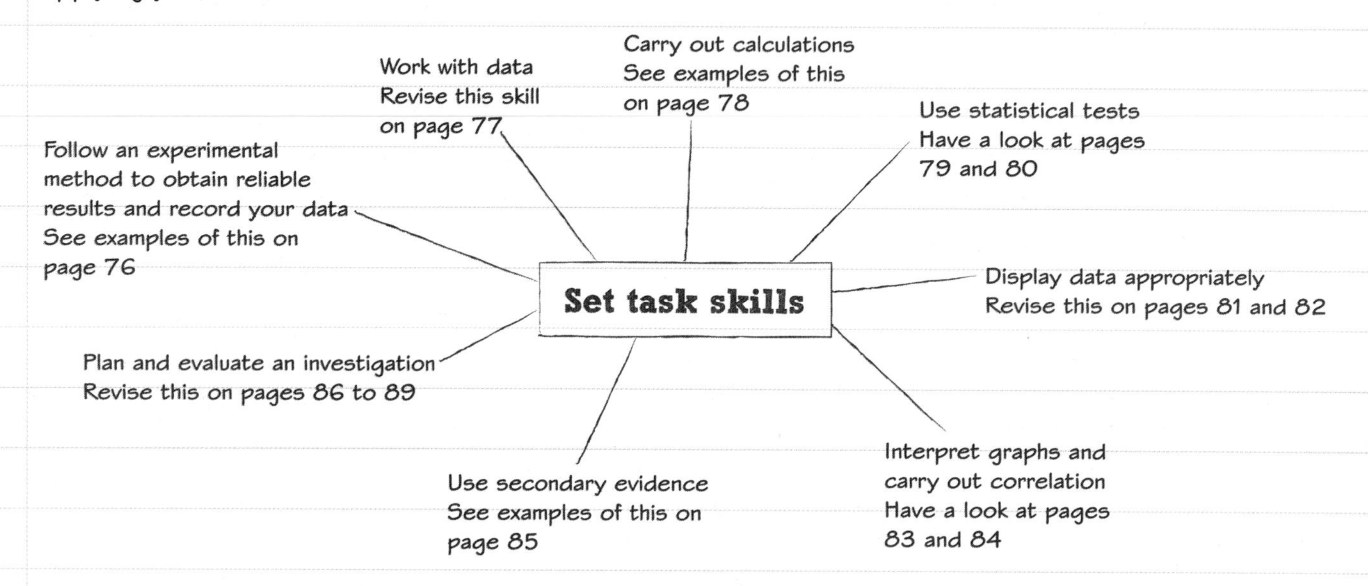

Checklist

For your assessed task, make sure you

- ✓ Have a black pen you find easy to write with and at least one spare
- ✓ Have a suitable calculator with you
- ✓ Follow any experimental method you are given
- ✓ Plan your time carefully
- ✓ Follow any safety instructions you are given
- ✓ Record your results carefully

Check the Pearson website

The questions and sample response extracts in this section are provided to help you to revise content and skills. Ask your tutor or check the Pearson website for the latest **Sample Assessment Material** and **Mark Scheme** to get an indication of the structure of the actual assessed task and what this requires of you. Details of the actual assessed task may change so always make sure you are up to date.

Now try this

Visit the Pearson website and find the page containing the course materials for BTEC National Applied Science.
Look at the latest Unit 3 Sample Assessment Material for an indication of:

- The structure of your set task, and whether it is divided into parts
- How much time you are allowed for the task, or different parts of the task
- What briefing or stimulus material might be provided to you
- The activities you are required to complete and how to format your responses

Types of task

You will complete a written task that is made up of two sections.

Section 1

This section is related to the practical work you will have carried out in advance. You will be given your observation sheets back to use. For this section, you will be asked to **process** your results in some way. This could include calculations, graphs and questions. You may be asked to analyse your results and use secondary data provided to support your conclusions.

Section 2

The second section involves writing a **plan** for an investigation that is not related to the one in the first section. You may also be asked to evaluate a given method, which will also not be related to the work you have done previously.

Data you will use

For both Sections 1 and 2, you may use:

- primary data – results you have practically obtained
- secondary data – results you have researched.

Links Primary and secondary data are described on page 57.

Command words

The questions you are asked will include verbs (command words), which give you instructions on what the question requires. For example:

Evaluate the method provided by the student for the experiment described.

Explain the results shown by the graph drawn. These verbs mean different things. Here are some examples:

Calculate – Work out a numerical answer. It is important to show your working and include the units if appropriate.

Convert – Change one unit to another e.g. g to kg.

Complete – Complete a missing part of a table or a diagram based on information provided.

Describe – Give an account of what the results of an experiment shows.

Evaluate – Review the infomation provided about the experiment and say what the strengths and weaknesses are. Suggest how it could be improved. Make a judgement about the quality of any data / data that will be obtained. Say whether the experiment is fully addressing the hypothesis.

Explain – You may be asked to say why the results of an experiment have been produced or to justify what could have caused an error to occur.

Identify – You may be asked to pick a piece of information out from a given resource, such as an image or a table.

Plot – You may be asked to produce a graph by marking data points accurately on a grid. A suitable scale and appropriately labelled axes must be included.

Now try this

Molly is carrying out an experiment to investigate the effect of changing the concentration of red dye on how fast the dye moves to the edge of a piece of filter paper. Molly starts by cutting the filter paper into 10 cm long strips, then she makes a mark on the paper and places it in the first concentration of red dye so that the red dye sits on the mark made. Molly then times how long it takes for the dye to move up to a second mark which she drew 5 cm up the paper. She repeats the same method for different concentrations of dye.

Evaluate this method:

(a) What are the good points (advantages) of this method?

(b) What are the bad points (disadvantages) of this method?

(c) Will the data collected be reliable?

Links You can revise evaluating an investigation on page 58.

Had a look ☐ Nearly there ☐ Nailed it! ☐

Recording data

Here are some examples of skills involved when you need to record experimental results in an appropriate format. These results could be made up of qualitative data, quantitative data, or both.

Links Look back at page 55. There is an example of a typical results table for quantitative data.

Worked example

Investigate the effect of temperature on enzyme activity.

Record your results/observations in the space provided. **3 marks**

Sample response extract

Temperature (°C)	Volume of oxygen given off (cm^3)
10	1
20	3
30	12
40	18
50	2

Clear headings with correct units.

All results recorded to the same precision. However, is this precise enough? Learners should consider how precise the measuring equipment is. It is unlikely that oxygen volume was always a whole number.

Only one set of results recorded? The result for 50°C looks unusual and yet no repeats have been recorded.

You have been given three compounds: A, B and C. Carry out flame tests on these compounds.

Record your results/observations in the space provided. **3 marks**

This learner was asked to complete a Part A method to conduct flame tests.

Sample response extract

Compound	Colour of flame
A	lilac
B	red
C	pink

Clear headings and no units are needed.

Several cations give a red flame. The learner should be more specific if possible, for example, brick red or crimson red.

You must be as specific as possible about your observations; this can include colour, smell, and appearance.

Now try this

You have been asked to investigate how the length of a wire affects the resistance of the wire.

Draw a results table you could use to record the results from this investigation.

Look at the results table on page 55 to give you ideas.

Dealing with data

You need to be confident to express data using decimal places, significant figures and standard form. This can be part of a calculation question during your assessment.

Decimal places

When carrying out calculations you may be asked to give your answer to a certain number of **decimal places**. To do this you:

- Look at the number after the one you are told to stop at. For example, if the question says three decimal places, look at the fourth number.
- If this number is 5 or above, then round the previous number up. If the number is 4 or below, the previous number stays the same.

Two significant figures

You have been asked to provide an answer to two decimal places and have calculated 11.679436.

The answer would be: 11.68

Here, you need to look at the third number, which in this case is 9. This number is above 5 so the previous number (7), is rounded to 8. Remember to include the units in your answer.

Significant figures

Significant figures can also be used to display numbers. When using significant figures, it is the **first non-zero** in the number that is important. If you are asked to give an answer to one significant figure, you look for the first non-zero number. If you are asked to give the answer to two significant figures, you look for the first two non-zero numbers, etc. All of the numbers after the significant ones are then written as zeros, or if the numbers come after a decimal place, they are ignored.

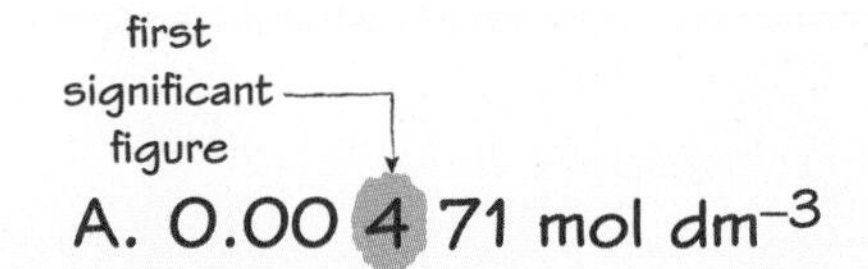

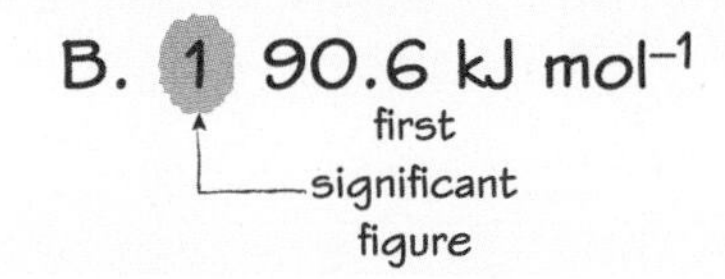

In the image above, 0.00471 mol dm^{-3} would be written as 0.005 to one significant figure. Can you work out what it would be to two significant figures? (0.0047)

190.6 kJ mol^{-1} would be written as 200 to one significant figure. Two significant figures would be 190. Make sure you don't forget the units.

Standard form

Numbers can be expressed using **standard form**. You use this when the number is very large (e.g. 19000) or very small (e.g. 0.000054). Standard form always has a number between 1 and 10, followed by a multiplication by a power of 10 (e.g. 5.4×10^{-5}). If the original number is big, the power is positive. The power shows the number of times the decimal place is moved to produce the number between 1 and 10.

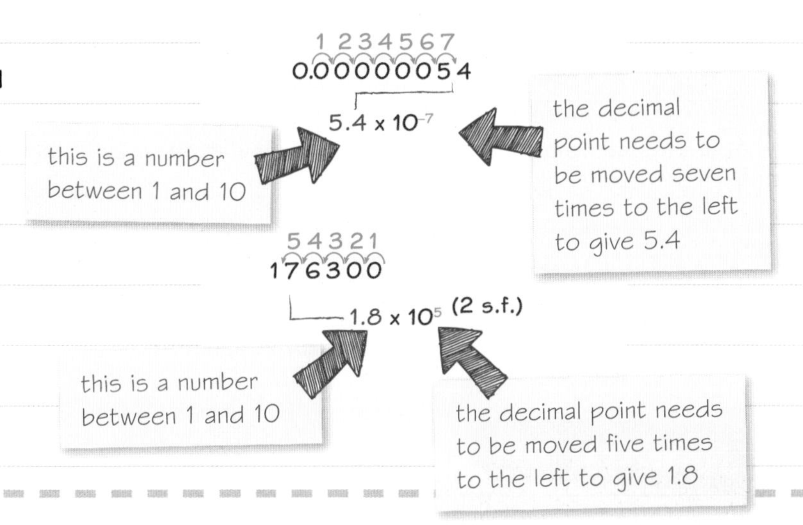

Now try this

1 Express the following to two decimal places and to two significant figures:
(a) 0.04942 (b) 168.472 (c) 0.1086 (d) 19464.6843

2 Express all of the above numbers using standard form and three significant figures.

Had a look ☐ Nearly there ☐ Nailed it! ☐

Calculations using data

You may be asked to carry out a variety of calculations based on your results or results given to you.

Calculating a mean

When you draw your results table, include a column for **mean values**.

Once you have recorded your data, you should calculate the mean result for each set of data.

To calculate the mean, you need to add together all of the repeated values and divide this by the number of repeats you carried out.

Fill the calculated means into a final column of your results table.

In your answers you should show your working for at least one example of calculating a mean.

Remember to ignore any **anomalous data** and do not include it in your calculation (unless you have repeated it and it is no longer anomalous).

Always make it clear in your answer which data is anomalous.

Standard deviation

This is a measure of how spread out the results collected are. **Standard deviation** is a measure of how reliable a set of data is. A smaller standard deviation indicates more reliable data.

Worked example

Calculate the standard deviation for the following data 10, 12, 14, 16. **6 marks**

Sample response extract

1 Calculate the mean $\frac{(10 + 12 + 14 + 16)}{4} = 13$.

2 For each number subtract the mean and square the result:
$10 - 13 = -3$, $-3^2 = 9$. $12 - 13 = -1$, $-1^2 = 1$. $14 - 13 = 1$, $1^2 = 1$. $16 - 13 = 3$, $3^2 = 9$.

3 Add up these values and divide by one less than your sample number $\frac{(9 + 1 + 1 + 9)}{4 - 1} = 6.67$.

4 The answer to this is the variance = 6.67.

5 Square root the variance to get the **standard deviation**, $\sqrt{6.67} = 2.58$.

Calculating the standard error

Calculating the **standard error** gives you the **accuracy** of your mean.

You can use your calculated standard deviation to calculate the standard error.

To calculate standard error, you divide the standard deviation by the square root of the sample number.

- The standard deviation of the sample 10, 12, 14, 16 is 1.29.
- Sample number is 4.
- Square root of 4 = $\sqrt{4} = 2$.
- 2.58/2 = 1.29; this is the **standard error**.

Now try this

An investigation was carried out into the energy released when ethanol is burnt. Each mass was tested 5 times.
The energy produced by burning 100 g of ethanol gave the following set of results:
3.7 kJ, 3.8 kJ, 3.9 kJ, 1.8 kJ and 3.7 kJ.

(a) Calculate the mean energy produced by 100 g of ethanol.
(b) Calculate the standard deviation and standard error for this set of results.

Remember to ignore anomalous results.

Statistical tests: Chi-squared

Chi-squared is a test that can be used to see if the difference between observed and expected numbers is significant or due to chance.

Null hypothesis

When you carry out chi-squared, you start by writing a **null hypothesis**. The null hypothesis states that there is **no significant difference** between observed and expected numbers.

$$\chi^2 = \sum \frac{(O - E)^2}{E}$$

This is the equation for chi-squared.

χ^2 = chi-squared O = observed
$\sum$ = sum of E = expected

Worked example

A researcher expects that there is no difference in the number of left-handed people compared to right-handed. The table shows the results of 400 people surveyed

Number of left-handed individuals	157
Number of right-handed individuals	243

If there is no difference, we would expect that there are 200 left-handed and 200 right-handed people.
Chi-squared would allow you see if the numbers observed are significantly different to these expected numbers.

Sample response extract

It is easier to carry out chi-squared calculations, using a table.

This is how a chi-squared calculation can be carried out on experimental data. A statistical test like this could form part of your assessment.

	1	2	3	4	5
	Observed	Expected	$O - E$	$(O - E)^2$	$\frac{(O - E)^2}{E}$
Left-handed	157	200	−43	1849	9.25
Right-handed	243	200	43	1849	9.25

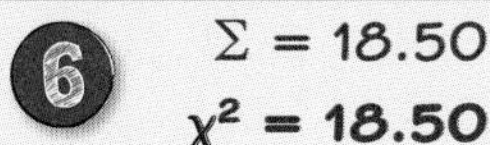

6 $\sum = 18.50$
$\chi^2 = 18.50$

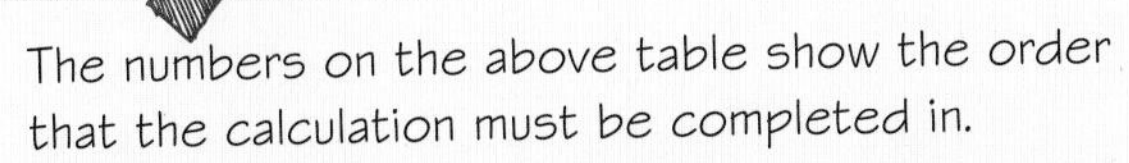

The numbers on the above table show the order that the calculation must be completed in.

Once you have your value of χ^2 you need to use a probability table to look up the critical value. You then compare your χ^2 to this critical value.

When do you use chi-squared?

- When you have data that has been counted and you know, or can calculate, what numbers you were expecting.

OR

- When you have expected numbers of organisms from genetic crosses and observed numbers from a sample of individuals.

Critical value

The critical value is the point at which the null hypothesis is rejected. To find the critical value we need to calculate the degrees of freedom. This is the number of items – 1. So, here, the degrees of freedom is (2 – 1) = 1.

Degrees of freedom	p = 0.05	p = 0.01
1	3.841	6.635
2	5.991	9.210
3	7.815	11.345
4	9.488	13.277
5	11.070	15.086

The critical value is 3.841, since we always use the 0.05 probability column.

If χ^2 is greater than the critical value, then the observed data is significantly different from the expected data.

If χ^2 is smaller than the critical value, then the observed data is not significantly different from the expected data. Any differences are due to chance alone.

Now try this

In the Worked Example question given, is there a significant difference between the number of left-handed and right-handed individuals observed and the number expected? Explain your answer.

Had a look ☐ Nearly there ☐ Nailed it! ☐

Statistical tests: t-test

A **t-test** can be used when you want to assess if there is a significant difference between the mean values of two sets of data. It is based on the null hypothesis that there is no significant difference.

Equation for carrying out a t-test for two sets of independent data

$$t = \frac{\bar{x}_1 - \bar{x}_2}{\sqrt{\left(\frac{s_1^2}{n_1} + \frac{s_2^2}{n_2}\right)}}$$

$\bar{x}_1$ is the mean of first dataset

$\bar{x}_2$ is the mean of **second** dataset

s_1^2 is the standard deviation of first dataset

s_2^2 is the standard deviation of **second** dataset

n_1 is the number of elements in the first dataset

n_2 is the number of elements in the **second** dataset.

We can use the t-test here, as the mean of heights of the two groups are independent of each other.

Worked example

You have been asked to compare the heights of a group of five German shepherd dogs and five Siberian husky dogs. Here are the heights of each of the dogs in the group:

Breed of dog	Height of individual dogs to the shoulder (mm)						
	1	2	3	4	5	Mean	Standard deviation
German shepherd	581	602	620	615	590	602	16
Siberian husky	585	580	610	585	590	590	12

Use the t-test to determine if there is any significant difference between the mean heights of the two groups. **5 marks**

Sample response extract

$$t = \frac{602 - 590}{\sqrt{\left(\frac{16^2}{5} + \frac{12^2}{5}\right)}}$$

$$= \frac{12}{\sqrt{80}}$$

$$= 1.342$$

Substitute the numbers into the above equation to calculate your t-test result. Use a scientific calculator and only round up when you have the final answer.

You then need to compare the t–test result to the critical value in a probability table.

1. Calculate the degrees of freedom. This is the number of measurements made – 2. In this example the degrees of freedom = (10 – 2) = 8.
2. Using this row of the probability table to find the critical value at 0.05 probability or 5% significant level.

Degrees of freedom	Significance level					
	20%	10%	5%	2%	1%	0.1%
1	3.078	6.314	12.706	31.821	63.657	636.619
2	1.886	2.920	4.303	6.965	9.925	31.598
3	1.638	2.353	3.182	4.541	5.841	12.941
4	1.533	2.132	2.776	3.747	4.604	8.610
5	1.476	2.015	2.571	3.365	4.032	6.859
6	1.440	1.943	2.447	3.143	3.707	5.959
7	1.415	1.895	2.365	2.998	3.499	5.405
8	1.397	1.860	2.306	2.896	3.355	5.041
9	1.383	1.833	2.262	2.821	3.250	4.781
10	1.372	1.812	2.228	2.764	3.169	4.587
11	1.363	1.796	2.201	2.718	3.106	4.437
12	1.356	1.782	2.179	2.681	3.055	4.318
13	1.350	1.771	2.160	2.650	3.012	4.221
14	1.345	1.761	2.145	2.624	2.977	4.140
15	1.341	1.753	2.131	2.602	2.947	4.073

Remember that if the calculated value of *t* is less than the critical value, there is no significant difference between the two means. If the calculated value of *t* is more than the critical value, there is a significant difference between the two means.

The calculated value of *t* is **smaller** than the critical value so there is **no significant difference** between the mean height of the German shepherd and the Siberian husky.

Now try this

Can you use the table to make a judgement on the probability that there is no significant difference between the mean height of the German shepherd and the mean height of the Siberian husky?

Displaying data (1)

You will be assessed on your ability to display data appropriately. This will include selecting and drawing the correct **graph** for your data.

Types of display

In your assessed task you will need to present data correctly.

You may need to draw:

- tables
- bar charts
- scatter graphs.

Categorical data

Categorical data is data in which the values fall into fixed categories, for example, favourite colour, blood group. This type of data can be discrete or continuous. Categorical sets of data are usually displayed as a bar chart. There should be should be spaces between the bars if the data is discrete, but no spaces if the data is continuous.

Numerical data

You will need to use different types of graph to display different types of data.

Numerical data can fall into one of the following types:

- **Discrete** – for this type of data, there are a fixed number of options, for example, the outcome of rolling a dice, the number of students.
- **Continuous** – this type of data can take any value within a range, for example, time, height, weight, volume.

Numerical sets of data are usually displayed using a scatter graph. Here, the independent variable is plotted on the x-axis and the dependent variable is plotted on the y-axis.

Worked example

The following set of data was collected:

Type of seed	Mean number of bird species feeding in 24 hour period
corn	4
thistle	5
sunflower	15
sorghum	9

Display the data correctly using a graph. 3 marks

Sample response extract

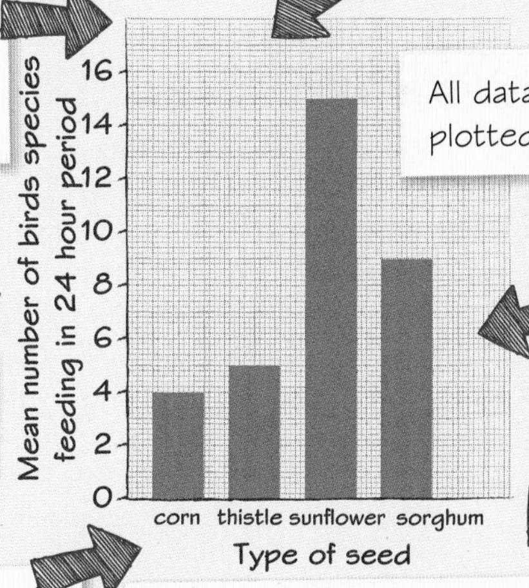

Mean number of birds species feeding in 24 hour period

0 2 4 6 8 10 12 14 16

corn thistle sunflower sorghum

Type of seed

Axis intervals are ascending and equidistant

Plotted area takes up more than 50% of graph paper

All data points are plotted correctly

Label on y-axis same as in table

Bars are all the same width

Label on x-axis same as in table with individual bars also labelled

Discrete categorical data so leave spaces of the same size between bars

Now try this

1 Using the information in the worked example and your knowledge, write a set of rules for drawing bar charts correctly.

2 Use these rules to display the following data

Type of bread	Diameter of mould growing after the use by date (mm)	
	48 hours	72 hours
white	6	9
wheat	2	5
rye	0	3

You will need two bars for each type of bread. Give a key to indicate which bars represent each time period.

Had a look ☐ Nearly there ☐ Nailed it! ☐

Displaying data (2)

When both your independent and dependent variables are numeric, you will need to display them using a line graph in order to assess the relationship between the two variables.

Marking

A graph will be marked against the following criteria:

- axes with correct labels with units
- axes with suitable, ascending and equidistant scales
- a plotted area that occupies more than 50% of the graph paper
- plots pointed correctly
- suitable line or curve of best fit.

Error bars

Error bars can be included on a graph to show uncertainty in the measurements. Error bars are usually the size of one standard deviation above and below the mean value. Error bars can also be drawn to represent the range of repeated readings taken for each plotted point.

The mean value is plotted on graph.

The top of the error bar is the standard deviation added to the mean value.

The botton of the error bar is the mean value minus the standard deviation.

Worked example

Plot a graph of volume of oxygen produced against distance of light source from plant using the graph paper provided. **3 marks**

Distance of light source from plant (cm)	Mean volume of oxygen produced (cm^3)	Standard deviation (cm^3)
10	33	1
20	27	2
30	20	2
40	7	2
50	1	0

Remember that you can use these standard deviations to plot error bars on a graph displaying the mean data.

Sample response extract

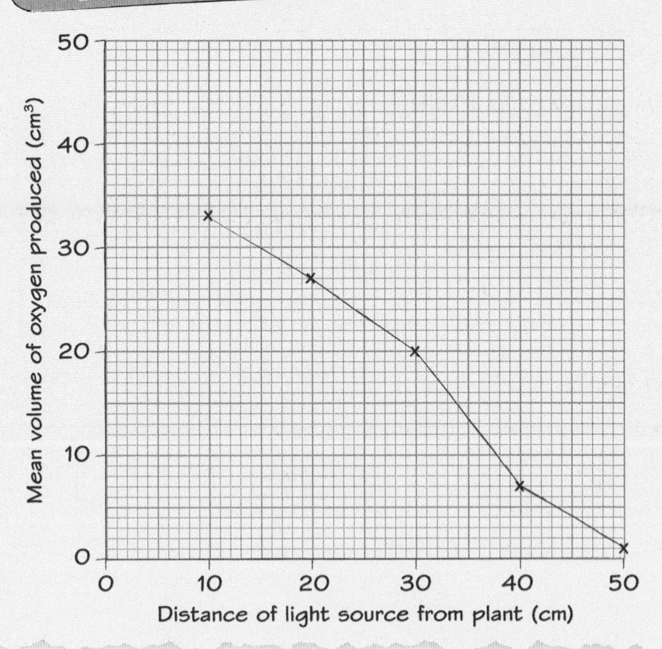

Now try this

Plot a graph of time taken to break down starch against temperature for the results in the table. Include error bars to show the standard deviation.

Temperature (°C)	Mean time taken to break down starch (seconds)	Standard deviation
20	240	6
30	180	4
40	12	1
60	86	3
80	200	8

Interpreting graphs

You will need to identify patterns and trends in your graph.

Worked example

Use the graph to describe the relationship between resistance of a thermistor and the temperature. **2 marks**

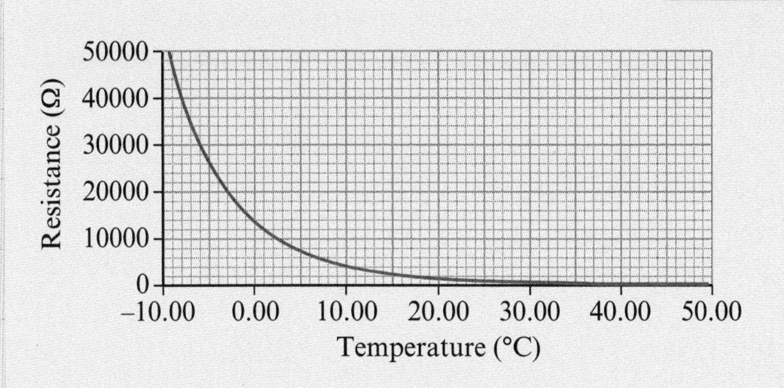

Sample response extract

The higher the temperature the lower the resistance, but as the temperature increases the difference in the resistance becomes less. Temperature and resistance are not directly proportional. Eventually the increase in temperature will make no difference to the resistance.

This is a good answer, as it describes what is happening at each part of the graph. It could be improved by using specific data from the graph.

Drawing a conclusion

When drawing a conclusion, you need to consider:

- The patterns in the graph.

When deciding if you can make a valid conclusion you need to consider:

- Do you have enough evidence?
- What are the percentage errors of your equipment?
- Did you deal with anomalies?

Error bars

Error bars can be drawn onto graphs.

They show the standard deviation at each plot.

These also give you evidence to state if your conclusion is valid or not.

Links For more information on error bars, see page 82.

Percentage errors

You need to calculate the percentage error for the equipment you used.

Worked example

Calculate the percentage error for the equipment you used to measure current. **1 mark**

Sample response extract

The ammeter measures in units of 1 amp (A).

Therefore the maximum error for each measurement is 0.5 (A)

$$\% \text{ error} = \frac{\text{maximum error} \times 100}{\text{measurement made}}$$

% error for a reading of 3 A = $\pm 0.5 \times 100/3$
= 17%

This answer is correct. You need to look at the divisions on the measuring tool. Half the distance between each division is as accurate as you can get.

Now try this

1. Look at the graph on page 56. Use the graph to describe the relationship between volume of oxygen produced and time.
2. Calculate the percentage error of a stopwatch that can measure to 0.01 seconds and records a time of 45.55 seconds. Give your answer to two significant figures.
3. What else would affect the error of a stopwatch?

Had a look ☐ Nearly there ☐ Nailed it! ☐

Correlation analysis

A **line of best fit** on a graph can show whether there is no **correlation** or a positive or negative correlation between variables. A **correlation coefficient** can be calculated to determine the **strength** of any correlation.

Types of correlation

When there is **no clear pattern** in your data, there is **no correlation**. This can be seen in graph (a).

If an **increase** in one variable (x-axis) causes an **increase** in the other (y-axis), then there is a **positive correlation**. This can be seen in graph (b).

If an **increase** in one variable (x-axis) causes a **decrease** in the other (y-axis), then there is a **negative correlation** as seen in graph (c).

A correlation is strong if most of the plotted data points are close to the line of best fit. Graph (d) shows a strong positive correlation compared to graph (b) where the positive correlation is weaker.

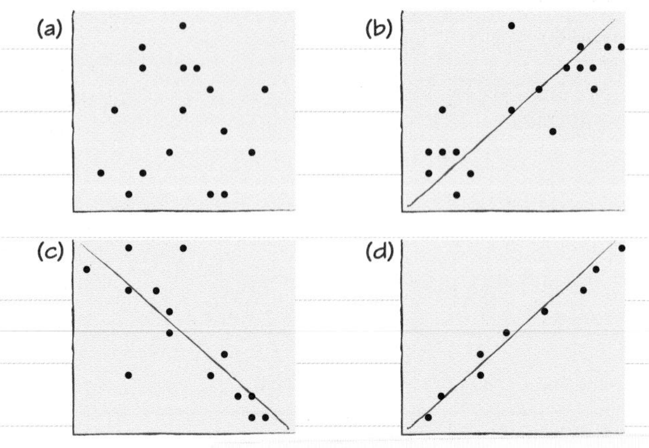

This shows the different types of correlation you may see in your data.

Correlation coefficient

A line of best fit tells you what correlation (if any) is present in your data. But to measure the strength of the correlation and to be sure of your conclusions, the correlation coefficient (r) can be calculated.

The formula for correlation coefficient is:

$$r_{xy} = \frac{\sum_{i=1}^{n}(x_i - \bar{x})(y_i - \bar{y})}{\sqrt{\sum_{i=1}^{n}(x_i - \bar{x})^2 \sum_{i=1}^{n}(y_i - \bar{y})^2}}$$

- Σ = sum of
- $(x_i - \bar{x})$ is each x-value minus the mean of x
- $(y_i - \bar{y})$ is each y-value minus the mean of y

You would probably use a computer program like Excel to calculate it for you but it is important you know how to interpret the results

1.00 r = 1.00. It is a strong positive relationship between the two variables.

0.50 r is greater than 0.00 but less than 1.00. It is a positive relationship between the two variables.

0.00 r = 0.00. There is no relationship between the two variables.

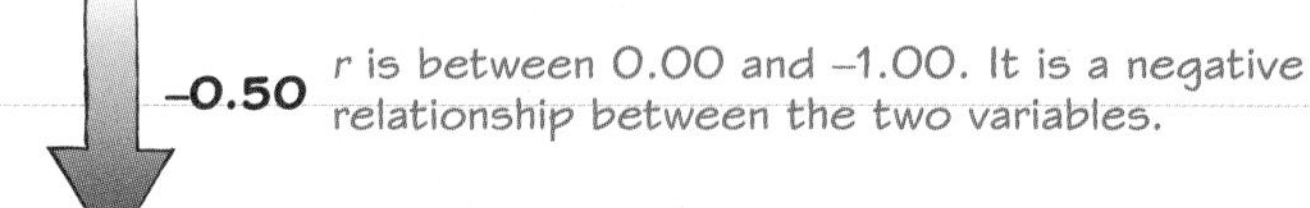

–0.50 r is between 0.00 and –1.00. It is a negative relationship between the two variables.

–1.00 r is –1.00. It is a strong negative relationship between the two variables.

Use this to help interpret correlation coefficient.

Worked example

In an experiment to investigate how distance from a light source affects the volume of oxygen produced by Elodea, the following graph was produced. Using the data obtained, the correlation coefficient determined was –0.98.

Comment on the correlation shown by the variable and explain your answer. **2 marks**

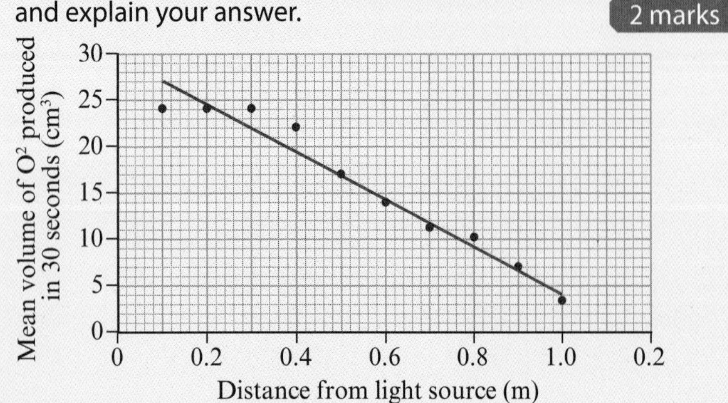

Sample response extract

The graph shows that there is a negative correlation between the distance from a light source and the volume of oxygen produced.

The correlation coefficient tells you that the relationship between the two variables is strong and you can therefore be confident in any conclusions.

Now try this

What do the following correlation coefficients tell you about the strength of the relationship between the variables investigated?

(a) Temperature and rate of reaction, r = 0.99.

(b) Length of index finger and intelligence, r = –0.13.

Using secondary evidence

You will need to use **secondary** evidence as well as **primary** evidence to back up any conclusions you make.

Examples of secondary evidence

- Melting point and boiling point values from data books
- Government reports
- Weather reports
- Manufacturer data
- Heat of combustion from data books.

Comparison of primary and secondary evidence

In order to compare primary and secondary data you need to know:

- If the practical investigations were carried out using similar or different methods.
- If the practical investigations were carried out using similar or different techniques/equipment.
- Who collected the data?
- Why was the secondary data collected?

Worked example

In an experiment, you have collected primary data on the effect of temperature on enzyme activity. You investigated the breakdown of hydrogen peroxide by the enzyme catalase at different temperatures. You measured the volume of oxygen produced in 30 seconds as a result of hydrogen peroxide break down. Your data showed that the enzyme activity increased with temperature, but that enzyme activity stopped completely after 40 °C as no oxygen was produced at this temperature.

You obtain some results from a scientific paper in which the effect of temperature on the breakdown of hydrogen peroxide by catalase was also investigated.

Give two reasons why there is a difference between your results and the scientific paper results for volume of oxygen produced. **2 marks**

Sample response extract

1. A less precise thermometer was used for the primary data.
2. The technique to measure enzyme activity was more accurate for the secondary data.

You could also say that the technique to raise the temperature and keep it at a steady temperature was more accurate for the secondary data.

Commenting on a conclusion

When you are commenting on a conclusion you should use all the **evidence**.

'Comment on' is a command word. It means you must synthesise (understand) a number of variables from data/information to form a judgement.

So, if you are commenting on a conclusion, you could use primary and secondary data to form a judgement.

Now try this

Plants grow faster when potassium and magnesium are present in the soil.

Find some secondary data showing the results of an investigation into the effects of magnesium and potassium on plant growth.

Comment on whether you think the statement is correct.

Writing a plan

Here are some examples of the skills involved in writing a plan for an investigation that you won't have seen before. Any new investigation you are given will have a title and some basic information to help you write the plan. Below is an example investigation.

Worked example

Effect of width on resistance of a wire

Materials with high resistance have low conductivity. Copper is a very good conductor and is used for circuits. It is useful for the copper wires to be thin so that the circuits can be small. However, it is important that enough electricity can flow through the wires for the equipment in the circuit to work.

You have been asked to write a plan for an investigation into the effect of different widths of copper wire on the resistance of the copper wire.

Your plan should include the following details:

- A hypothesis
- Selection and justification of equipment, techniques or standard procedures
- Health and safety associated with the investigation
- Methods for data collection and analysis to test the hypothesis including:
  - the quantities to be measured
  - the number and range of measurements to be taken
  - how equipment may be used
  - control variables
  - brief method for data collection analysis.

12 marks

Links Look back at page 50 to help you write your hypothesis.

Variables

Remember to consider:

- independent variable
- dependent variable
- control variables.

A sample response extract for this question can be found on page 87.

Links There is more information on variables on page 53.

Sample notes extract

Hypothesis

Thinner the copper wire = higher the resistance. So wider the wire = lower resistance. This is because current is affected by the flow of electrons. A reduced area (thinner wire) will mean less electron flow so lower current.

Equipment

Copper wire of at least 5 different widths to give good range of results. Keep length the same.

Ruler to ensure all the same length

Ammeter to measure current

Voltmeter to measure voltage

Bulb to check circuit is working

Variable resistor to keep current constant

Power pack

Circuit wires

Crocodile clips

Health and safety

Be careful of hot wires and electrical current.

Method

1 Set up circuit (give description)

2 Ensure ammeter in series and voltmeter in parallel with wire to be tested

3 Use the bulb to check circuit is working ...

This is part of a good initial **plan** for the answer. The full answer would need to give a detailed hypothesis, ensuring that the method has enough information to be followed.

Now try this

Try to answer the question on this page before comparing your answer with the one on page 87.

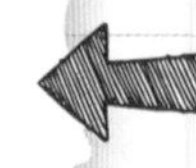

Make sure you answer the question fully. Do not miss any of the details that have been asked for in the question.

Constructing an answer

Below is one learner's answer to the question on page 86.

Sample response extract

Hypothesis

The thinner the copper wire the higher, the resistance. This is because current is a flow of electrons. In a thin wire, there is not much space for the current to flow so the resistance is higher.

The learner has tried to use science to explain their hypothesis. This is good practice, but could be more detailed.

Equipment

Copper wire of 5 different widths – this will allow me to see a trend in resistance as the width changes. Each wire will be 50 cm long as I am only changing width and not length.

Ruler – to measure length of copper wire

Ammeter in series with copper wire – to measure current

Voltmeter in parallel with copper wire – to measure voltage. I can use the voltage and current to work out resistance $V = IR$.

Variable resistor – to control the current and keep it constant

Power pack – to provide electricity for circuit

Circuit wires – to build circuit

Crocodile clips – to connect wire into circuit

It is clear what equipment is to be used and how. For this experiment, a circuit diagram could have helped. It is often a good idea to draw the arrangement of the equipment, as it is easier to see what you are planning.

Health and safety

There are few risks, but take care not to leave the circuit on too long as the wires can get hot and burn you.

It is important to cover health and safety. Some experiments have fewer hazards than others.

Method

I will set up a circuit containing the copper wire, ammeter, voltmeter, variable resistor, power pack and bulb. I will change the setting on the variable resistor to keep the current the same all the time. I will connect the first width of copper wire and ensure that the circuit is complete by making sure the bulb lights up. I will then record the ammeter and voltmeter readings. I will then disconnect the copper wire and allow the copper wire time to cool down as temperature can also affect resistance.

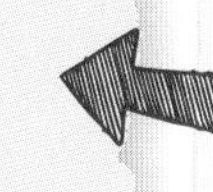

It would be better to write the method as a set of instructions. Think about a recipe in a cookery book. You should use the same style.

I will take these readings three times to ensure there are no anomalies and then I will repeat this for each different width of wire.

Once I have got all my results, I will work out the mean current and voltage reading for each width. I will then work out resistance for each width and then plot a graph of the width of the wire against the mean resistance.

Now try this

Mark your answer from page 86 using the suggested answers in the back of this revision guide. How could you improve your answer?

Why not mark the response by the student above against the same answers, too?

How could you improve this answer? It will also be helpful to check the mark scheme in the latest Sample Assessment Materials on the Pearson website to find out.

Had a look ☐ Nearly there ☐ Nailed it! ☐

Evaluation questions

Evaluate is a command word. It tells you what you are required to do in this type of question.

Evaluate

You must review information, then bring it together to form a conclusion, drawing on evidence including strengths, weaknesses, alternative actions, relevant data or information. You must come to a supported judgement.

Many learners find this a very difficult skill, but it can carry a lot of marks so it is worth practising. Read through the example evaluation question below and answer the Now Try This question at the bottom of the page.

Worked example

A learner investigates the effect of length of carbon chain in five alcohols on energy released during combustion.

The learner follows the following method:

- Measure the mass of a spirit burner filled with ethanol.
- Place it on a heat-proof mat under a calorimeter filled with water.
- Light the wick.
- Stir the water in the calorimeter at regular intervals.
- Extinguish the flame.
- Record the temperature of the water.
- Measure the new mass of the spirit burner and ethanol.

The learner repeats the steps for four other alcohols.

The diagram shows the setup of the equipment.

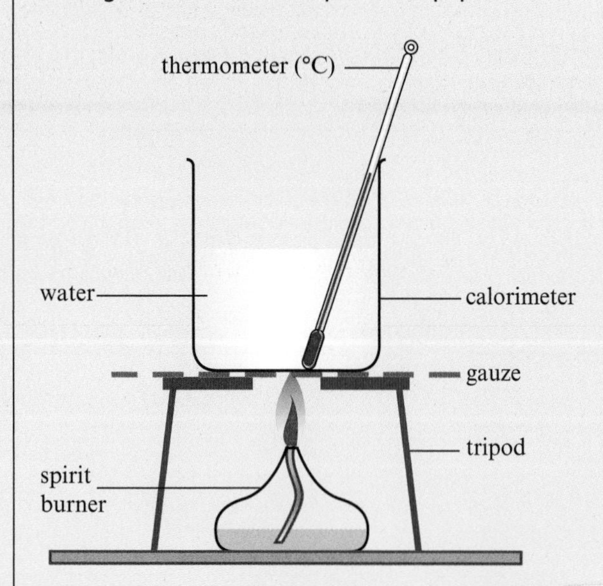

The results of the learner's investigation are shown in the table.

Alcohol	Mass used up during combustion (g)
methanol	1.98
ethanol	1.75
propan-1-ol	1.68
butan-1-ol	1.64
pentan-1-ol	1.54

The learner concludes that the longer the carbon chain in an alcohol the more energy it gives off per 1 g.

Evaluate the learner's investigation.

Your answer should include reference to:

- the method of the experiment
- the results collected
- the conclusion made.

8 marks

A sample response extract for this can be seen on page 89.

Now try this

Try answering the question on this page. Evaluate the method provided.

Read the information on page 58 about evaluating an investigation to help you.

Answering evaluation questions

Below is one learner's answer to the question on page 88. To gain high marks you must link ideas and explain and justify your comments.

Sample response extract

There are problems with the method. It is not clear how much water is in the calorimeter. A volume should be chosen and this should be measured and kept the same for each alcohol. It is not clear when the learner should stop heating. Is it after a set time or when the temperature has gone up by the same amount? I think they should say stop heating when the temperature has gone up by 10 degrees. They need to measure the temperature of the water before and after heating. At the moment, they cannot get results that they can compare.

The learner is evaluating the method. They have stated the problems with the method. They have also given alternative actions to improve the method. They haven't said what is good about the method and this would gain more marks.

They have only taken one set of results for each alcohol and so do not know if these results are correct. They should repeat each result at least once to check it and more often if they observe any anomalous results.

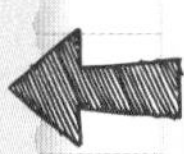

They have commented briefly on the results and said how these could be improved. It would be useful to say what is good, if anything, about the results.

At the moment they cannot make a conclusion. If they had heated the water to the same temperature rise each time then their conclusion would be correct based on the results that they have. However, even with this information there is not enough data to definitely state that their conclusion is true.

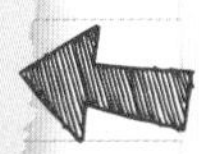

Again they have said what is weak about the conclusion and given a suggestion to improve it. They should list exactly what information they would need to make this conclusion.

They also haven't said in their method how they will prevent heat loss to the air. And they haven't controlled the variables. They need to make sure the flame is the same distance from the calorimeter for each alcohol and they could put some sort of cover round the experiment to stop heat being lost to the atmosphere. The flame should be as close as possible to the calorimeter to prevent heat loss.

This section of the answer might be better with the evaluation of the method above. However, some good points have been made especially about controlling variables.

Now try this

Evaluate the following method produced by a student.

'To investigate the effect of pH on amylase, I will use iodine to test if starch agar has been broken down into maltose by the amylase. I will put filter paper discs containing amylase at different pHs onto agar plates containing starch agar. I will leave these for a bit and then add some iodine. I will see how big the yellow/brown circles are around each disc as the yellow/brown circles will show where starch has been broken down. If the circles are blue/black the starch has not been broken down and therefore the enzyme hasn't worked.'

Links Use what you have learned about evaluation from pages 58, 88 and 89.

The heart

The heart consists of four chambers (pumps). The right and left sides are completely separate, but there are valves between the top chambers (atria) and bottom chambers (ventricles).

A simple model of the four-chambered heart

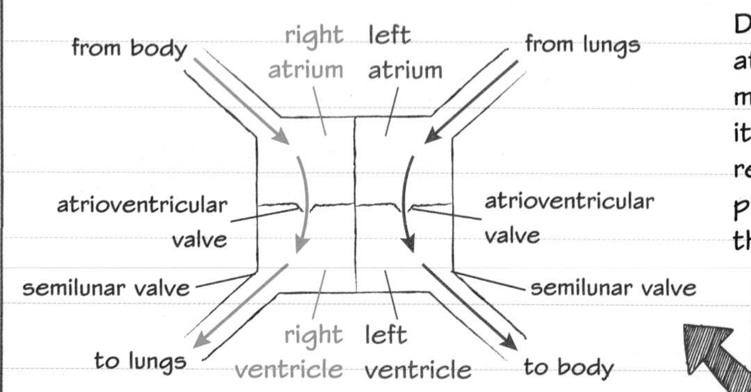

Deoxygenated blood (blue) enters the right atrium from the body. It is then pushed, by muscles in the right ventricle, to the lungs where it is oxygenated. The oxygenated blood (red) returns from the lungs to the left atrium, is pushed into the left ventricle and then out along the main blood vessel of the body, the aorta.

You need to know the **structure** of the heart and understand how the structure relates to how it **functions**. It might help you to think of the heart as a box with four chambers and four pipes: two going in and two coming out. In addition, the top two chambers are connected to the bottom two by valves.

The anatomical structure of the heart

The heart has a more complicated structure than the four-chambered box at the top of the page.

The heart muscles are myogenic, meaning that they contract spontaneously, without being stimulated by nerve cells.

right pulmonary artery (carries deoxygenated blood to right lung)

aorta carries oxygenated blood to the body

superior vena cava carries deoxygenated blood from the upper body

left pulmonary artery (carries deoxygenated blood to left lung)

right pulmonary vein (carries oxygenated blood from right lung)

left pulmonary vein (carries oxygenated blood from left lung)

sinoatrial node (SAN), heart's pacemaker, responsible for the regular contraction of the heart muscle

left atrium

semilunar valve

semilunar valve (valves prevent backward flow of blood)

right atrium

atrioventricular (tricuspid) valve

atrioventricular (bicuspid) valve

bundles of His – transmits impulses from the AV node to the ventricles

inferior vena cava carries deoxygenated blood from the lower and middle body

thinner cardiac muscle of **right ventricle**

aorta

Purkinje fibres (send nerve impulses to the ventricles of the heart)

thick cardiac muscle of **left ventricle**

septum divides left and right side of the heart

Key
→ flow of oxygenated blood
→ flow of deoxygenated blood

Cardiac output

Cardiac output is the volume of blood pumped by the heart per minute. The heart rate is the number of heart beats per minute. The stroke volume is the volume of blood in millilitres, pumped out of the heart at each beat.

For example:

cardiac output = heart rate × stroke volume

cardiac output = 70 × 132

= 9240 ml/minute

Now try this

Explain why the mammalian heart has two sides.

You need to make sure you relate structure to function.

Blood vessels and types

The mammalian circulatory system is a transport system. The blood vessels are like roads and the blood is like the vehicles.

There are many blood type systems, which are inherited.

Links For the structure of blood vessels, see page 8. For more on the heart, see page 90.

Structure and function in arteries, veins and capillaries

Vessel	Structure	Function
arteries	• relatively thick wall • smooth muscle • elastic fibres • lined with smooth layer of endothelial cells • narrow lumen	• withstand high blood pressure • alter diameter of lumen to vary blood flow • allow walls to stretch when blood is pumped into the artery and then recoil, smoothing blood flow • low friction surface to ease blood flow
capillaries	• very thin wall (just one cell thick) • small lumen	• allow rapid exchange between blood and tissues • link arteries and veins
veins	• relatively thin wall • very little smooth muscle or elastic fibres • wide lumen • valves	• blood under low pressure • no pulse of blood so no stretching and recoiling • large volume acts as blood reservoir • valves stop backflow, ensuring a one-way flow of blood toward the heart

Valves in veins

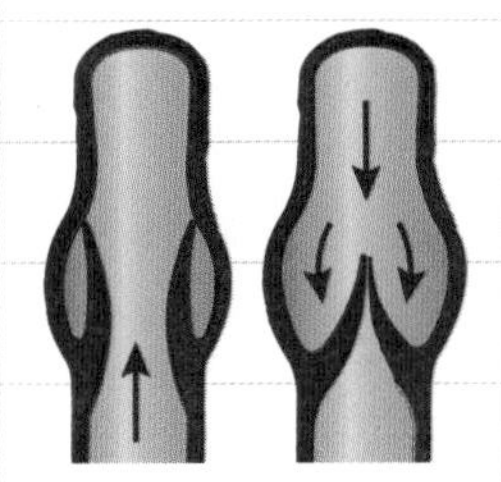

Veins have valves, which are not found in arteries or capillaries (except where arteries leave the heart).

Universal donor

People with O Rh-negative are called universal donors. They can donate to all other blood types. People with this blood group have no A, B or rhesus antigens, so the recipient's blood cell antibodies will not cause the donor's, blood cells to clump together, or agglutinate, which can be fatal.

Transfusions using O Rh-negative blood only happen in emergencies, as there is still a risk involved. Normally, patients will receive blood with the same Rh and ABO group as their own.

Blood type systems

The ABO type system is the most important. A and B are types of antigens on red blood cells (O type is when neither A nor B antigens are present).

	Group A	Group B	Group AB	Group O
Red blood cell type	A	B	AB	O
Antibodies in plasma	Anti-B	Anti-A	None	Anti-A and Anti-B
Antigens in red blood cell	A antigen	B antigen	A and B antigens	None

universal recipients (Group AB) — universal donors (Group O)

The Rhesus system (Rh)

This system is linked to the ABO system and classifies blood as either Rh-negative (rhesus factor, an inherited blood protein not present on red blood cells, first discovered in the Rhesus monkey) or Rh-positive (present). People with Rh-positive blood can receive Rh-negative blood, but people with Rh-negative blood cannot receive Rh-positive blood.

Now try this

Describe **two** differences between the structure of a capillary and the structure of a vein.

Had a look ☐ Nearly there ☐ Nailed it! ☐

The cardiac cycle and the heartbeat

The heart goes through a cyclical process about 60 to 70 times a minute, receiving oxygenated blood from the lungs and pumping it around the body.

The cardiac cycle in the left side of the heart

Blood drains into left atrium from lungs along the pulmonary vein. → Raising of the blood pressure in the left atrium forces the left tricuspid valve open. → Contraction of the left atrial muscle (left atrial systole) forces more blood through the valve. → As soon as left atrial **systole** (muscle contraction) is over, the left ventricular muscles start to contract. This is called left ventricular systole. → This forces the left tricuspid valve to close and opens the valve in the mouth of the aorta (semilunar valve). Blood then leaves the left ventricle along the aorta.

Cardiac diastole is when the heart refills with blood. Ventricular diastole is when the ventricles are refilling and relaxed.

The same steps are repeated on the right side at the same time.

Electrocardiogram

The electrical changes in the heart can be measured and presented as an **Electrocardiogram** (ECG).

If disease disrupts the heart's normal conduction pathways there is a disruption of the expected ECG pattern. This can be used for diagnosis of cardiovascular disease.

The QRS complex is the time of ventricular systole.

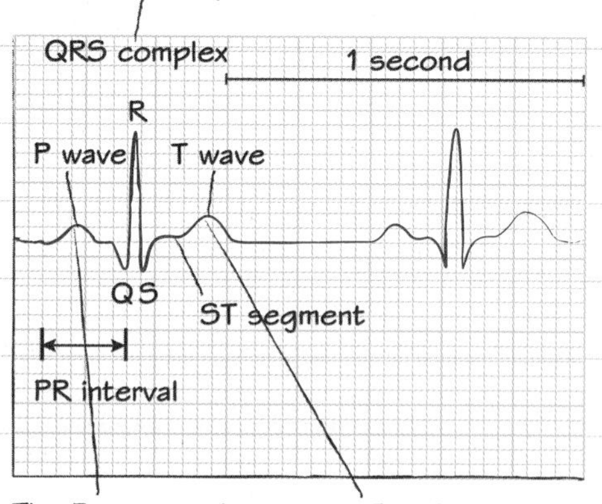

The P wave is the time of atrial systole.

The T wave is caused by repolarisation of the ventricles during diastole.

Electrical changes in a normal heart during one cardiac cycle

ECG waves

Normal rhythm: between 60 and 100 beats per minute.

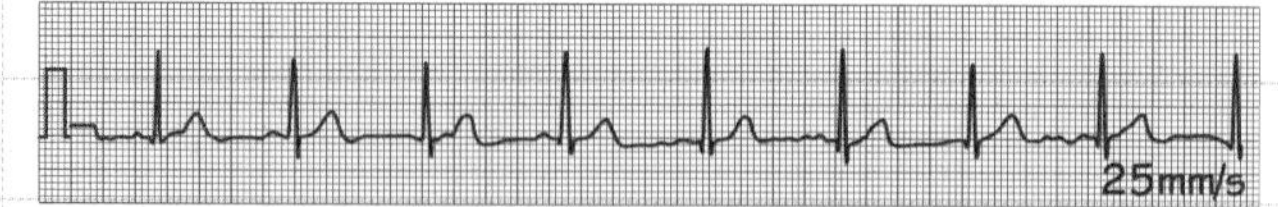

Sinus arrhythmia: normal beats but triggered at an irregular interval.

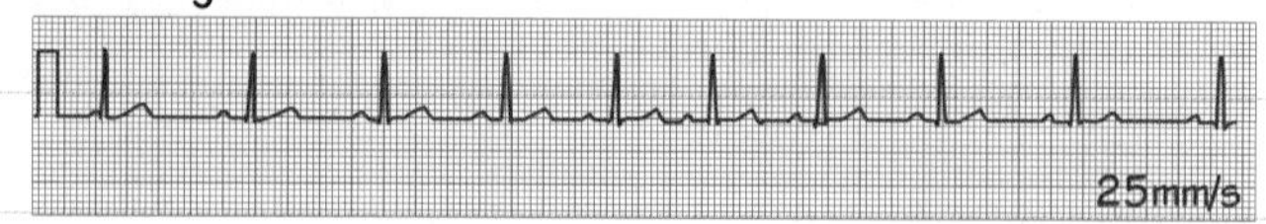

Bradycardia: less than 60 beats per minute.

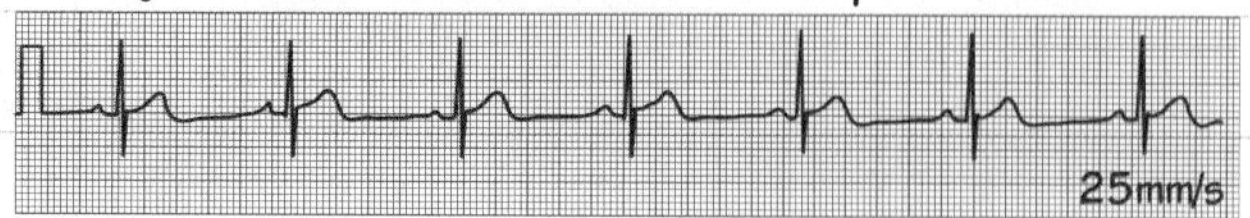

Tachycardia: more than 100 beats per minute.

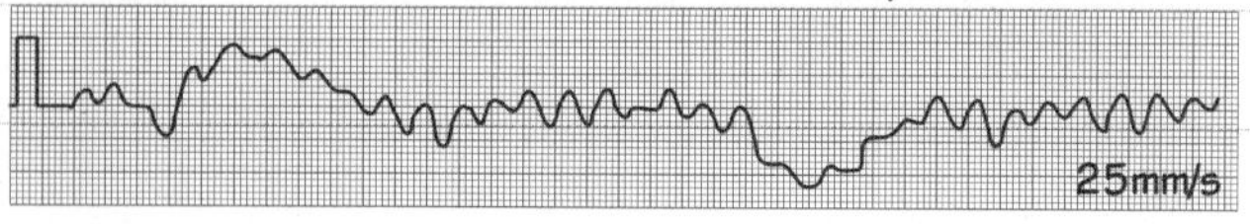

Ventricular fibrillation: irregular ventricular rate.

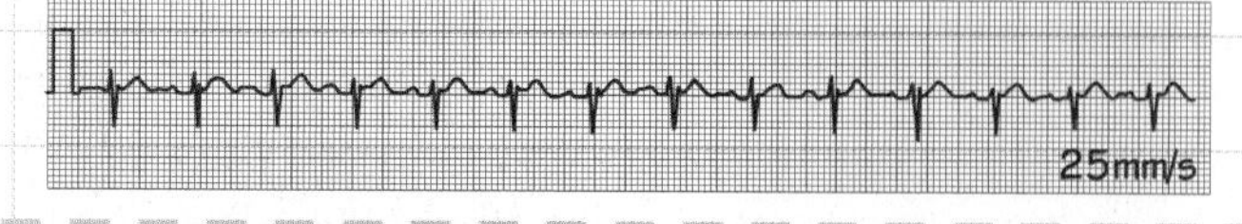

A flat line: there is no signal and indicates that resuscitation is needed, or death results.

Now try this

Describe the path of a red blood cell through the heart and lungs from when it enters the heart from the vena cava.

Cardiovascular disease (CVD): Risks and treatment

Lifestyle choices and factors, such as age, gender and genetics, can increase the risk of cardiovascular disease (CVD).

Factors that increase the risk of CVD

risk factors for CVD:

- **diet** – many correlations between dietary habits and levels of CVD, e.g. lipoprotein and salt levels; some might be causal, particularly for blood cholesterol levels
- **high blood pressure** – very important – should not be above 140 mm Hg systolic and 90 mm Hg diastolic
- **genetics** – inherit tendency to: high blood pressure; poor cholesterol metabolism; arteries that are more easily damaged; relative HDL : LDL levels in blood
- **smoking** – correlation and causation shown because chemicals in smoke physically damage artery linings and also cause them to constrict
- **inactivity** – regular vigorous exercise reduces the risk of CVD by reducing blood pressure and raising HDL (good cholesterol) levels
- **gender** – oestrogen gives women some protection from CVD before the menopause
- **age** – elasticity and width of arteries decrease with age

Treatments for CVD

Treatment	Benefit	Risk
anti-hypertensive (diuretics, calcium channel blockers)	reduces high blood pressure	occasional dizziness, nausea, cramps
statins	reduces LDL (low density lipoprotein cholesterol) by inhibiting enzyme in the liver	tiredness, disturbed sleep, nausea, diarrhoea, headache, muscle weakness also people may depend wholly on statins and neglect to eat a healthy diet
transplantation and immunosuppressants	properly functioning heart	risk of rejection, and immunosupressants reduce immunity so increasing risk of infections

Lifestyle impact

High blood cholesterol leads to fatty deposition in artery walls. This causes the lumen of the coronary arteries to narrow.

High blood pressure damages the lining of the arteries and increases the risk of a blood clot blocking coronary arteries.

Smoking increases blood pressure and increases the risk of aneurysm which is a swelling in the wall of a blood vessel.

Links For more on cholesterol, see page 8.

Now try this

There are two types of cholesterol: high density lipoprotein (HDL) and low density lipoprotein (LDL). The Framingham study investigated the link between each type of cholesterol and coronary heart disease (CVD). CHD is a form of CVD that affects the coronary arteries.

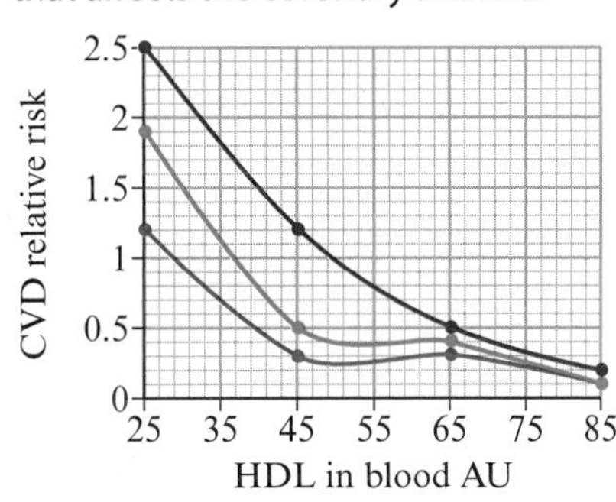

(a) Analyse the data in the chart to explain the interaction between LDL cholesterol and CVD and HDL cholesterol and CHD relative risk.

(b) State how you could reduce the relative risk of CHD through your intake of cholesterol.

Remember that, to compare LDL and CVD, you need to compare the different levels when HDL is the same, so look at one point on the *x* axis (e.g. 25AU). To compare HDL with CVD, LDL should be the same, so look at one line at a time (e.g. Medium LDL).

Daphnia heart rate

Some dietary components, such as caffeine, may have an effect on the heart, but this is not easy to study in humans for ethical reasons. Instead, animals such as the water flea (or *Daphnia*) can be used to assess the effects of dietary components.

Ethical issues

Daphnia	Human
simple nervous system	complex nervous system
no need for dissection, as transparent and can see heart beating	needs dissection
abundant in nature	cannot kill for experiment
bred for fish food	
cannot give consent	can give consent

Why use *Daphnia* for experiments?

- ✓ abundant
- ✓ easily obtained
- ✓ transparent, so heart can be seen
- ✓ simple nervous system, so ethically less of an issue than a mammal.

Some limitations

Although simple animal models such as *Daphnia* are useful to get ideas about how human body systems work, it must be remembered they are not human. Conclusions drawn from such experiments should be considered carefully when extended to humans.

Ensuring a successful experiment

Step in experiment	Method
immobilise the *Daphnia*	use strands of cotton wool in a small dish of the experimental solution to trap the *Daphnia*
control other variables, such as water temperature and *Daphnia* size	difficult to maintain a constant temperature but it should be monitored with a thermometer in the water; *Daphnia* of similar size and age should be used for all experiments
accurate measurement of heart rate	dots are put on a piece of paper (in an S shape to avoid putting one dot on top of another) or repeatedly press a button on a calculator
repeatability	ensure that variables other than caffeine concentration are controlled

Variables and safety

The variables to be considered are:

- ✓ temperature
- ✓ age, size and sex of *Daphnia*
- ✓ aspects of pretreatment such as type of water, length of time out of natural habitat.

The safety measures to be taken are:

- ✗ do not mix water and electricity
- ✗ don't forget to wash your hands after handling the *Daphnia*.

Choosing a temperature

If an experiment is investigating the effect of a substance, such as caffeine or ethanol (alcohol), on heart rate in *Daphnia*, it is important to choose a sensible temperature. This needs to be kept constant, as it avoids the heart rate changing due to temperature change. 25 °C is a good choice as it ensures a good level of Daphnia acivity without leading to enzyme denaturation.

Now try this

Explain ways in which the accuracy of this experiment might be improved.

The human lungs

Most living organisms gain their energy from aerobic respiration. This produces adenosine triphosphate. This process needs oxygen to be brought into the body and carbon dioxide to be excreted. This happens by the ventilation system, which includes the lungs.

Links For more on epithelial tissue in the lungs, see page 7.

The lungs and ventilation system

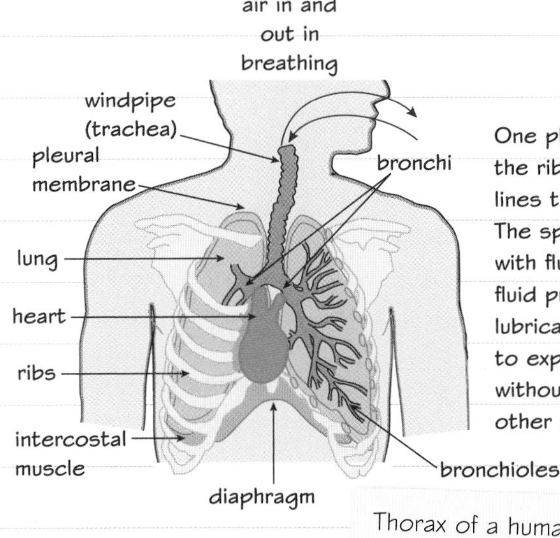

One pleural membrane lines the rib cage and the other lines the outside of the lungs. The space between is filled with fluid. The membranes and fluid provide protection and lubrication, allowing the lungs to expand and contract easily without being damaged by other organs.

Thorax of a human showing lungs

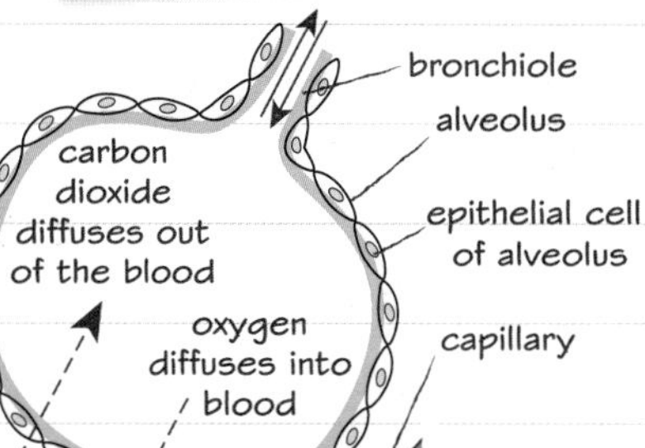

A single alveolus showing gas exchange by diffusion

Features of the gas exchange system

1. Very large surface area: adult human has about 500–700 million alveoli (70–100 m^2)
2. Very thin gas exchange surface: alveoli walls and capillary walls are both only one cell thick and made up of squamous epithelium, which are flat cells.
3. A good blood supply: alveoli are well supplied with a network of capillaries, so oxygen is continually taken away from the alveoli by red blood cells in the blood and carbon dioxide is continually being brought to the alveoli. This maintains a high **concentration gradient** of both gases.

Ventilation of the lungs by inspiration and expiration (breathing) means that oxygen is brought into the lungs and carbon dioxide excreted.

Gas exchange

In the alveoli, there is a high concentration of oxygen and a low concentration of carbon dioxide. Oxygen dissolves in the layer of moisture, so it can diffuse through cell membranes. In the blood is a low concentration of oxygen, so the oxygen diffuses into the blood (high to low). The blood has a high concentration of carbon dioxide, so diffuses from the blood into the alveoli (high to low).

Now try this

The diagram shows a small part of the lungs where gas exchange occurs.

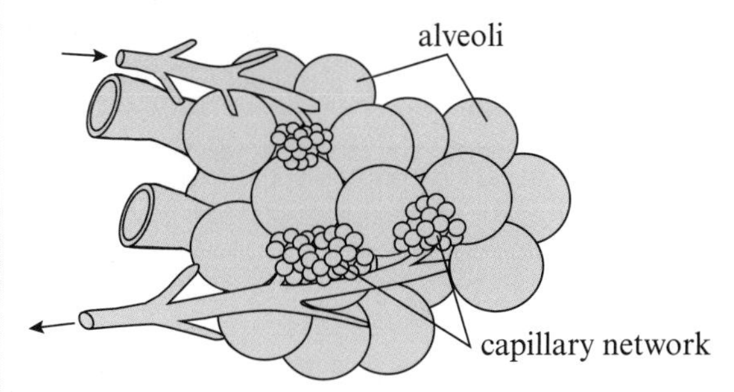

Describe how the structures shown are adapted to this function.

Had a look ☐ Nearly there ☐ Nailed it! ☐

Lung ventilation

Air gets into and out of the lungs by inspiration (breathing in) and expiration (breathing out).

Inspiration

inspiration
intercostal muscles
diaphragm

1. Diaphragm contracts and moves down to become flat.
2. If more oxygen required, for example, when exercising, external intercostals contract and move rib cage upwards and outwards.
3. Volume of thoracic cavity (chest) increases.
4. Pressure decreases in lungs compared to outside.
5. Air rushes in.

Expiration

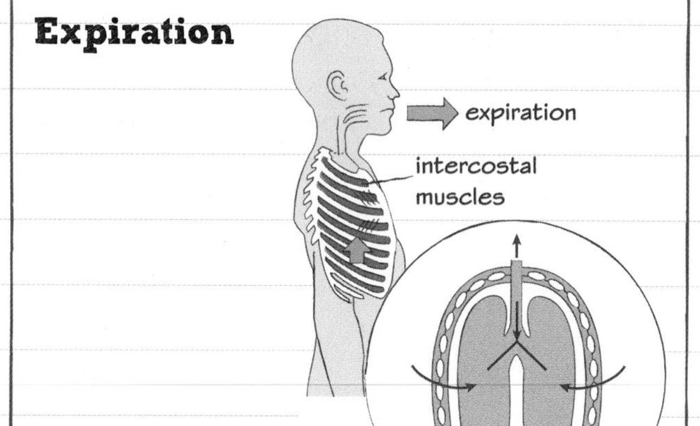

The elastic fibres in the alveoli shrink, increase pressure and air is squeezed out.

1. When we are exercising, internal intercostals contract and move rib cage inwards and downwards.
2. Abdominal muscles contract and the stomach and liver push diaphragm back so it is domed again.
3. Volume of thoracic cavity (chest) decreases.
4. Pressure increases.
5. Air rushes out.

Mechanical ventilation

If a person is not able to ventilate their lungs themselves (breathe), they are assisted by using a mechanical ventilator. Mechanical ventilators usually require power and are controlled by a computer.

Air is pumped into the patients' trachea through an endotracheal tube (inserted through the mouth) or a tracheostomy tube, which is inserted through the skin straight into the trachea. This increases the pressure (positive pressure) allowing air to flow into the airways until the end of the ventilator breath. The pressure then drops to zero. The chest and lungs contract and push the air in the lungs out through passive expiration.

Pressure changes in the lung

If a lung is punctured, there is a hole in the thoracic cavity. Air would enter the thoracic chamber and increase the pressure outside the lungs. The pressure also increases inside the lung. This means air no longer flows in and out.

Now try this

Which of the following statements is true?

A During inspiration, the diaphragm contracts.
B During inspiration, the internal intercostals contract.
C During expiration, the abdominal muscles relax.
D During expiration, the external intercostals contract.

Measuring lung volumes

A device called a spirometer is used to measure different aspects of how the lungs are working.

The spirometer

A person using a **spirometer** breathes in and out of an airtight chamber, causing it to move up and down.

These movements can be recorded on a revolving drum or to a computer.

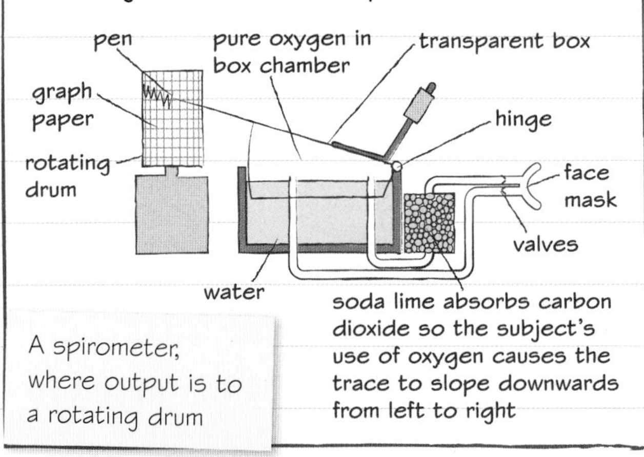

A spirometer, where output is to a rotating drum

The terminology

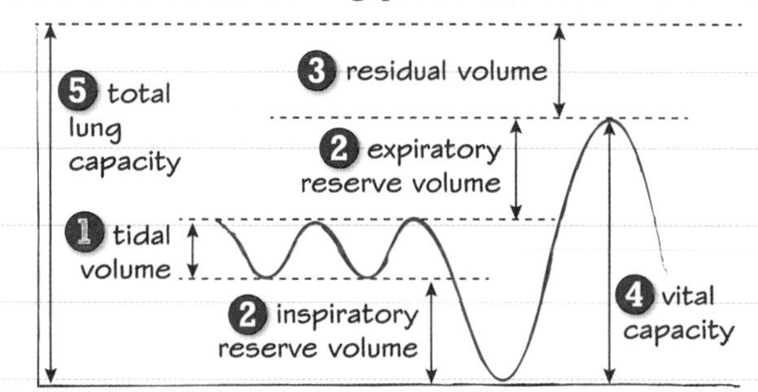

1. is the normal breathing volume, about 0.5 dm^3
2. are the maximum amount you can breathe in and out
3. is the air that is in the lungs after you have fully breathed out
4. is the total volume from fully breathed in to fully breathed out
5. is the vital capacity plus the residual volume.

Peak expiratory flow test

This is a measure of how fast you can breathe out. A person takes a full breath in and blows out as fast as they can into a peak flow meter, which is a small hand-held plastic device. Peak flow is measured in $dm^3\ min^{-1}$.

Forced vital capacity (FVC)

You use a spirometer to measure the amount of air that can be forcibly exhaled from the lungs after taking the deepest breath possible. A lower than normal FVC indicates a range of respiratory diseases.

Reading a spirometer trace

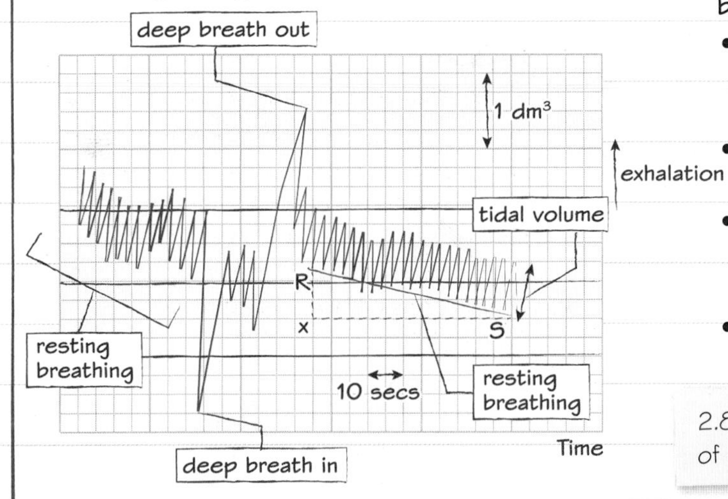

From the spirometer trace the following can be calculated:

- **Tidal volume** is the height of the peaks after the deep breath out.
  $2.8 \times 0.25 = 0.7\ dm^3$
- **Breathing rate** = 22/55 s
  = 0.4 breaths sec^{-1} = 24 breaths min^{-1}
- **Respiratory minute ventilation** is tidal volume × breathing rate
  $= 0.7 \times 24 = 16.8\ dm^3\ min^{-1}$
- **Oxygen consumption** = xR/xS
  $= 0.6 \times 60/55 = 0.65\ dm^3\ min^{-1}$

2.8 is the number of squares and 0.25 is the volume of oxygen per square – remember to use the scale.

Now try this

The subject then does 5 minutes of vigorous exercise. These are his measurements immediately after the exercise:

tidal volume	1.3 dm^3
breathing rate	42 breaths min^{-1}
respiratory minute ventilation	47.5 $dm^3\ min^{-1}$
oxygen consumption	3.6 $dm^3\ min^{-1}$

Explain the effect of exercise on breathing.

Had a look ☐ Nearly there ☐ Nailed it! ☐

Structure of the kidney

An important role of the kidney is to remove **urea** from the blood. Urea is produced in the **liver** when excess amino acids are broken down. Osmoregulation is another important function.

The urinary system

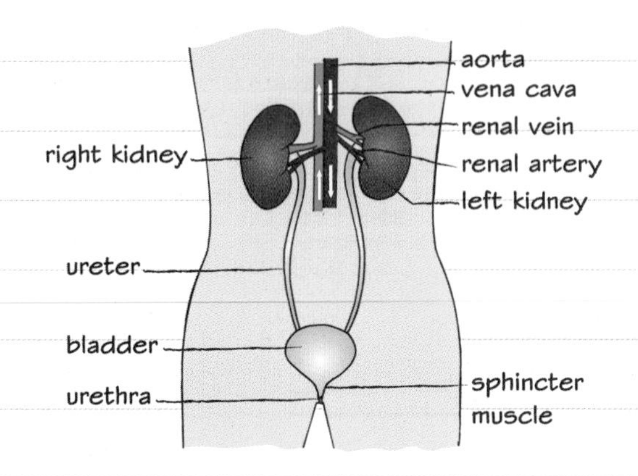

Gross structure of the kidney

Mammals have two kidneys that sit either side of the aorta at the rear of the abdomen. There are about a million **nephrons** in each kidney, which give the kidney its distinctive layers.

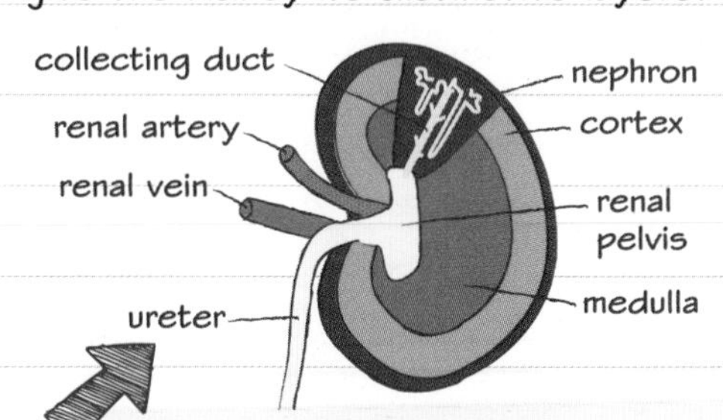

Blood is filtered in the cortex and the ureter carries urine to the bladder.

The nephron

The nephron is the functional unit of the kidney. There are 1–2 million per kidney.

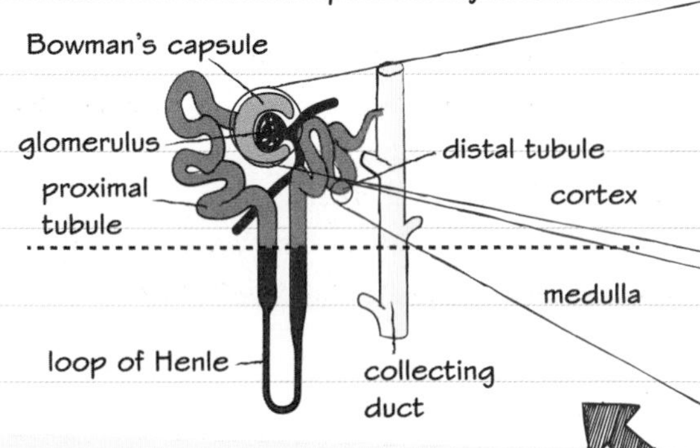

The structures in the cortex are mostly concerned with filtering toxins such as urea from the blood. The tubules extending into the medulla are mostly concerned with **osmoregulation**.

Microscopic structure of the nephron

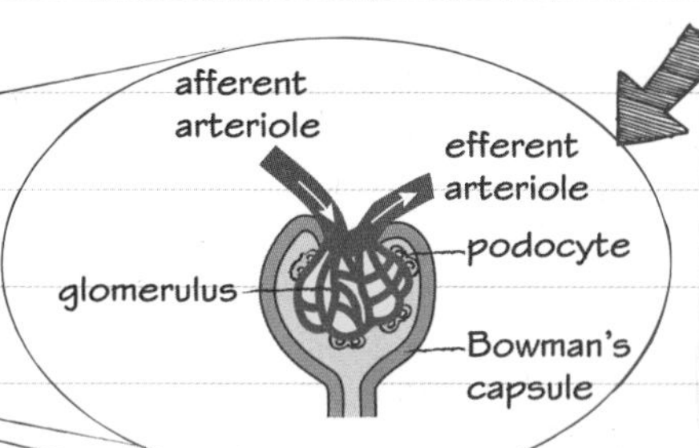

The **Bowman's capsule** has little resistance to fluid leaving blood but prevents cells and large proteins from passing into the tubule.

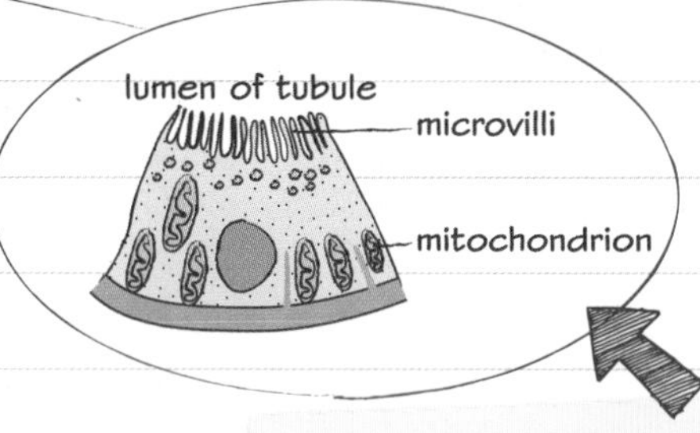

The wall of the **proximal tubule** has adaptations for rapid absorption, including a large surface area, due to **microvilli**, and ATP readily available from many mitochondria.

Blood supply of the nephron

The kidneys filter around 180 dm^3 of fluid out of the blood each day. The vast majority of this returns to the blood; a complex network of blood vessels around the kidney tubules makes this possible.

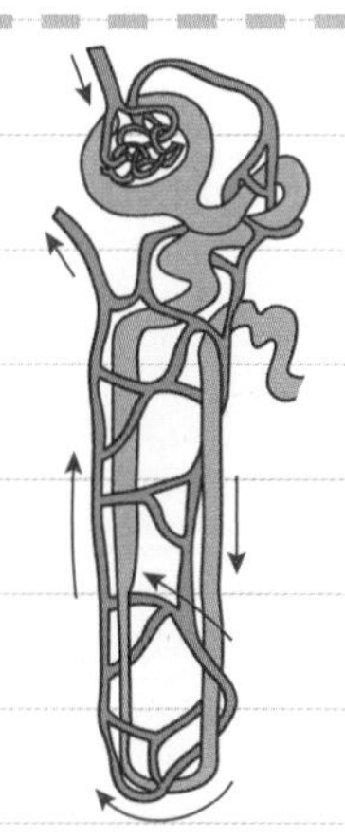

Now try this

Describe the function of the renal vein and the renal artery.

Kidney function

The kidneys remove urea from the bloodstream by **ultrafiltration**.

Ultrafiltration

The first stage in producing urine is the filtering of blood in the Bowman's capsule. The filtration is based on particle size and the filtrate produced has a composition very similar to plasma. Most proteins are too big to pass into the tubules.

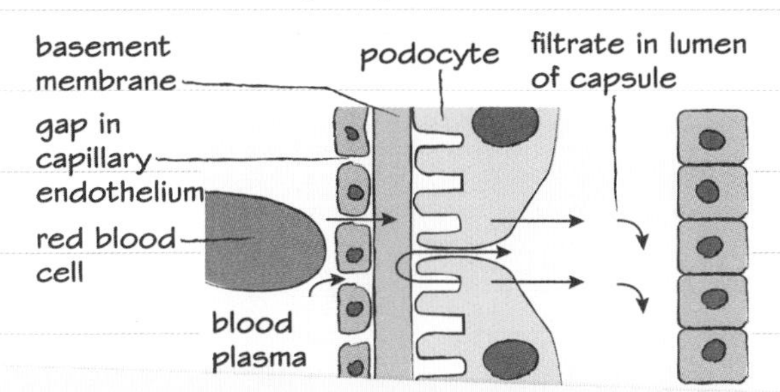

There is a very thin barrier between blood in the capillary and the tubule lumen as there are many gaps between the capillary endothelium cells.

Filtrate formation is promoted, because:

- blood enters the glomerulus under pressure
- the **afferent arteriole** entering the glomerulus is wider than the **efferent** vessel leaving.

Loop of Henle

The loop of Henle produces a very high concentration of solutes in the medulla of the kidney so it has a very low water potential. This allows mammals to produce urine more concentrated (hypertonic) or less concentrated (hypotonic) than plasma. The differences in permeability of regions of the loop account for changes in the filtrate.

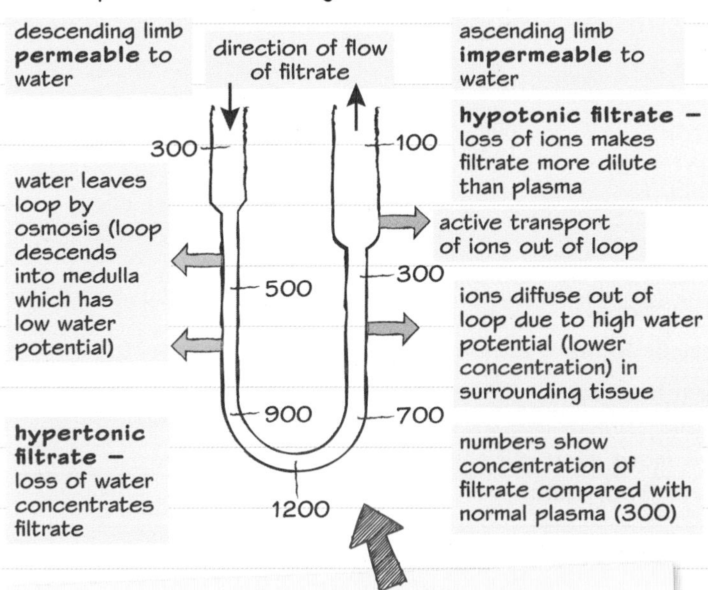

The loop of Henle is a countercurrent multiplier. This means that the flow of filtrate in the two limbs is in opposite directions and the longer the loop the greater the effect on the concentration of the filtrate.

Selective reabsorption

The majority of filtrate is returned to the blood in the **proximal convoluted tubule (PCT)** but much less urea returns.

	Filtered/day	Reabsorbed to blood/day	Present in urine/day
Water	180 dm^3	179 dm^3	1 dm^3
Protein	2 g	1.9 g	0.1 g
Glucose	162 g	162 g	0 g
Urea	54 g	24 g	30 g

Useful substances such as glucose are fully reabsorbed. Waste such as urea is left in high concentration in the filtrate. Water reabsorption is variable.

Active transport recovers glucose, amino acids and proteins, vitamins and hormones. Many mitochondria supply the ATP needed.

Sodium ions are also actively reabsorbed. Water and other ions follow passively.

The **distal convoluted tubule (DCT)**:

- secretes waste chemicals such as **creatinine** into the filtrate
- pumps ions to control blood pH
- helps to control blood volume and therefore concentration of urine (see page 100).

Urine tests

Urine is often tested for proteins or glucose.

Finding plasma protein would suggest damage to kidney membranes, as plasma proteins are too big to cross the filtration barrier in the Bowman's capsule – often caused by high blood pressure.

Finding glucose might suggest that the mechanism for reabsorption was damaged, as all glucose is normally removed from filtrate in the PCT. It is also possible that the concentration of glucose in blood is too high for it all to be reabsorbed, as occurs in diabetes.

Now try this

Explain how filtrate in the loop of Henle is hypertonic to blood plasma in some places and hypotonic in others.

Had a look ☐ Nearly there ☐ Nailed it! ☐

Osmoregulation

Mammalian plasma concentration is controlled by the **pituitary gland** and **osmoreceptors** in the **hypothalamus** of the brain. **Antidiuretic hormone (ADH)** is the hormone released by the pituitary that affects the permeability of the kidney tubules to water.

Osmoregulation

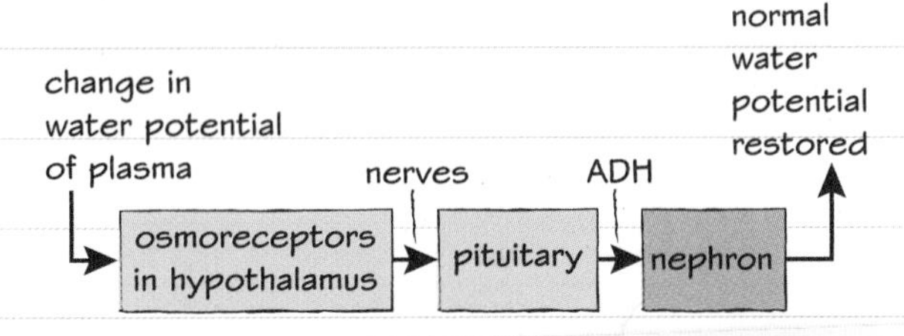

Controlling the water potential of body fluids is an example of a **negative feedback** mechanism. If the concentration of the blood changes, the amount of ADH released changes to maintain a dynamic equilibrium. Nervous stimulation from the hypothalamus controls the amount of ADH released by the pituitary.

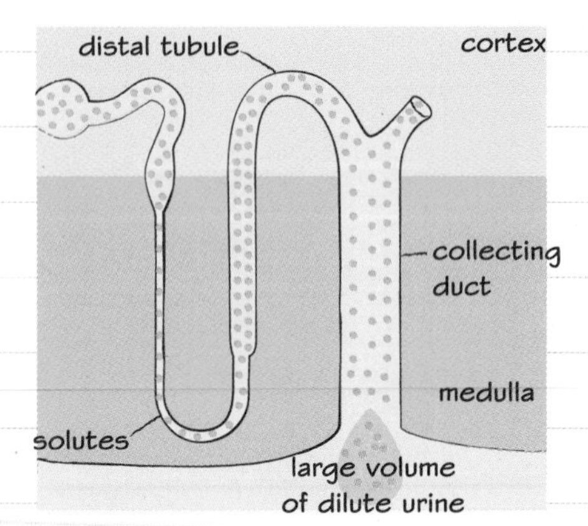

Without ADH, the hypotonic filtrate passes through the collecting duct unchanged, producing high volumes of dilute urine.

Effects of ADH

The quantity of ADH released by the pituitary gland controls the concentration and volume of urine produced by the kidneys. This helps to maintain the water potential of body fluids within a very narrow range.

The effect of ADH on urine concentration is dependent on two properties of the fluids in the kidney caused by the countercurrent multiplier in the loop of Henle:

 the filtrate as it reaches the end of the loop has a water potential higher than that of plasma (it is more dilute)

 the medulla has a very low water potential.

ADH causes channels to open in the collecting duct, which allows water to pass through. Because the medulla has a very low water potential, water will leave the filtrate by osmosis as it passes through. This produces small quantities of concentrated urine.

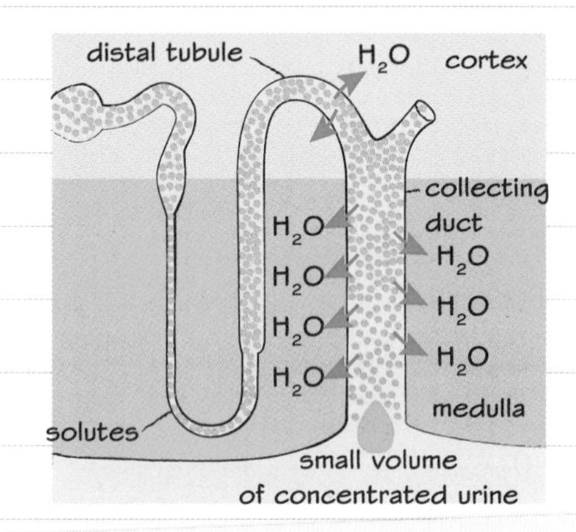

ADH makes the walls of the collecting duct permeable to water. Water leaves by osmosis and is carried away by capillaries.

Kangaroo rats

These animals live in places so arid that they are never able to drink water. They obtain water from food and biochemical reactions.

They need to conserve water by producing very little, very concentrated urine and their kidney structure reflects this.

They have nephrons with very long loops of Henle to produce a very low water potential in the medulla. The tubules have more microvilli and more mitochondria than in most animals, for efficient reabsorption.

Now try this

Describe how a very salty meal affects the concentration of urine.

Maintaining balance

The kidney helps maintain balance in the body (homeostasis). The kidney:

- regulates the amount of water in the plasma
- maintains blood pressure and electrolyte balance using the angiotensin-aldosterone mechanism
- regulates the pH of the plasma.

Links For more on kidney function, see page 99.

Regulation of blood pressure and electrolytes

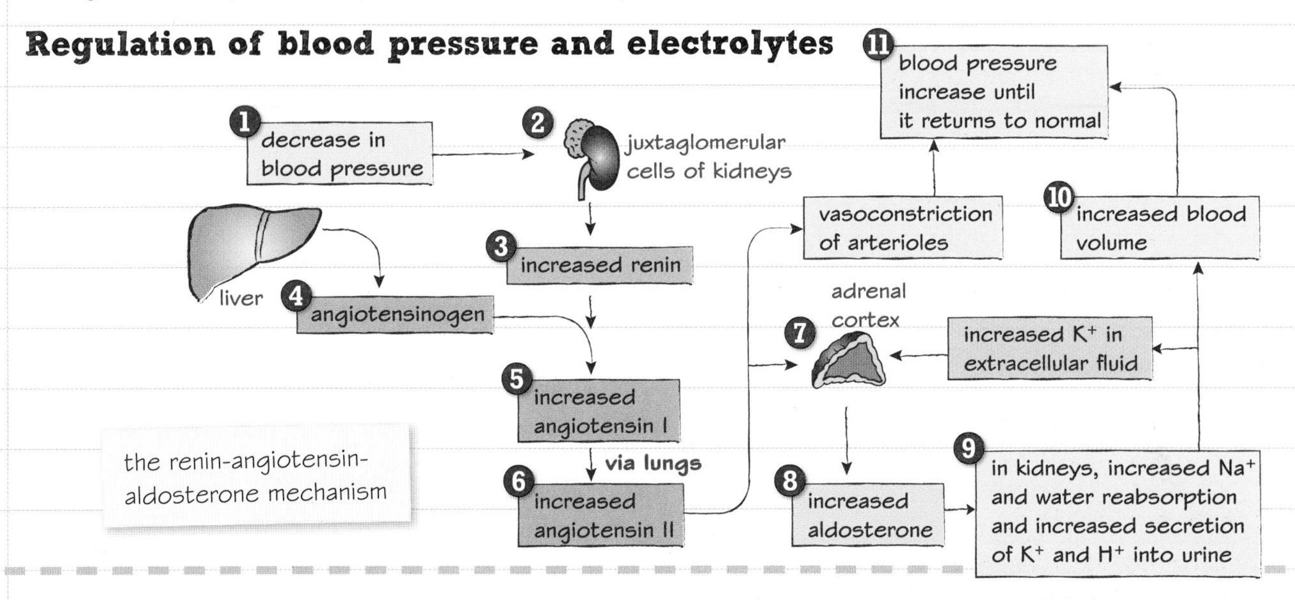

pH

pH stands for potential of hydrogen. A low pH means that a solution is acidic. A high pH means a solution is alkaline. The scale is from 0 to 14. Neutral pH is 7. The greater the concentration of hydrogen (H^+) ions the lower the pH and the more acidic a solution is. A base is a substance that produces hydroxide (OH^-) ions in solution and/or accepts hydrogen ions.

Acid–base balance

Fluids, such as blood in the human body, have to be between pH 7.35 and 7.45. Otherwise proteins, such as enzymes, are denatured.

The kidneys respond slowly to imbalance, but if tissue fluid is too acidic, hydrogen ions are absorbed into the tubular fluid and bicarbonate ions are excreted. If the tissue fluid is too alkaline, hydrogen ions are excreted and bicarbonate ions are absorbed.

Dialysis

If a person's kidneys fail, they stop removing toxic waste from the blood, such as urea. These toxins would build up and kill the person.

A dialysis machine takes over the job of the kidneys and the person's blood passes through the machine. It removes waste products and balances the levels of water and salts in the blood. The person would need dialysis three to four times a week and each time it takes 4–5 hours.

Kidney transplants

If both kidneys fail, a single healthy kidney from a donor can be joined to the blood vessels and the bladder in the groin of the recipient.

If the transplant is successful, it can then clean and balance the blood and the recipient can lead a normal life.

The antigens on a donor kidney may be different from a patient's. The recipient's immune system detects the difference and may reject the new kidney, so it is very important to find a donor with a tissue type as close to the recipient as possible. Immunosuppressant drugs are used to suppress the immune system to prevent rejection.

Now try this

Describe how the kidney regulates blood pressure.

Had a look ☐ Nearly there ☐ Nailed it! ☐

Surface area to volume ratio

The efficiency of transport of substances by diffusion in an organism depends on the size of the organism – specifically the **surface area to volume ratio**. The higher this ratio is, the more efficient diffusion is.

Modelling surface area to volume ratios

The surface area (SA) to volume (V) ratio is given by SA/V.

Small objects have a larger $\frac{SA}{V}$.

Look at this cube.

$V = 1 \times 1 \times 1 = 1\,\text{cm}^3$

$SA = 6 \times 1 \times 1 = 6\,\text{cm}^2$

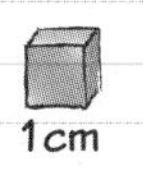

So $\frac{SA}{V} = \frac{6}{1} = 6$

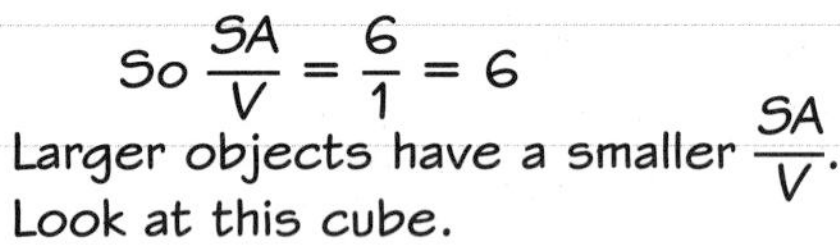

Larger objects have a smaller $\frac{SA}{V}$.

Look at this cube.

$V = 4 \times 4 \times 4 = 64\,\text{cm}^3$

$SA = 6 \times 4 \times 4 = 96\,\text{cm}^2$

So $\frac{SA}{V} = \frac{96}{64} = 1.5$

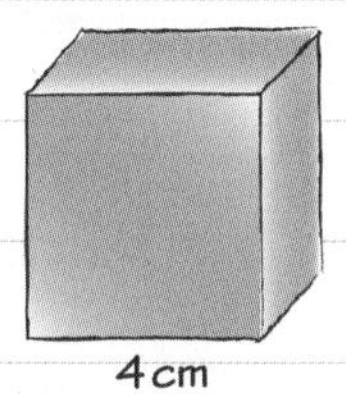

Specialised gas exchange surfaces

Diffusion rate is affected by three related factors:

- area of diffusion
- difference in concentration
- thickness of diffusion surface.

Diffusion is faster when surfaces have a large area and are thin. In addition, the higher the difference in concentration the faster diffusion will be.

Small organisms

Small single-celled organisms, such as bacteria, and very small multicellular organisms, such as some worms, can obtain all they need, and get rid of all the CO_2 and waste they produce, by diffusion because:

- the diffusion distance is short
- the $\frac{SA}{V}$ ratio is high
- metabolic demands are low.

Large organisms

Large organisms have a low $\frac{SA}{V}$ ratio. Diffusion is not adequate to supply needs and remove waste, due to the large distances involved. Materials would arrive less quickly than they are used and vice versa for waste.

Large organisms have evolved specialised **organs** for gas exchange, such as tracheae in insects.

Read more about breathing mechanisms on pages 95–97.

These structures each have a large $\frac{SA}{V}$ ratio.

Large organisms have also evolved transport systems for carrying O_2, CO_2 and food molecules quickly around the body.

For more about transport, see pages 90–92.

Now try this

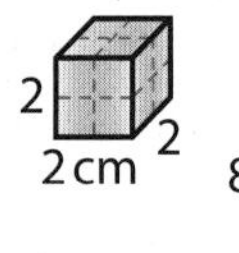

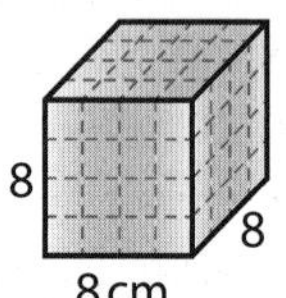

Work out the $\frac{SA}{V}$ of these two cubes.

Set your calculation out step by step on paper so you do not go wrong. This is good advice with any calculation. Also, if you get the answer wrong you can still get marks for the working, but only if it is there and correct!

The cell surface membrane

The cell surface membrane contains the cell, controls movements of substances in and out of the cell, maintains the osmotic balance of the internal environment and allows the cell to be recognised by, for example, hormones.

The chemical structure of the cell membrane

- lipids (in the form of phospholipids and cholesterol)
- proteins
- carbohydrate (in the form of glycoproteins and glycolipids).

Properties of cell membranes

- flexible
- fluid
- selectively permeable.

Developing models of the cell membrane

Finding out how these components are arranged took many years, requiring the development of techniques such as electron microscopy and the use of radioactive isotopes. The resulting data were used to build models such as the **fluid mosaic model** to provide a scientific explanation of the structure and properties of cell membranes.

The fluid mosaic model

The fluid mosaic model has a fluid lipid bilayer with a mosaic of proteins, glycoproteins and glycolipids floating in it.

Facts and evidence for the model

- Phospholipids are both hydrophilic (water-loving) and hydrophobic (water-hating), so form bilayers in an aqueous environment.
- Microscope images of cell surfaces show proteins sticking out.
- Some water-soluble substances pass into and out of cells.
- Ionic and polar molecules do not pass easily through membranes, but lipid-soluble substances do.

Other models

The Davson–Danielli model (below), proposed in 1935, accounts for protein and lipid, but it does not explain many of the known facts about membranes.

There are both similarities and differences between the two models:

- Both have a phospholipid bilayer with protein, but the fluid mosaic model has proteins within the phospholipid layer, while the Davson–Danielli model has protein layer on the outside of the membrane.
- The fluid mosaic model has glycolipid, glycoprotein and cholesterol.

Now try this

Describe the structure of a cell membrane.

Had a look ☐ Nearly there ☐ Nailed it! ☐

Passive transport

Membranes control the movement of materials in and out of cells and organelles. Substances can move across membranes either **passively** (sometimes with help, but energy is never needed) or **actively**, which needs energy. How exactly they move depends on their properties.

Passive movement

There are three kinds of passive movement: **diffusion**, **facilitated diffusion** and **osmosis**.

In all of these, when there are more particles in one area (high concentration) than another (low concentration), more move away from the high concentration area than from the low concentration area. So the **net movement** (the balance of movement) is **away** from the high concentration **to** the low.

Equilibrium

Particles move across cell membranes in **both directions all the time**. If the concentration is the same on both sides of a membrane, particles are still moving across in both directions but the **net** movement is zero.

Diffusion

The concentration of a particle may be higher on side A than side B, but there will **only** be movement if:

- the membrane is **permeable** (pores big enough)
- the particle and/or pore is **not charged**
- the particle is soluble.

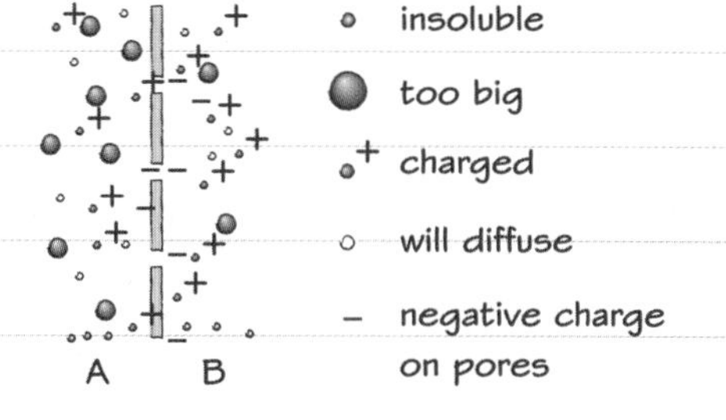

Facilitated diffusion

Small, non-charged molecules, like O_2 and CO_2, can pass through the lipid bilayer. Bigger (glucose) or charged molecules (sodium ions) need a channel in the form of a **channel protein** in order to pass through.

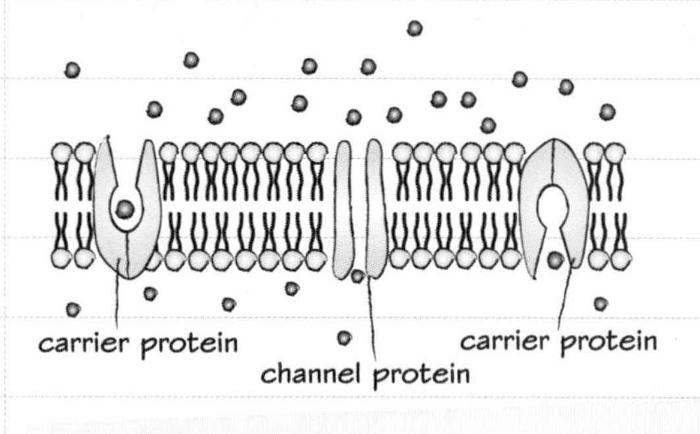

Membrane proteins allowing molecules through a pore (channel protein) or a 'flip-flop' system for specific molecules (carrier protein). Here, the molecule causes the protein to change shape, thus carrying it across the membrane. Carrier proteins are passive or active when transporting molecules down a concentration gradient (high to low concentration).

Osmosis

Osmosis is the diffusion of free water molecules. It involves:

- net movement of water molecules from a solution with high water potential (low solute concentration) to a solution with a low water potential (higher solute concentration)
- movement through a partially permeable membrane (permeable to water but not the solute).

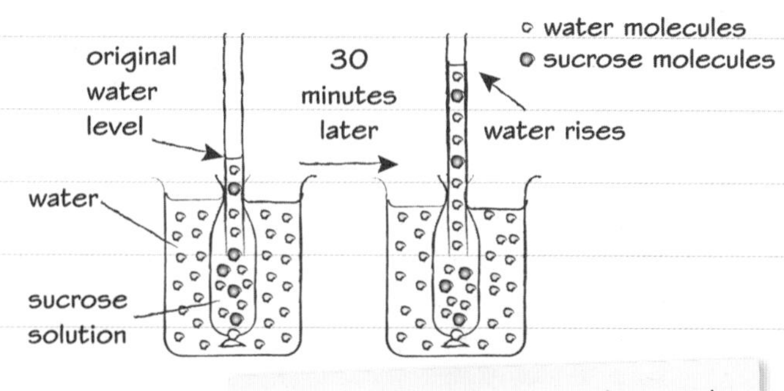

A simple demonstration of osmosis

Now try this

Explain how drugs could be designed to enter a cell rapidly.

The drugs would be entering by facilitated diffusion – osmosis is just the movement of water molecules and simple diffusion just relies on concentration gradient. So think about the factors that would enable this.

Active transport, endocytosis and exocytosis

Living things can move substances in and out of their cells even if the concentration gradient is in the wrong direction (from low to high concentration).

Movement against a concentration gradient

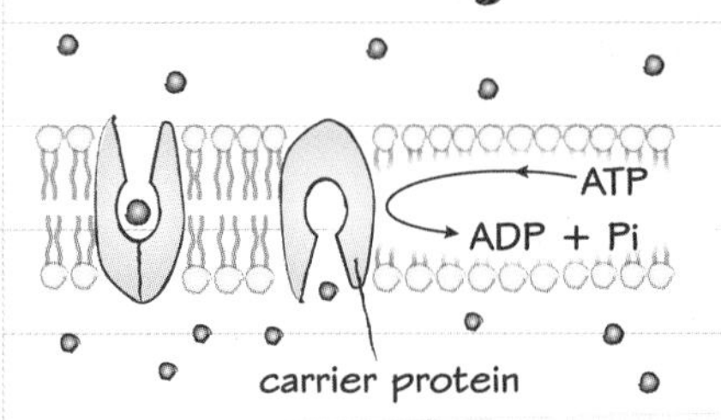

When some channel proteins are given energy (from ATP) they can move molecules **against** a concentration gradient. This is called **active transport**.

Adenosine triphosphate (ATP), the energy currency of the cell

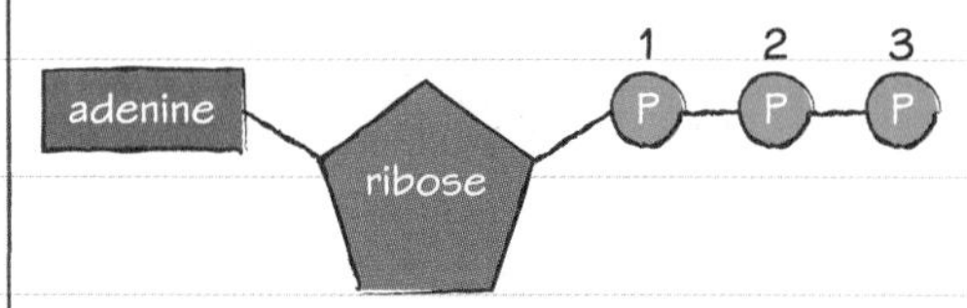

The ATP molecule is made of a ribose, an adenine base and three phosphate groups (P).

When the bond between the third and second phosphate is broken by hydrolysis, energy is released. This can be used in energy-requiring processes taking place within the cell.

Energy is required to add a third phosphate bond to adenosine diphosphate (ADP) to create ATP again.

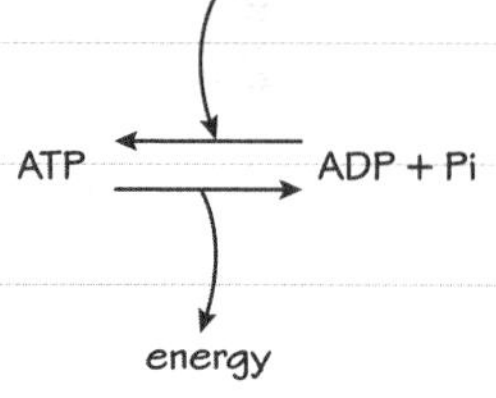

Endocytosis and exocytosis

Vesicles transport large molecules into and out of cells. This is the process of **endocytosis** (movement into the cell) and **exocytosis** (movement out of the cell). Both of these processes need energy in the form of ATP. They are possible because of the fluid nature of the cell membrane.

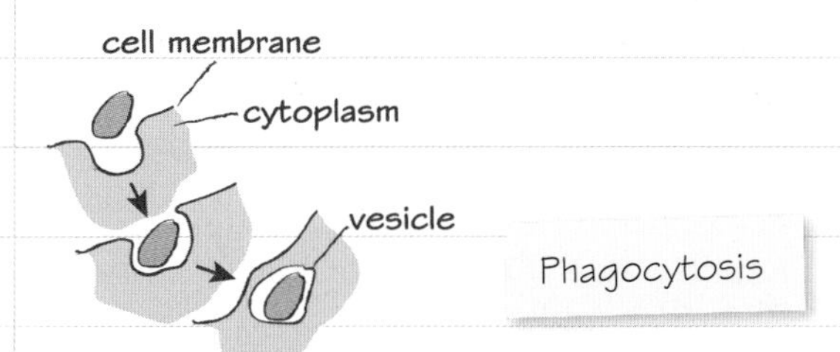

Phagocytosis

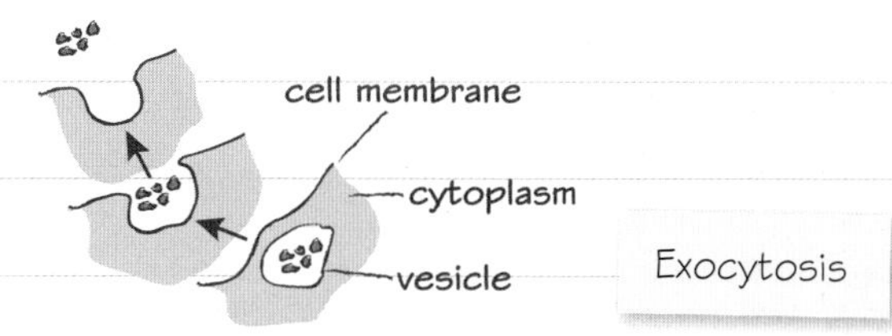

Exocytosis

Phagocytosis ('cell-eating') is a kind of endocytosis. A large structure, such as a bacterial cell or a molecule, is surrounded by the cell membrane and engulfed in a vesicle, which moves into the cell. If the material being taken in is a liquid it is called 'cell-drinking' (**pinocytosis**).

To remove large particles from inside the cell, the cell performs exocytosis. This is a process in which large molecules in the cell are engulfed in a vesicle. The vesicle then moves to the cell membrane where it fuses with the membrane. The contents that were engulfed are then released out of the cell.

Now try this

Explain how insulin, made in the β cells of the pancreas, is moved out of the cell into the pancreatic duct.

Think about how substances are packaged and moved in and out of cells.

Had a look ☐ Nearly there ☐ Nailed it! ☐

Metal oxides and hydroxides

Many metal oxides and hydroxides behave as bases and neutralise acids.

Basic behaviour

Bases are substances that accept hydrogen ions, H^+. These ions are often referred to as protons, so bases are sometimes called **proton acceptors**. Both metal oxides and metal hydroxides can behave as bases.

Metal oxides

The oxide ion in metal oxides can accept protons and neutralise acids. For instance, calcium oxide will neutralise sulfuric acid to form calcium sulfate:

$$CaO(s) + H_2SO_4(aq) \rightarrow CaSO_4(s) + H_2O(l)$$

$$O^{2-} + 2H^+ \rightarrow H_2O$$

Calcium oxide, sometimes called lime, is used in agriculture to neutralise acidic soil.

This is an important reaction as if the pH of soil is too low, the plant will tend to absorb toxic ions that affect root growth.

Metal hydroxides

Metal oxides will form hydroxides when they react with water:

$$MgO(s) + H_2O(l) \rightarrow Mg(OH)_2(s)$$

The hydroxide ions can accept protons and neutralise acids. For instance, magnesium hydroxide is used as an antacid to relieve the symptoms of indigestion and heartburn, by neutralising hydrochloric acid that has entered the oesophagus.

$$Mg(OH)_2(s) + 2HCl(aq) \rightarrow MgCl_2(aq) + H_2O(l)$$

$$OH^-(aq) + H^+(aq) \rightarrow H_2O(l)$$

Calcium hydroxide is used to neutralise acidic effluent in the chemical industry, like a dilute solution of sulfuric acid in the waste water from the manufacture of polymers.

$$Ca(OH)_2(s) + H_2SO_4(aq) \rightarrow CaSO_4(s) + 2H_2O(l)$$

Amphoteric oxides

Amphoteric oxides can behave as either acids or bases. Elements that form such oxides tend to be in the middle of a period, for example, aluminium.

Aluminium oxide will **not dissolve in water**, but it will **react with acids** to form a salt and water in a similar way to other metal oxides. For instance, it will react with HCl(aq) in warm conditions:

$$Al_2O_3(s) + 6HCl(aq) \rightarrow 2AlCl_3(aq) + 3H_2O(l)$$

It will also **react with bases to form aluminates**. For instance, it will react with NaOH, to form sodium tetrahydroxoaluminate, which is used in the Bayer process to convert the mineral bauxite into aluminium oxide:

$$Al_2O_3 + 2NaOH + 3H_2O \rightarrow 2NaAl(OH)_4$$

Alumina

Alumina is a form of aluminium oxide found in the mineral bauxite. It is extracted and purified using the Bayer process. The stages in the process are:

1. Crush the bauxite.
2. React with NaOH(aq) at 170 °C to form $NaAl(OH)_4(aq)$.
3. Filter out solid impurities.
4. Allow to crystallise to form $Al(OH)_3$.
5. Heat in rotary kiln to form Al_2O_3.

Most of the alumina separated in this way is used in the Hall-Héroult process to form aluminium by electrolysis.

Some is used as a **refractory material** in kilns. These are materials that retain their strength and are chemically stable at high temperatures.

Now try this

Nitric acid is used in the steel industry to clean oxides from the surface of steel. Effluent from the industry will often contain nitric acid.

Describe, including an equation, how the effluent can be neutralised.

For more on the Hall-Héroult process, see page 107.

Aluminium and titanium

Both these elements are low-density and corrosion-resistant metals that form strong alloys.

Extraction of aluminium

Aluminium ore, bauxite, is mined and then processed to form alumina, which is aluminium oxide. Molten alumina is then electrolysed using the **Hall-Héroult** process.

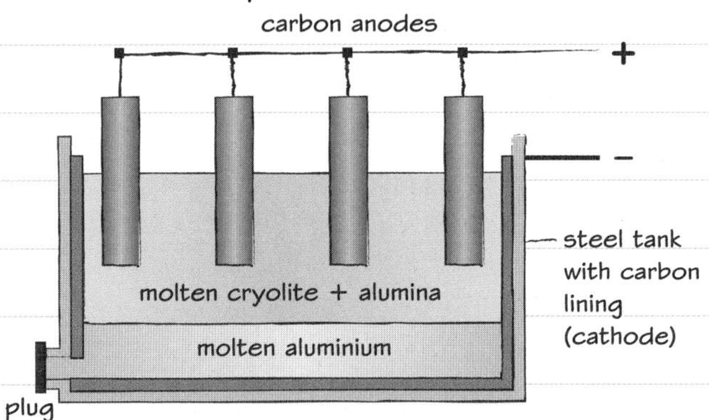

Cryolite is added to the alumina to lower the melting point and save energy. The lining of the steel tank is made from carbon, which acts as the negative electrode. Here, aluminium ions are reduced to form molten aluminium:

$Al^{3+} + 3e^- \rightarrow Al$

The molten aluminium can then be drained off and cast into ingots.

The positive electrodes are made from carbon. Here, oxide ions are oxidised to form oxygen gas:

$2O^{2-} \rightarrow O_2 + 4e^-$

The carbon electrodes have to be replaced regularly as they react with the oxygen as it forms.

Extraction of titanium

The main titanium ore is rutile, which contains titanium (IV) dioxide, TiO_2. Although titanium has a similar reactivity to aluminium, it is not generally extracted by electrolysis as the titanium formed often has 'tree-like' crystals, which can affect the electrodes. Also, side-reactions with titanium ions at both electrodes can lead to impurities.

Most titanium is extracted using the **Kroll process**. In the process:

 Titanium(IV) dioxide, coke (a form of carbon) and chlorine are heated together at about 900°C, to form titanium(IV) chloride:

$TiO_2(s) + 2C(s) + 2Cl_2(g) \rightarrow TiCl_4(g) + 2CO(g)$

 Magnesium is used as a reducing agent to form titanium:

$TiCl_4(g) + 2Mg(l) \rightarrow Ti(s) + 2MgCl_2(l)$

The process is expensive due to the large amounts of energy needed to create the very high temperatures involved. Also the magnesium used in step 2 is produced by an energy-intensive electrolysis process. The process is time consuming as it is a batch process. This means the reactor has to be shut down and cooled before the titanium can be removed for purification. The impurities are then removed by further heating.

Choosing between aluminium and titanium

When considering the choice of material for a particular use, the properties of the material must be evaluated in comparison to the desired use. For instance, both aluminium and titanium could be used to manufacture bicycle frames. Some of the properties of the metals are summarised in the table.

	Stiffness (GPa)	Strength (MPa)	Density ($g\,cm^{-3}$)	Cost (£ $tonne^{-1}$)	Resistance to corrosion
titanium	114	434	4.51	46000	high
aluminium	69	276	2.70	1000	high

An aluminium frame will be durable, quite strong, lightweight and quite rigid, so unlikely to flex in use. It is also much cheaper than titanium. Titanium is about 1.6 times stronger than aluminium, so although denser, a smaller volume of titanium is needed to produce a frame of similar strength. It is nearly twice as stiff as aluminium, so the frame would be more rigid. Titanium is corrosion resistant. Less titanium is needed for a frame, but the frame will be more costly as titanium is 46 times more expensive per tonne. It is likely that aluminium would be chosen, as the benefits of titanium are not offset by the extra cost in this context.

Now try this

Carbon fibre is a strong, rigid, low density, non-metal material used in bike frames.

Suggest a reason, other than cost, that explains why fewer frames are made from carbon fibre than aluminium or titanium.

Had a look ☐ Nearly there ☐ Nailed it! ☐

Useful products from electrolysis of brine

The chlor-alkali industry produces useful materials for other chemical processes by the electrolysis of brine, NaCl(aq).

Electrolysis of brine

A solution of NaCl(aq) produces chlorine, hydrogen and sodium hydroxide when electrolysed.

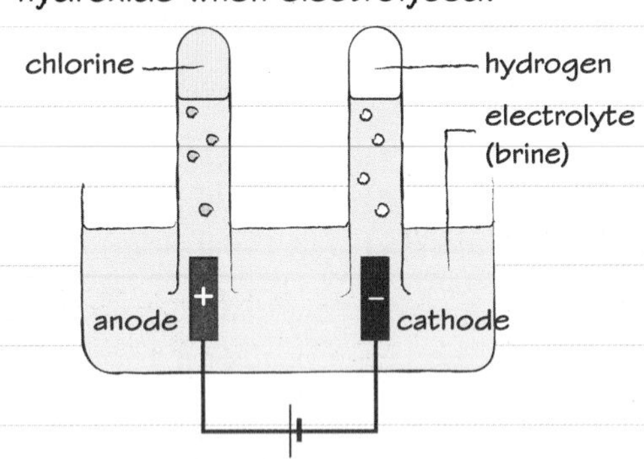

Reactions at each electrode

Chloride ions react at the positive electrode (anode), to form chlorine gas:

$$2Cl^-(aq) \rightarrow Cl_2(g) + 2e^-$$

Each chloride ion has lost an electron so the reaction is **oxidation**.

Hydrogen ions, from the water, react at the negative electrode (cathode), to form hydrogen gas:

$$2H^+(aq) + 2e^- \rightarrow H_2(g)$$

Each hydrogen ion has gained an electron so the reaction is **reduction**.

The remaining sodium ions and hydroxide ions stay in solution forming sodium hydroxide solution, often referred to as caustic soda.

Diaphragm cell

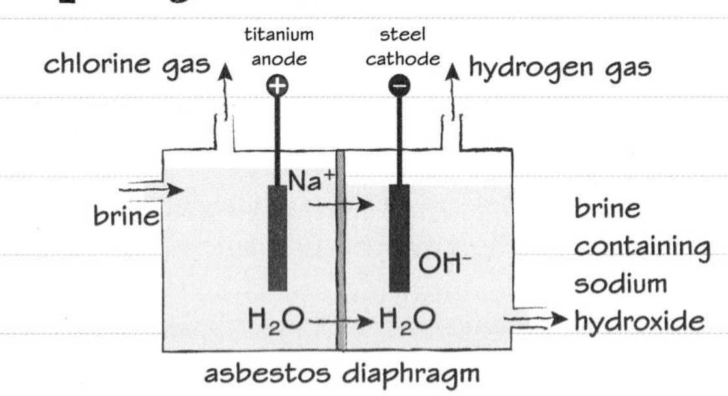

This cell, used in the chlor-alkali industry, has a **diaphragm** dividing the cell. It is porous, so allows brine to pass from one side to the other but prevents chlorine and hydrogen gas from passing through, so the gaseous products are kept separate. The sodium hydroxide formed is still mixed with brine. The sodium chloride is less soluble than sodium hydroxide so can be recrystallised to remove most of it.

Membrane cell

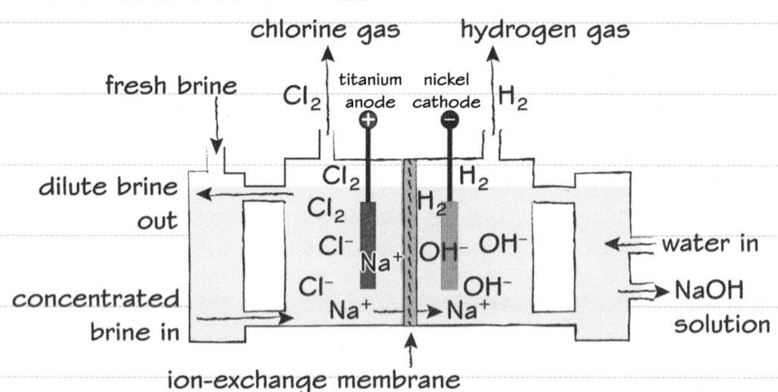

This cell, used in the chlor-alkali industry, has an **ion-exchange membrane**, dividing the cell. It allows the positive sodium ions, but will not allow negative chloride ions, to pass through from the anode side to the cathode side. As the brine only enters the cell from the anode side, the sodium hydroxide left over at the cathode side is not contaminated by sodium chloride.

Pros and cons of the diaphragm cell

The diaphragm cell is cheaper to construct, but the diaphragm needs replacing regularly. The purity of the sodium hydroxide formed is lower and the cell uses slightly more energy per tonne of chlorine produced.

Pros and cons of the membrane cell

The membrane cell is more expensive to construct, but the membrane needs little maintenance. The purity of the sodium hydroxide formed is high and the cell uses slightly less energy per tonne of chlorine produced.

Now try this

Explain why the sodium hydroxide formed in the membrane cell is of greater purity than that formed in the diaphragm cell.

Formulae in organic chemistry

Each type of formulae provides different levels of information about the molecule they represent.

Molecular formulae

Molecular formulae tell you the number and type of each atom present in a molecule. For instance:

- Butane, C_4H_{10} contains 4 carbon atoms and 10 hydrogen atoms.
- Phenol, C_6H_6O contains 6 carbon atoms, 6 hydrogen atoms and an oxygen atom.

In each case, the formulae give you no information as to how the atoms are arranged within the molecule.

Displayed formulae

These show all the bonds present in a molecule, so it is clear to see how each joins to others.

For instance, the hydrocarbon propene has the molecular formula C_3H_6. Its displayed formula is:

```
H       H
|       |
C = C — C — H
|   |   |
H   H   H
```

Structural formulae

These show how all the atoms are arranged, but do not require all the bonds to be shown. The displayed formula of butane is:

```
    H   H   H   H
    |   |   |   |
H — C — C — C — C — H
    |   |   |   |
    H   H   H   H
```

Its structural formula is shown as $CH_3CH_2CH_2CH_3$.

The displayed formula of 2-methylbutane is:

```
          H
          |
      H — C — H
    H     |   H   H
    |     |   |   |
H — C  —  C — C — C — H
    |     |   |   |
    H     H   H   H
```

In the structural formula any group branching off the main carbon chain is shown in a bracket to the right of the carbon atom to which the group is attached. So the structural formula here is $CH_3CH(CH_3)CH_2CH_3$.

Skeletal formulae

These are simplified formulae with any hydrogen atoms removed and the carbon chain reduced to a 'skeletal' line:

```
    H   H   H   H   H
    |   |   |   |   |
H — C — C — C — C — C — H        /\/\
    |   |   |   |   |
    H   H   H   H   H
```

pentane displayed — pentane skeletal

```
          H
          |
      H — C — H
    H     |   H   H
    |     |   |   |
H — C  —  C — C — C — H
    |     |   |   |
    H     H   H   H
```

2-methylbutane displayed — 2-methylbutane skeletal

Remember the 'end of the line' and 'corners' in a skeletal formula represent a carbon.

Once you have practised using skeletal formulae they save a lot of time as they do not require you to draw and check all of the hydrogen atoms.

3D representations of formulae

These use dashes and wedges to emphasise the 3-D nature of many molecules.

This bond from the carbon to the methyl group is facing away from you.

Bond in the plane of the paper.

This bond from the carbon to the hydrogen is facing towards you.

```
        Cl
        |
$H_3C$ ''''C
        ▼   \
        H    Br
```

Now try this

A component of petrol, the hydrocarbon octane C_8H_{18}, has the structural formula $CH_3CH_2CH_2CH_2CH_2CH_2CH_2CH_3$.

Draw the displayed and skeletal formulae of octane.

Alkanes

Alkanes are a series of hydrocarbons with the general formula C_nH_{2n+2}.

Bonding in alkanes

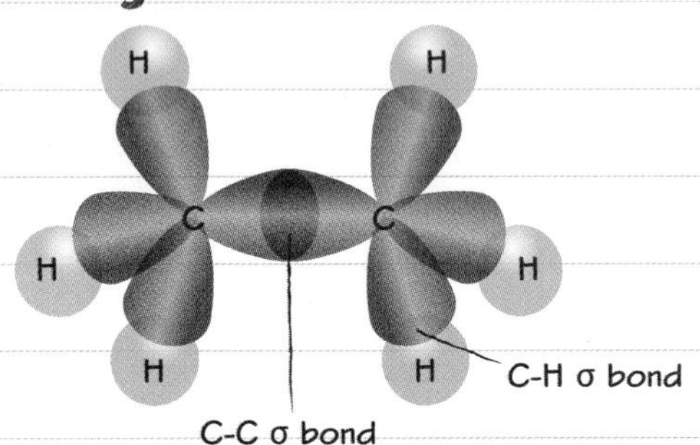

The overlap of orbitals in alkanes results in the formation of sigma (σ) bonds between carbons and hydrogens, and between adjacent carbons. These single bonds are free to rotate.

The carbon orbitals involved in the bonding are called **sp^3 hybrid orbitals**. They are formed when the 2s orbital and three 2p orbitals in carbon's outer shell rearrange themselves into four identical orbitals, each containing one unpaired electron. This process is called **hybridisation**.

Types of alkanes

Alkanes can be classified according to their structure. Alkanes with a single chain of carbon atoms are called **straight chain alkanes**, for example, hexane, $CH_3CH_2CH_2CH_2CH_2CH_3$.

hexane

Alkanes with one or more carbon atoms attached to a carbon in the main chain are called **branched alkanes**, for example, 2-methylpentane, $CH_3CH(CH_3)CH_2CH_2CH_3$.

2-methylpentane

Alkanes with carbon atoms joined together in a ring are called **cyclic alkanes**, for example. cyclohexane, C_6H_{12}.

cyclohexane

Boiling points of alkanes

The longer the carbon chain of a straight chain alkane, the higher the boiling point.

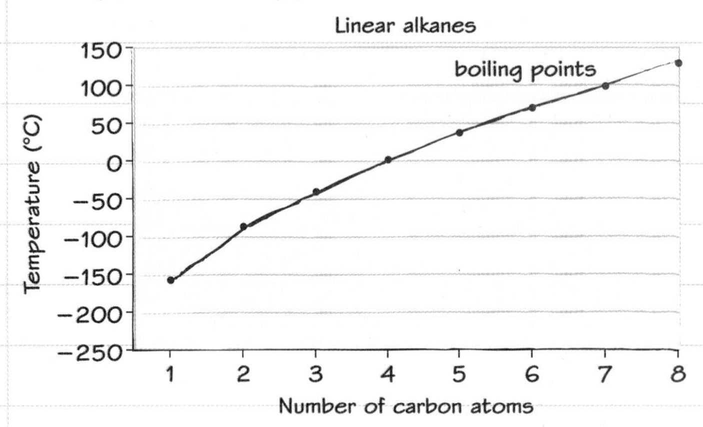

Longer carbon chains have stronger London forces as they have more electrons. The electron density of the larger electron clouds fluctuates more readily, so instantaneous dipoles are stronger in magnitude. Hence, more energy is required to break the London forces.

Structural isomers

Structural isomers are molecules with the same molecular formula but a different structural formula. Hydrocarbons such as alkanes can have a variety of isomers due to a range of possible branches on a carbon chain. For instance, there are three isomers with the molecular formula C_5H_{12}.

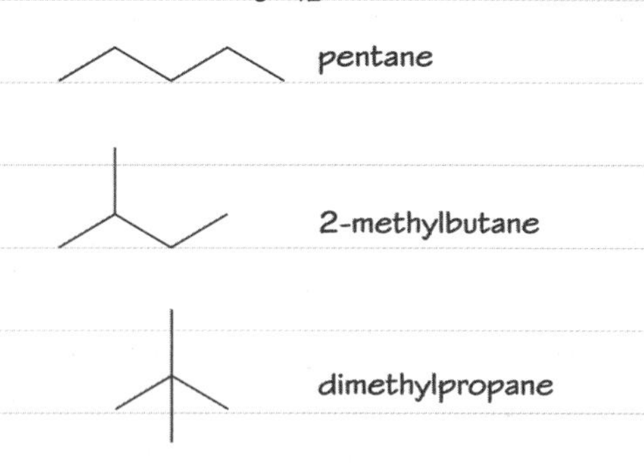

These isomers are chemically very similar. However, physical properties, such as boiling points, may differ.

Now try this

Two fractions found in crude oil are diesel, boiling point range 250–350 °C, and kerosene, boiling point range 175–250 °C. Each fraction consists mainly of alkanes.

Suggest reasons that explain why boiling point ranges are different and yet overlap.

Alkenes

Alkenes are a series of hydrocarbons with the general formula C_nH_{2n}.

Bonding in alkenes

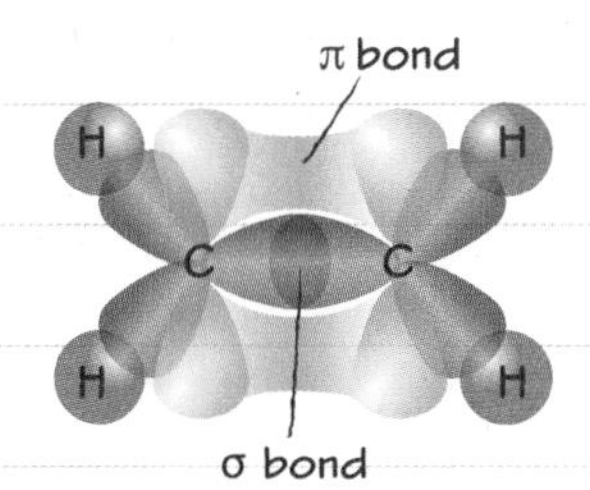

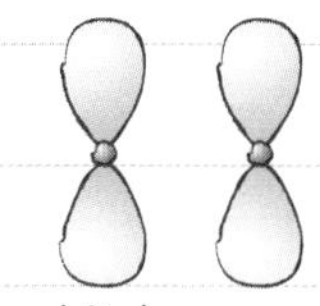

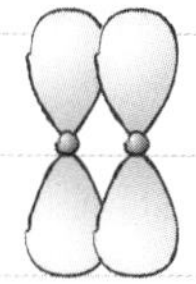

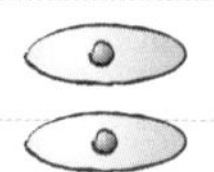

The overlap of orbitals in alkenes results in the formation of sigma (σ) bonds between carbons and hydrogens, and forms one of the bonds between adjacent carbons.

The carbon orbitals involved in the bonding are called **sp^2 hybrid orbitals**. They are formed when the 2s and **two** of the 2p orbitals in carbon's outer shell rearrange themselves into three identical orbitals, each containing one unpaired electron. The **remaining p orbitals** in each carbon overlap sideways to form a pi (π) bond.

The pi bond has two significant effects:

1. It restricts rotation around the double bond. This means groups attached to the carbons in the double bond are locked in position, which can lead to the formation of **stereoisomers**.
2. The high electron density influences many of the chemical reactions of alkenes. Electron deficient species (**electrophiles**) will be attracted to and often react with the double bond in **addition reactions**.

Links For more on the reactions of alkenes, see page 114.

Stereoisomerism in alkenes

Like alkanes, alkenes can form structural isomers. However, restricted rotation about the carbon–carbon double bond in alkenes can lead to stereoisomers. These are compounds with the **same structural formula**, but their atoms have a **different arrangement in space**. The haloalkene $CHClCBrCH_3$ has two possible spatial arrangements.

Links For more on alkanes, see page 110.

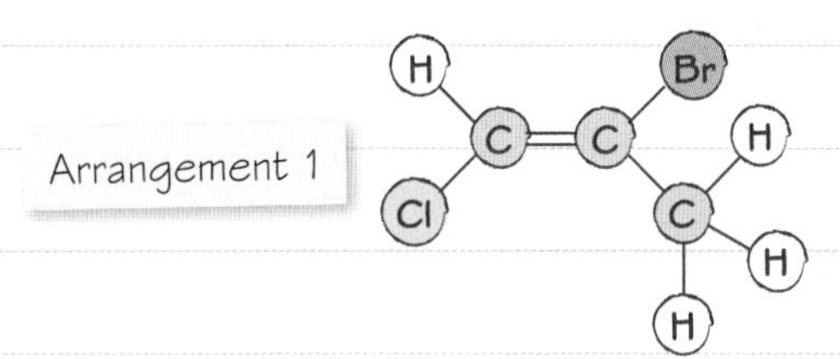

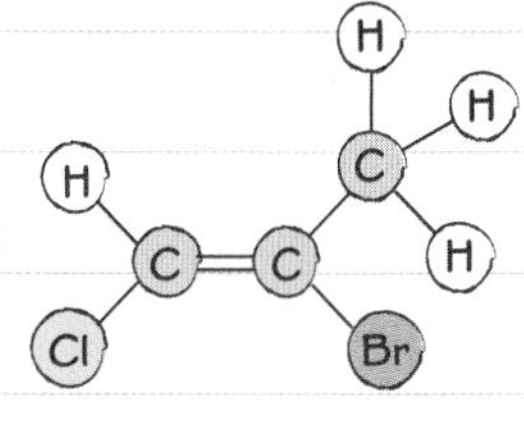

Bond lengths and strength in hydrocarbons

Figure 1

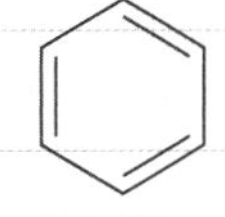

Figure 2

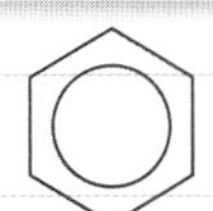

The double carbon–carbon bonds in hydrocarbons are shorter and stronger than carbon–carbon single bonds, as the attraction between the nucleus and the shared pairs of electron is greater. They are not twice as strong, as the p orbital overlap is less effective than overlap by hybrid orbitals. The hydrocarbon benzene, C_6H_6, can be shown as in Figure 1. If correct, this would mean benzene would have alternate long and short carbon–carbon bonds of different strengths. Measurements show that all the carbon–carbon bonds are equivalent length and strength, so Figure 2 is a more accurate model. The p-orbitals overlap over the entire structure, shown by the ring. The electrons in the orbitals are **delocalised**.

Now try this

The chemical industry uses hydrocarbons as starting materials.

Explain why alkenes are more reactive than alkanes.

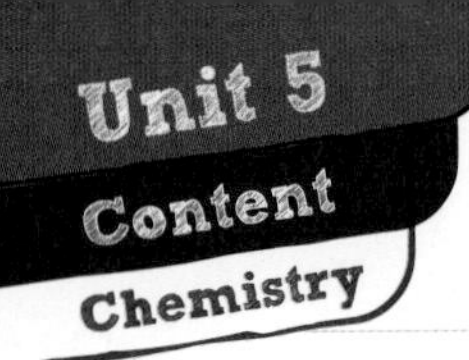

Naming hydrocarbons

The huge number and variety of organic compounds mean a systematic way of naming them, the IUPAC system, is required. This ensures all scientists have a consistent approach to naming.

Key rules for naming hydrocarbons

1. Find the longest unbranched chain of carbon atoms. This determines the **stem** of the name.
2. Identify any side chains, often alkyl groups, coming from the longest unbranched chain. These determine any **prefixes** in the name, added in front of the stem.
3. Identify the key functional groups. For hydrocarbons, these will be **alkanes** or **alkenes**. These determine the **suffix** in the name, added after the stem.
4. Number the position of any side chains or functional groups other than alkanes, if necessary. Count in the direction that gives the lowest possible numbers.

The naming rules in action

The compound below is named **2-methylbutane**:

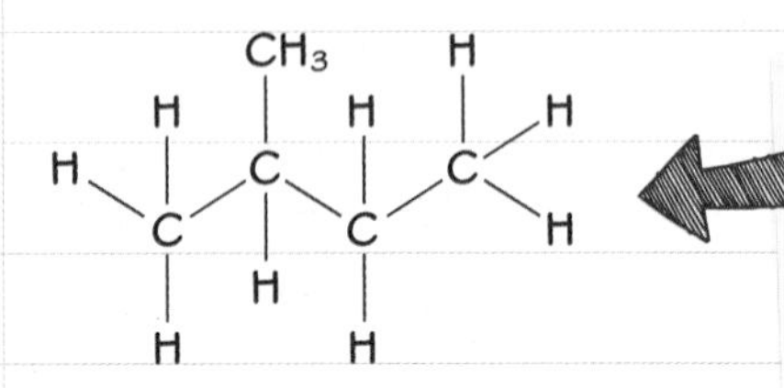

Four carbons in the longest unbranched chain so the stem is **but-**.
There is a methyl group on the **second** carbon so the prefix is **2-methyl**.
The compound is an alkane, so the suffix is **–ane**.
Hence, the name is **2-methylbutane**.

The compound below is named **2,3-dimethylpentane**:

Five carbons in the longest unbranched chain so the stem is **pent-**.
There are methyl groups on the **second and third** carbons so the prefix is **2,3-dimethyl**. The '**di**' indicates two methyl groups; the numbers show their position on the unbranched chain.
The compound is an alkane, so the suffix is **–ane**.
Hence, the name is **2,3-dimethylpentane**.

The compound below is named **pent-1-ene**:

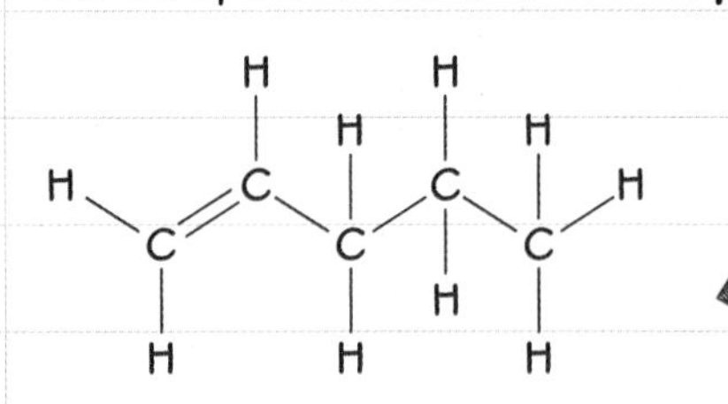

Five carbons in the longest unbranched chain so the stem is **pent-**.
No side chains, so no prefix required.
The compound is an alkene so the suffix is **–ene**.
The position of the double bond has to be stated using the lowest numbered carbon in the double bond, in this case, **-1-**.
Hence, the name is **pent-1-ene**.

The compound below is named **4-ethyl-3-methyloctane**:

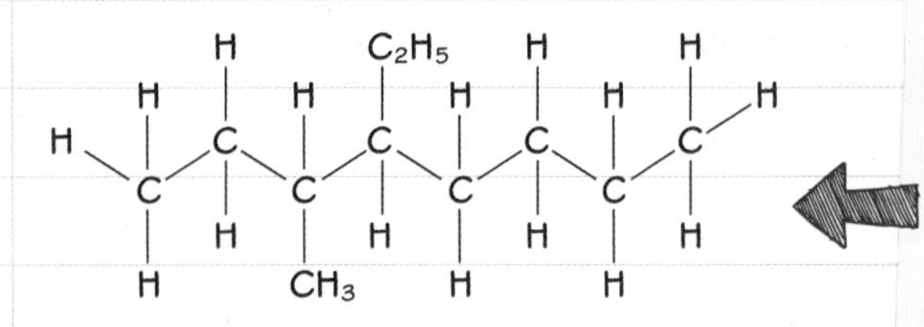

Eight carbons in the longest unbranched chain so the stem is **oct-**.
Ethyl group on fourth carbon, methyl group on third. Different side chains go in alphabetical order so prefix is **4-ethyl-3-methyl**.
The compound is an alkane, so the suffix is **–ane**
Hence, the name is **4-ethyl-3-methyloctane**.

Now try this

Many different alkanes and alkenes are found in crude oil.

Draw the skeletal formulae of these alkanes and alkenes: (a) 2,2,3-trimethylheptane; (b) 2-methylbut-2-ene; (c) 2,5-dimethylhept-2-ene.

Reactions of alkanes

The most important reactions of alkanes are combustion and the reaction with halogens.

Combustion of alkane fuels

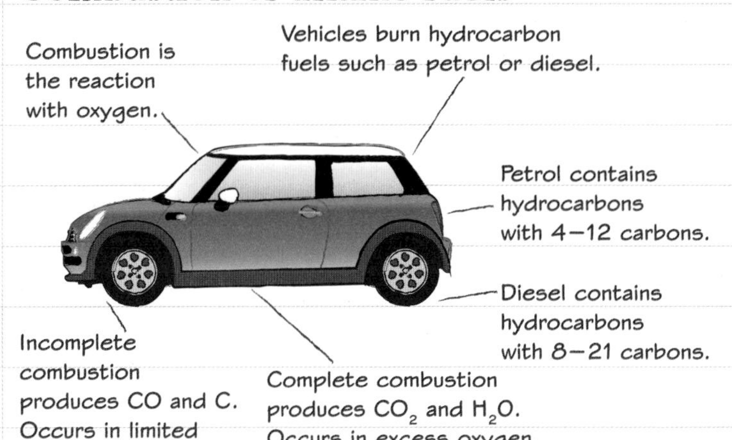

Writing combustion equations

A useful order to follow for this process is:

- ✓ Write the formulae of all the reactants and products.
- ✓ Balance carbon atoms using the carbon dioxide molecules.
- ✓ Balance the hydrogen atoms using the water molecules.
- ✓ Balance the oxygen atoms. As oxygen is diatomic, using halves is acceptable.

For example, complete combustion of octane:

$C_8H_{18}(l) + 12\frac{1}{2}O_2(g) \rightarrow 8CO_2(g) + 9H_2O(l)$

Reaction with halogens

This reaction is called a **free radical substitution**. The **mechanism** of a reaction shows what occurs at each stage of the process. The three stages in this mechanism are:

Initiation – homolytic bond fission of a halogen to form halogen free radicals. Uses UV light or heat.

Homolytic bond fission means when a bond breaks and each fragment formed retains one of the electrons from the bond pair.

Propagation – steps where one free radical reacts and forms a different free radical.

Termination – combination of any two free radicals to form a stable product.

The equations for each step in the reaction of ethane with bromine

Step	Equations
initiation	$Br_2 \xrightarrow{UV} 2Br^\bullet$
propagation	$C_2H_6 + Br^\bullet \rightarrow {}^\bullet C_2H_5 + HBr$
	${}^\bullet C_2H_5 + Br_2 \rightarrow C_2H_5Br + Br^\bullet$
termination	$2Br^\bullet \rightarrow Br_2$
	$2{}^\bullet C_2H_5 \rightarrow C_4H_{10}$
	${}^\bullet C_2H_5 + Br^\bullet \rightarrow C_2H_5Br$

The termination steps shown produce several products, and further substitutions also occur, so C_2H_5Br may not be the only haloalkane formed. As a result, this reaction is of limited use in the chemical industry, as it is difficult to control to produce only one desired product.

A **free radical** is a reactive atom or group of atoms, that contains an unpaired electron, often formed by homolytic bond fission.

Further free radical substitution

When ethane reacts with bromine, another haloalkane formed is $C_2H_4Br_2$. This is formed because:

- C_2H_5Br formed initially may collide with a bromine free radical and form a new radical, in a propagation step:

  $C_2H_5Br + Br^\bullet \rightarrow {}^\bullet C_2H_4Br + HBr$
- The new radical formed may collide with another bromine free radical, in a termination step:

  ${}^\bullet C_2H_4Br + Br^\bullet \rightarrow C_2H_4Br_2$

Other similar reactions take place producing many other by-products.

Now try this

MOT tests check the carbon monoxide levels of car emissions.

Write a balanced equation for the incomplete combustion of cetane, $C_{16}H_{34}$, which is found in diesel.

Reactions of alkenes

Most reactions of alkenes are addition reactions – two reactants combine to form one product.

Key ideas about the addition reactions of alkenes

- The electron-rich C=C bond in alkenes is susceptible to attack by electrophiles.
- Electrophiles are species that can accept a pair of electrons.
- Electrophiles are either positive ions or polar molecules with partial charges (δ+).
- Some species can become an electrophile by passing near to a C=C double bond, as its high electron density is enough to induce a dipole.

Key reactions

You need to know the reactants used and the products formed in each of these reactions.

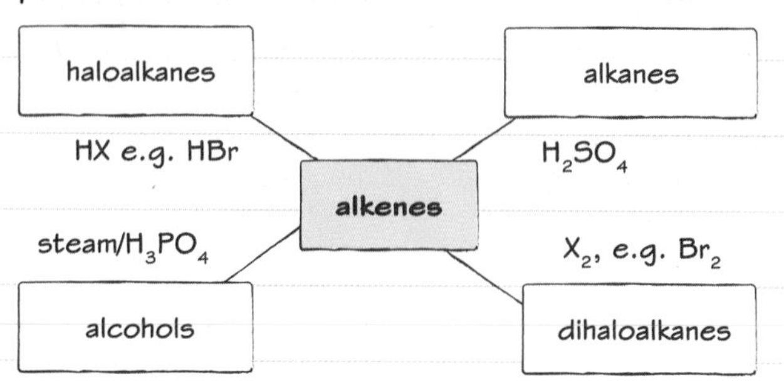

Addition reaction mechanism

This is the series of steps showing how ethene undergoes an addition reaction with bromine. A **curly arrow** shows the direction in which a **pair of electrons moves**. Although the electrophile or alkene will change, the basic steps of the mechanism are the same. The electrophilic ends in each of the other reactants are shown in red:

- with hydrogen halides, H–X
- with sulfuric acid, $H–OSO_3H$
- with steam, H–OH.

Electron pair from double bond moves to $Br^{\delta+}$.

Intermediate carbocation forms.

Product forms.

Bond pair moves onto $Br^{\delta-}$ to form bromide ion.

Lone pair moves from bromide ion to carbocation, forming a bond pair.

Carbocation intermediates formed in the first steps tend to be very unstable, and can influence the products formed.

Addition to asymmetric alkenes

Alkenes like ethene, which have the same two groups either side of the C=C bond, are symmetric. Alkenes like propene, $CH_2=CHCH_3$, have **different groups on either side of the double bond**, so are **asymmetric**. When asymmetric alkenes react with asymmetric electrophiles more than one product forms:

propene + HBr → 2-bromopropane (major product) / 1-bromopropane (minor product)

The major product is more likely to form as the **carbocation intermediate** formed is **more stable**.

route 1

route 2

The two CH_3 groups, attached to the positive carbon in route 1, push electrons towards the positive carbon in the intermediate to stabilise the carbocation, so it can form faster. In route 2 there is only one C_2H_5 group attached to the positive carbon in the intermediate, so it is less stable.

Now try this

The alkene, but-1-ene, $CH_2=CHCH_2CH_3$, is used in the manufacture of adhesives.

Name the products it forms in the reaction with hydrogen chloride, HCl. Explain which is the major product.

Hydrocarbon reactions of commercial importance

Hydrocarbons are involved in many industrial reactions to produce commercially viable products.

Addition polymerisation of alkenes

Addition polymerisation involves the joining together of many alkene monomers to form a large saturated molecule called a polymer. For example, ethene can form poly(ethene):

ethene monomers → double bonds open → and join with neighbours → to form a polymer

$$nCH_2 = CH_2 \rightarrow \left(CH_2{-}CH_2 \right)_n$$

During these reactions, **free radicals** are formed during an **initiation** step, often by reaction with an oxygen or peroxide catalyst. During **propagation** steps, free radicals react with a monomer forming another free radical with a longer carbon chain. This process continues forming longer and longer free radicals. Two long free radicals then combine to form the polymer in a **termination** step.

Cracking of alkanes

In the cracking of alkanes, larger hydrocarbons with lower commercial demand are broken down by thermal decomposition into smaller, higher demand alkanes and alkenes. The commercial reaction can be replicated in the laboratory using liquid paraffin, which contains decane, $C_{10}H_{22}$.

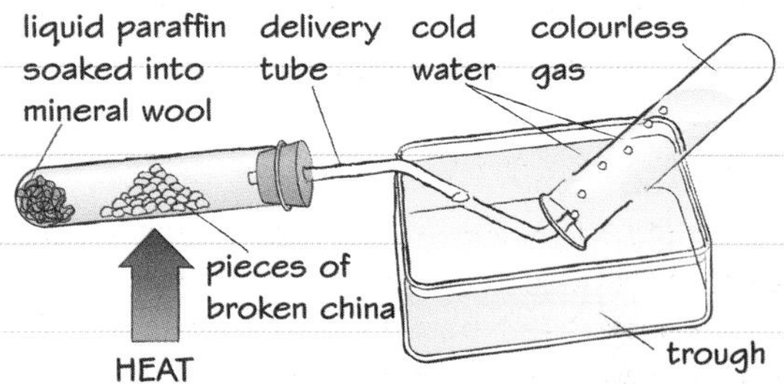

The pieces of china act as a catalyst and the products are octane and ethene:

$$C_{10}H_{22}(l) \rightarrow C_8H_{18}(l) + C_2H_4(g)$$

An example of cracking in industry uses the fraction diesel, with hydrocarbon chains in the range 14–20 carbons. This is cracked using heat (around 700 K) and a zeolite catalyst. The major products are ethene and propene, used to make polymers, and petrol, used as a fuel.

Hydration of ethene

Ethene, produced from the cracking of crude oil fractions, can be converted into ethanol by direct hydration with steam:

$$C_2H_4(g) + H_2O(g) \rightarrow CH_3CH_2OH(g)$$

The reaction is exothermic, so higher temperatures produce a lower yield. However, to increase the rate of reaction to a suitable level, the process is carried out at 500 K with a phosphoric acid catalyst. A high pressure of around 60 atm is used. This gives a relatively high yield, but even higher pressures would compromise safety, be expensive and cause some of the ethene to start to polymerise.

Combustion of alkanes

The commercial use of alkanes in combustion reactions is linked to the length of the carbon chain.

Shorter hydrocarbons such as butane and propane, can be liquefied under pressure and stored in tanks or canisters for portable use. Hydrocarbons such as octane are found in the fuel petrol, which is a volatile liquid fuel, often used in cars. Longer hydrocarbons are found in fuels such as diesel, kerosene (aviation fuels) and fuel oil, used in ships.

Read more about the combustion reactions of alkanes on page 113.

Now try this

The polymer polystyrene is used in insulation and packaging. It is formed from the monomer styrene, $CH_2{=}CH(C_6H_5)$.

Write an equation to show three styrene monomers joining together to make part of the polymer chain.

Issues with hydrocarbon use

- Crude oil, the main source, is finite.
- Polymers can be difficult to dispose of as they do not biodegrade.
- Combustion produces pollutants such as CO_2, solid particulates, acid rain from NO_x and SO_2.

Had a look ☐ Nearly there ☐ Nailed it! ☐

Enthalpy changes in chemical reactions

Enthalpy change, ΔH, in a chemical reaction is the change in energy, normally heat, at constant pressure.

Reaction profiles for exothermic and endothermic reactions

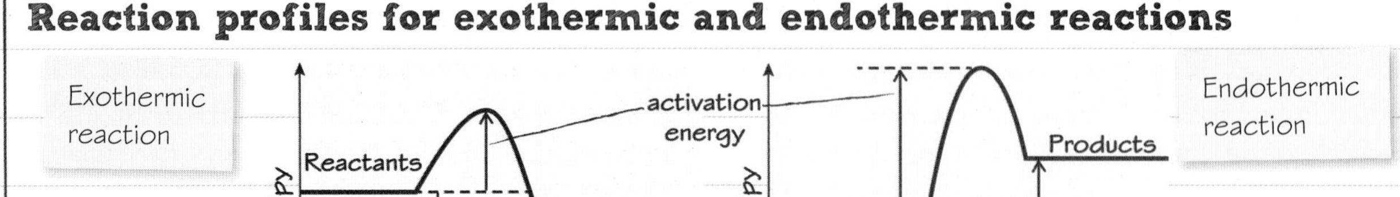

In an **exothermic** reaction, energy is transferred by the system (the reaction) to the surroundings (the rest of the universe), so the temperature of the surroundings increases. In an **endothermic** reaction, energy is transferred to the system from the surroundings, so the temperature of the surroundings decreases.

An enthalpy change can be defined as $\Delta U + p\Delta V$, where ΔU = change in internal energy, p = pressure, and ΔV = change in volume, within a system.

ΔH for an exothermic reaction has a negative sign as the enthalpy of the products is lower than the enthalpy of the reactants. ΔH for an endothermic reaction has a positive sign as the enthalpy of the products is higher than the enthalpy of the reactants.

Activation energy is the minimum collision energy needed for a reaction to occur. Usually only a small proportion of the particles in a reaction will have enough energy to react.

More about activation energy

Catalysts can speed up reactions by providing an alternative reaction mechanism that has a **lower activation energy**.

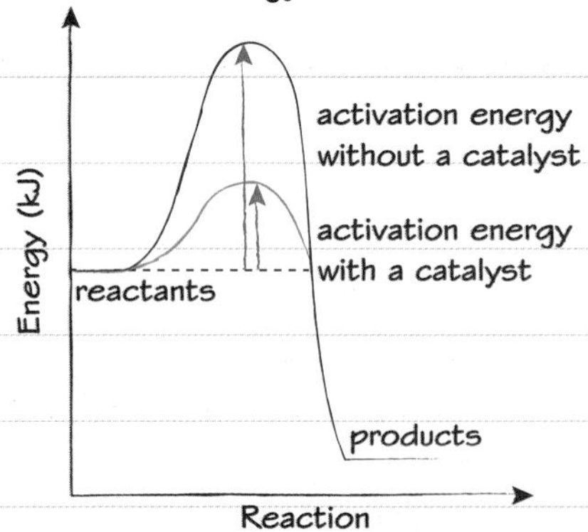

This means a greater proportion of the particles in the system have an energy greater than or equal to the activation energy, so the reaction is faster.

Transition metal complexes, a central metal ion with molecules or ions attached by a dative covalent bond, can act as catalysts in reactions.

Effective use of heat energy in industry

As energy costs for reactions are expensive, heat released by exothermic reactions in industry is transferred to other parts of a factory. This heat is then not wasted, as it can be used to speed up other reactions.

The use of **catalysts** is also vital as it means reactions can be carried out at an acceptable rate but with lower temperatures, so saving energy.

Iron catalyses the production of ammonia in the **Haber process**. The reactants adsorb onto the surface of the catalyst, react, then the products leave the catalyst surface (desorb).

Vanadium(V) oxide catalyses the formation of SO_3 in the production of sulfuric acid (**contact process**). It catalyses the reaction between SO_2 and O_2 to form SO_3, by oxidising SO_2, and then reforming:

$$SO_2 + V_2O_5 \rightarrow SO_3 + V_2O_4$$

$$V_2O_4 + \frac{1}{2}O_2 \rightarrow V_2O_5$$

Now try this

Platinum is a much more effective catalyst for the contact process than vanadium(V) oxide. It works in a similar way to iron in the Haber process. Suggest reasons that explain why, despite its effectiveness, it is not widely used in the process.

Measuring enthalpy changes in chemical reactions

The enthalpy change of a reaction in aqueous solution can be found using the temperature change.

Standard conditions for an enthalpy change of reaction

The standard enthalpy change of a reaction, $\Delta_r H^\circ$, is defined as the enthalpy change that occurs when the reaction is carried out using the amount of reactants shown by the equation, under standard conditions. Standard conditions are:

- pressure = 100 kPa
- concentration of solutions = 1 mol dm^{-3}
- temperature = 298 K
- shown by the symbol $^\circ$.

The Kelvin temperature scale

Kelvin is a unit of measure for temperature named after the physicist William Thomson, Lord Kelvin. It has the same magnitude as Celsius, but its zero point is at absolute zero, the lowest possible temperature. This means, unlike Celsius, it does not have negative values. When comparing the two scales you can see that:

- 0 K = −273 °C
- 273 K = 0 °C
- to convert from Kelvin to Celsius, you add 273
- to convert from Celsius to Kelvin, you subtract 273.

Simple calorimetry

Simple calorimetry is an experimental technique to find the temperature change during a chemical reaction. This set of data can then be used to calculate ΔH_r° for the reaction. In this reaction 50 cm^3 of 0.50 mol dm^{-3} $CuSO_4$ reacts with excess Zn. A maximum temperature rise of 24 K is recorded.

copper sulfate + zinc → zinc sulfate + copper

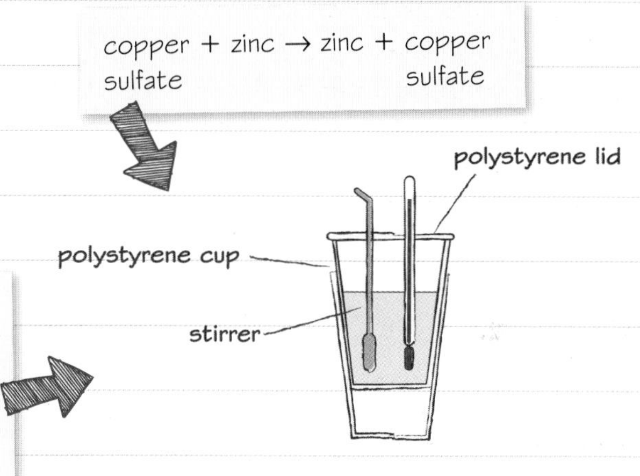

The polystyrene cup is used to prevent heat loss by conduction, the lid to prevent heat loss by convection. The stirrer ensures an even temperature distribution as well as making sure the Zn and $CuSO_4$ react quickly before too much heat is lost. An excess of Zn makes sure all the $CuSO_4$ reacts.

$CuSO_4(aq) + Zn(s) \rightarrow ZnSO_4(aq) + Cu(s)$

Calculating $\Delta_r H^\circ$

The equation used to calculate the energy change is $Q = mc\Delta T$. To calculate $\Delta_r H^\circ$ for the reaction between Zn and $CuSO_4$ you have to calculate:

1. $Q = 50 \times 4.18 \times 24 = 5016\,J = 5.016\,kJ$.
2. The amount of copper sulfate $= 50/1000 \times 0.50 = 0.025$ mol.
3. $\Delta_r H^\circ = -5.016/0.025 = -201\,kJ\,mol^{-1}$.

This standard enthalpy change has a negative sign as the reaction releases heat.

In the equation:

- Q is the energy change (J)
- m is the mass of water (g)
- c is the specific heat capacity of water (4.18 J K^{-1} g^{-1})
- ΔT is the temperature change (K).

As the density of water = 1 g cm^{-3}, we can assume the volume of the solution in cm^3, is the same as the mass, in grams.

Now try this

Calcium oxide reacts with water to release energy in self-heating cans.

The heating section of a self-heating can contained 50 cm^3 of water and the reaction released 7400 J of energy. Calculate the increase in temperature in the heating section of the can.

Had a look ☐ Nearly there ☐ Nailed it! ☐

Enthalpy changes of formation and combustion

Standard enthalpy changes of formation and combustion are used in enthalpy calculations.

Standard enthalpy change of formation

The standard enthalpy change of formation, $\Delta_f H^\circ$, is defined as the enthalpy change that occurs when one mole of a substance is formed from its elements in their standard state, under standard conditions.

For example, in this reaction, one mole of calcium carbonate is formed:

$$Ca(s) + C(s) + 1.5O_2(g) \rightarrow CaCO_3(s)$$

If the reaction is carried out under standard conditions, the enthalpy change will be the standard enthalpy change of formation of $CaCO_3$.

Standard enthalpy change of combustion

The standard enthalpy change of combustion, $\Delta_c H^\circ$, is defined as the enthalpy change that occurs when one mole of a substance, in its standard state, is burnt completely in excess oxygen, under standard conditions.

For example, in this reaction, one mole of ethanol is burned:

$$C_2H_5OH(l) + 3O_2(g) \rightarrow 2CO_2(g) + 3H_2O(l)$$

If the reaction is carried out under standard conditions, the enthalpy change will be the standard enthalpy change of combustion of $C_2H_5OH(l)$.

Measuring enthalpies of combustion

By transferring the heat released during a combustion reaction to water, the standard enthalpy of combustion can be calculated. The calorimeter needs to be made from a material with high thermal conductivity, for example, copper, to ensure the maximum amount of heat passes into the water. The pieces of data recorded are:

- mass of fuel burnt (g)
- maximum temperature change (K)
- mass of water in calorimeter (g).

In this experiment, 0.400 g of ethanol was burned and the temperature of 100 g of water increased by 25 K.

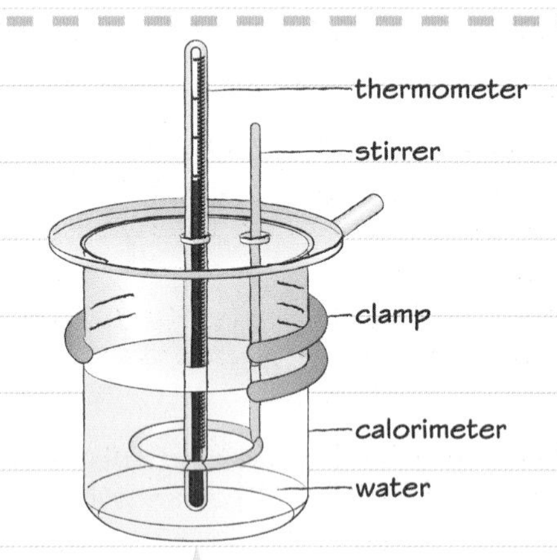

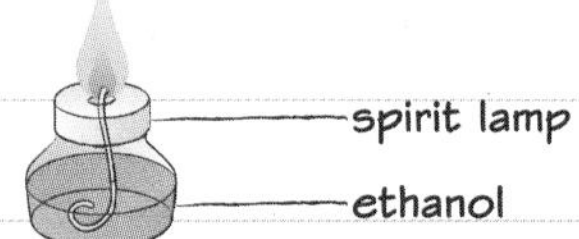

Calculating $\Delta_c H^\circ$

The equation used to calculate the 3 heat energy change is $Q = mc\Delta T$. To work out $\Delta_c H^\circ$ for ethanol you have to calculate:

1. $Q = 100 \times 4.18 \times 25 = 10450\,J = 10.45\,kJ$.

2. The amount of ethanol $= 0.400/46.0 = 0.00870\,mol$.

3. $\Delta_c H^\circ = -10.45/0.00870 = -1200\,kJ\,mol^{-1}$.

Standard enthalpy changes of combustion always have negative signs as they always release heat.

The true value of $\Delta_c H^\circ$ of ethanol, sometimes called the **literature value**, is $-1371\,kJ\,mol^{-1}$. The experimental value calculated is smaller in magnitude as:

- Heat from the combustion of the ethanol is lost to the atmosphere.
- Some unburnt ethanol may be lost when the spirit lamp is extinguished.
- Some of the ethanol may undergo incomplete combustion, producing carbon monoxide or carbon.

Now try this

A butane camping gas burner is used to heat 500 cm^3 of water from 20 °C to 80 °C. The change in mass of the burner is 2.80 g. Calculate the enthalpy of combustion for butane under these conditions. Is your calculated value likely to be higher or lower than the true value? Justify your answer.

Enthalpy change of hydration

When ionic solids dissolve in water, ions interact with water molecules in a process called hydration.

Dissolving ionic solids

When soluble ionic solids, such as sodium chloride, dissolve in water, two processes occur. The ions break away from each other in the ionic lattice. This process requires energy, so is endothermic. The newly separate ions then form electrostatic interactions with the slightly charged water molecules. This process releases energy so is exothermic. The energy released in the second process compensates for the energy requirement when the lattice breaks, meaning water is a good solvent for many ionic compounds.

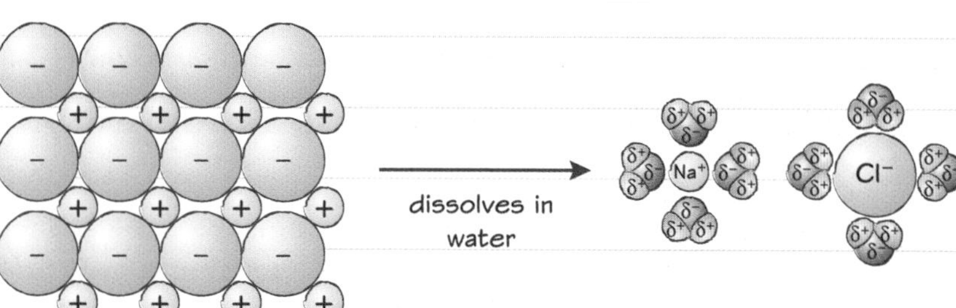

The positive sodium ions form strong interactions with the oxygen atom in water, as the oxygen has a slight negative charge. The negative chloride ions form strong interactions with the hydrogen atoms in water, as the hydrogens have a slight positive charge.

Standard enthalpy change of hydration

The standard enthalpy change of hydration, $\Delta_{hyd}H^{\circ}$, is defined as the enthalpy change that occurs when one mole of gaseous ions is dissolved in water to make an infinitely dilute solution. This is a solution with enough excess water to make sure it does not affect the overall enthalpy change.

Literature values for $\Delta_{hyd}H^{\circ}$ can be found in data books. For example:

$Na^+(g) \rightarrow Na^+(aq)$, $\Delta_{hyd}H^{\circ} = -406\,kJ\,mol^{-1}$

$Cl^-(g) \rightarrow Cl^-(aq)$, $\Delta_{hyd}H^{\circ} = -378\,kJ\,mol^{-1}$

Factors that affect the standard enthalpy change of hydration

The two key factors are:

Charge on the ion – the greater the charge on the ion, the stronger the electrostatic interactions with water. This gives a more exothermic enthalpy of hydration.

Size of the ion – the larger the ion, the more the charge is spread out over the ion. This means the electrostatic interactions with water will be weaker, giving a less exothermic enthalpy of hydration.

Using standard enthalpy change of hydration to calculate the enthalpy change when ionic solids dissolve

By comparing the energy required to break the lattice and the energy released when the ions interact with water, the enthalpy change of solution, $\Delta_{sol}H$ can be calculated:

$$\Delta_{sol}H = -\Delta_{LE}H + \sum\Delta_{hyd}H^{\circ}$$

So for sodium chloride:

$$\Delta_{sol}H = -(-787) + (-406 - 378) = +3\,kJ\,mol^{-1}$$

$\Delta_{LE}H$ is called the lattice enthalpy and is the energy released when the gaseous ions combine to form the lattice. Hence, $-\Delta_{LE}H$ is the energy needed to break apart the lattice. For sodium chloride, $\Delta_{LE}H = -787\,kJ\,mol^{-1}$

'$\sum$' means 'the sum of' as enthalpy changes of hydration for both ions in the compound have to be added together.

The positive value for $\Delta_{sol}H$ means when sodium chloride dissolves in water, the enthalpy change is slightly endothermic.

Now try this

Ammonium nitrate, NH_4NO_3, is mixed with water in 'Instant Cold Packs', used to treat sports injuries.

Use the data to calculate the enthalpy change of solution, when ammonium nitrate dissolves.

[$\Delta_{LE}H$ (NH_4NO_3) = $-646\,kJ\,mol^{-1}$, $\Delta_{hyd}H^{\circ}$ (NH_4^+) = $-307\,kJ\,mol^{-1}$, $\Delta_{hyd}H^{\circ}$(NO_3^-) = $-314\,kJ\,mol^{-1}$]

Had a look ☐ Nearly there ☐ Nailed it! ☐

Calculations using enthalpy changes

Standard enthalpy changes of formation and combustion can be used in a variety of enthalpy calculations.

Calculating $\Delta_r H^\circ$, the standard enthalpy of a reaction

$\Delta_r H^\circ$ is calculated using the relationship below:

$\Delta_r H^\circ = \sum \Delta_f H^\circ$ (products) $- \sum \Delta_f H^\circ$ (reactants).

For instance, $\Delta_r H^\circ$ for the breaking down of sodium hydrogencarbonate can be calculated by:

1. Writing the balanced equation:
   $2NaHCO_3(s) \rightarrow Na_2CO_3(s) + H_2O(l) + CO_2(g)$
2. Looking up values for $\Delta_f H^\circ$ given to you in the question.
3. Processing the data to find $\Delta_r H^\circ$
   $= (-1131 - 286 - 394) - (2 \times -951)$
   $= +91\,kJ\,mol^{-1}$.

$\Delta_f H^\circ$ data

Compound	$\Delta_f H^\circ$ ($kJ\,mol^{-1}$)
$NaHCO_3(s)$	−951
$Na_2CO_3(s)$	−1131
$H_2O(l)$	−286
$CO_2(g)$	−394

Note that the $\Delta_f H^\circ$ value for $NaHCO_3(s)$ is multiplied by two as the data in the table is per mole, but there are two moles of $NaHCO_3$ in the equation. The units for enthalpy changes are $kJ\,mol^{-1}$.

Calculating $\Delta_f H^\circ$, the standard enthalpy of formation

$\Delta_f H^\circ$ of a compound can be calculated using the relationship below:

$\Delta_f H^\circ$(compound) $= \sum \Delta_c H^\circ$ (elements) $- \Delta_c H^\circ$(compound).

For instance, $\Delta_f H^\circ$ for the formation of methane can be calculated by:

1. Writing the balanced equation for formation of one mole of methane from its elements:
   $C(s) + 2H_2(g) \rightarrow CH_4(g)$
2. Looking up values for $\Delta_c H^\circ$ given to you in the question.
3. Processing the data to find $\Delta_c H^\circ$
   $= [-394 + (2 \times -286)] - (-890)$
   $= -76\,kJ\,mol^{-1}$.

$\Delta_c H^\circ$ data

Compound	$\Delta_c H^\circ$ ($kJ\,mol^{-1}$)
$C(s)$	−394
$H_2(g)$	−286
$CH_4(g)$	−890

Note that the $\Delta_c H^\circ$ value for H_2 is multiplied by two as the data in the table is per mole, but there are two moles of $H_2(g)$ in the equation. Standard enthalpies of formation are often found using enthalpies of combustion, as experimentally it is easier to find the enthalpy change when elements and compounds burn, than it is when they combine directly.

Now try this

Hydrogen gas can be used as a fuel in internal combustion engines or fuel cells. It can be made from methanol in the reaction:

$$CH_3OH(l) + H_2O(l) \rightarrow CO_2(g) + 3H_2(g)$$

The methanol used can be produced from trees, and the combustion of the hydrogen produces no carbon dioxide, making this a possible source of more sustainable energy for transport. Scientists need to find out $\Delta_r H$ for reactions like this so they can choose conditions to get the optimum rate and yield for the reaction.

Use the standard enthalpy of formation data from the table above and the standard enthalpy of formation of methanol, $\Delta_f H^\circ(CH_3OH) = -238\,kJ\,mol^{-1}$, calculate $\Delta_r H^\circ$.

$\Delta_f H^\circ$ for elements is always zero.

The industrial process is carried out at about 600 K and 200 kPa. Explain why your calculated value may differ from the actual value for the enthalpy change of the industrial reaction.

Energy and work

To do work you need to transfer energy. It might be from potential energy (due to position and a force – as in hydro power) or kinetic energy (due to a moving mass – like a flywheel) or it could be from other kinds of energy. The energy used equals the amount of work done.

Force, F, and displacement, s

- Force and displacement (distance moved from a reference point) are vectors – that is, they each have a direction as well as a size.
- So, direction matters – the force and the distance moved must be in the same direction.

work done = force × distance moved in the same direction

$$w = F\Delta s$$

Remember from your GCSE studies

These two important energy formulae come from the work done by a force:

- **gravitational potential energy = $mg\Delta h$** (only vertical distance moved, Δh, counts because the force of gravity, mg, is vertical)
- **kinetic energy = $\frac{1}{2}mv^2$** (velocity, v, is the result of acceleration, a, caused by a force, $F = ma$).

Some equations will be given to you in the exam. Check the latest SAMS on the Pearson website to see which equations appear on the formula sheet.

Pressure, p, and volume, V

- A liquid or gas exerts pressure in all directions.
- When it changes volume by flowing or expanding it does work, pushing against its surroundings.

work done = pressure × volume change

Expanding hot gases are what move a rocket.

You don't have to worry about directions!

So this formula for work is very useful in thermal physics:

$$W = p\Delta V$$

'Delta' Δ always means **'final value minus starting value'** – that is, 'change in'.

Pressure is measured in **Pascal (Pa)**, which is the same as Newton per square metre (Nm^{-2}).

Volume is in **cubic metres (m^3)**.

Remember to convert units:

$1\,m^3 = 1000$ litres
$= 10^6\,cm^3$
$= 10^9\,mm^3$.

Power

Power is the rate of using energy or doing work.

The unit for energy is joule (J).

The unit for power is watt (W) = J/s.

You need to recognise the following big multiples:

- kilowatt (kW) = 1000 watts
- megawatt (MW) = 10^6 watts
- gigawatt (GW) = 10^9 watts.

pedal force
force on brake
master cylinder
brake fluid
slave cylinder

A hydraulic braking system

Other forms of energy

- Chemical energy (chemical bonding)
- Nuclear energy (nuclear reactions)
- Electrical energy
- Thermal energy (associated with temperature).

Now try this

The diagram in the box above shows how energy is transferred in a car from a foot pedal to the braking system. The diameter of a master cylinder is 10 mm and its piston is depressed by 25 mm using a pedal force of 12 N.

1 Calculate (a) the volume of fluid displaced, (b) the pressure of the fluid, and (c) the work done.

2 Calculate what force a slave cylinder piston of diameter 50 mm would deliver.

First calculate the piston area.

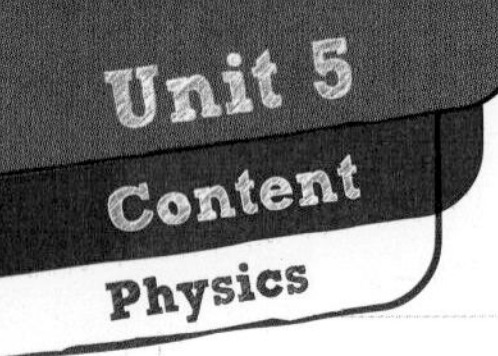

Thermal energy

All matter is made up of atoms and molecules. Thermal energy is the energy you cannot see or measure that is stored in their vibrations, rotations or other random movements.

Random molecular motion

The kinetic theory of matter describes:

- solids as atoms having fixed positions but able to vibrate
- liquids as molecules flowing past one another and colliding, while inside the molecule atoms still vibrate
- gases as molecules separated and travelling fast between collisions. The atoms still vibrate and the molecule can spin.

Thermal energy is stored in these motions. It is **added to whenever work is done on materials** – for example, when they are stirred, drilled, sawn, hammered, electric currents pass through them or fluids are pumped.

Mixing up pastry dough in a food processor gets it warm. To make crusty pies you need to cool it in a fridge before rolling, because rolling will also make it warmer.

Heat flow, *Q*

If two bodies are **in thermal contact**:

- ✓ the **quantity** of thermal energy that is transferred between the two bodies is called **heat**
- ✓ **heat transfer** happens by conduction, convection and radiation
- ✓ heat flows from the hotter to the cooler, and will keep flowing until they reach **thermal equilibrium**.

Temperature, *T*

The **Zeroth Law** of Thermodynamics states:

- ✓ there exists a physical quantity called temperature, such that
- ✓ any two bodies at the same temperature will be in thermal equilibrium with one another.

A **thermometer** can be anything with a measurable physical property that varies directly with its temperature.

A thermometer has to come to thermal equilibrium with its surroundings before it can be read.

Change of state

Heat transfer into a material will cause:

1. either temperature change: **sensible heat** (that is, you can sense its effect), or ...
2. a change of state (solid to liquid to gas/vapour, or vice versa): **latent heat** ('latent' means 'lying hidden' as binding energy of the material – you sense no temperature change).

During a change of state, latent heat is either absorbed or given out without a temperature change.

Now try this

1. *Work* and *heat* are both amounts of energy transferred. Explain what is different about the type of energy transferred as heat that distinguishes it from useful work.
2. A thermometer is placed into a child's mouth. Describe what happens to the temperature shown on the readout.
3. A beaker of water is left out on a frosty night and eventually begins to freeze. Its temperature is monitored at regular intervals. Describe and explain the shape of the temperature vs. time graph.

Gases and thermometry

To make valid measurements with a thermometer you first need to calibrate it against a temperature scale, but to establish the scale you need a chosen standard thermometer.

Ideal gases – pressure and volume

The ideal gas equation, $pV = NkT$, comes from kinetic theory, where we assume that gas molecules behave like:

- point particles, with
- elastic collisions, and
- no significant binding forces between them.

Some equations will be given to you in the exam. Check the latest SAMs on the Pearson website to see which equations appear on the formula sheet.

N is the number of particles. k is Boltzmann's constant, 1.38×10^{-23} J/K. T in this equation is proportional to the average kinetic energy of the gas molecules – an ideal concept of temperature.

So, an ideal gas would make a perfect standard thermometer.

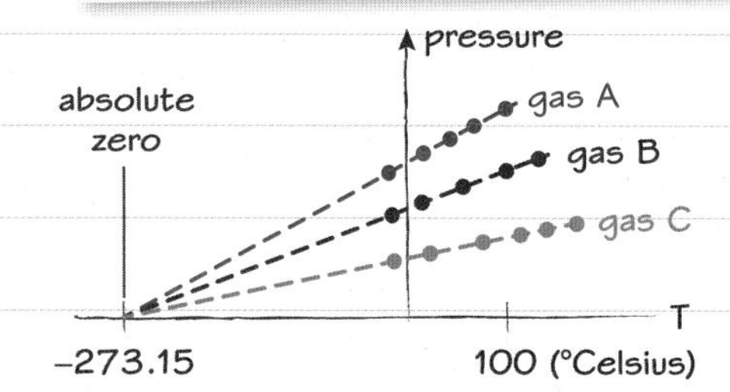

Real gases behave very like an ideal gas when they are well away from their condensation point – higher temperature, lower pressures.

Results for three different quantities of gas in a gas pressure thermometer show why you need to use absolute temperature, not °Celsius, in the ideal gas equation.

Absolute zero

0 Kelvin (K) = −273.15 °Celsius (°C):

- is where all thermal energy has been removed
- all molecules are in their ground (lowest possible energy) state
- cannot be practically achieved, though we can get close to it.

°C → add 273.15 → K

°C ← subtract 273.15 ← K

Temperature difference, $\Delta T = T_2 - T_1$, is the same either in Kelvin or in °Celsius.

Whenever you see T in an equation you must use Kelvin.

Practical thermometers you might use

Thermometer type	As temperature rises
liquid in glass	liquid expands
thermistor	electrical resistance decreases
platinum resistance	electrical resistance increases
infrared	increase in IR radiation emitted
thermocouple (measures temperature difference)	voltage difference increases

Now try this

1 (a) Liquid nitrogen boils at −195 °C. Restate that temperature in Kelvin. (b) 1 litre of liquid nitrogen contains approximately 1.74×10^{25} N_2 molecules. Calculate what volume it will initially displace at atmospheric pressure (1.01×10^5 Pa) when it boils to form a gas, and (c) by what factor this volume will increase as the nitrogen warms to room temperature, 20 °C? (d) Use these facts to explain why a spill of liquid nitrogen in an enclosed space can be hazardous.

2 Combustion in domestic fires produces a mixture of gases that has to be carefully ducted away. Detectors fitted to detect any leakage of poisonous carbon monoxide into the living space must be placed high up since it tends to rise.

Density is mass per unit volume: $\rho = m/V$.

Explain, using the ideal gas equation, why the different densities of gases at a given temperature and pressure are always in proportion to their relative molecular masses.

Had a look ☐ Nearly there ☐ Nailed it! ☐

Energy conservation

A helpful idea in thermal physics or chemistry is to draw an imaginary boundary line around each **system** that you are investigating, and to measure the transfers of energy in and out.

Energy transfers – work and heat

The **system** might be: hot gases inside an engine; a chemical reaction chamber; a single living cell or a whole ecosystem.

- Heat transfers, Q, into the system count as positive; flows out count as negative.
- Work done **by** the system (for example, a hot gas expanding) counts as positive.
- The actual direction of heat transfers depends on the temperatures outside the system – in external 'temperature reservoirs', which might be specially designed heat exchangers or just the system's surroundings.

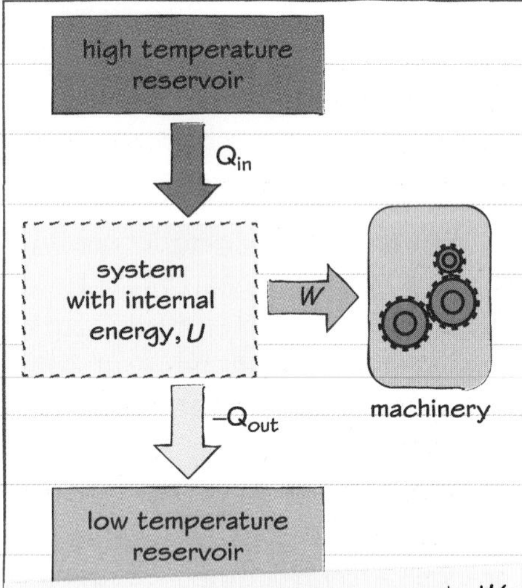

Transfers of heat, Q, **into** or work, W, **done by** the system count as positive.

Some equations will be given to you in the exam. Check the latest SAMS on the Pearson website to see which equations appear on the formula sheet.

Internal energy, U

Systems are complex, always with huge numbers of molecules and often with many parts. The 'internal energy' of a system includes all its:

- thermal energy (heat capacity)
- chemical and nuclear bond energies
- energy associated with physical states of matter (latent heat)
- kinetic or potential energy.

So U is far too complex to ever measure in total. We only calculate its changes:

$$\Delta U = U_2 - U_1$$

The First Law of Thermodynamics

This is just a statement of the Principle of Conservation of Energy as it applies to a thermal system:

$$Q = \Delta U + W$$

heat in = energy change in the system + work out

– or in the case of the system shown above with two heat reservoirs at different temperatures:

$$Q_{in} - Q_{out} = \Delta U + W$$

This equation prepares you to think about heat-driven engines and about chemical reactions that involve gases.

Specific heat capacity, c

The heat required to increase the temperature of 1 kg by 1 Kelvin (or °C)

$$Q = mc\Delta T$$

For water:

$c = 4.18\,\text{kJ}\,\text{kg}^{-1}\,\text{K}^{-1}$

Specific latent heat, L

The heat required to change the physical state of 1 kg of a material:

$$Q = mL$$

L has separate values for:

✓ Fusion (melting)
For ice to water:
$L = 333.6\,\text{kJ}\,\text{kg}^{-1}$

✓ Vaporisation
For water to steam:
$L = 2.26\,\text{MJ}\,\text{kg}^{-1}$

Now try this

A 3 kW electric kettle is filled with water at 15 °C and switched on. It then takes 2 minutes to come to the boil.

(a) Calculate the amount of heat input.
(b) Estimate the mass of the water in the kettle, and
(c) if it were left to continue boiling, how long it would take before it boiled dry.

Save time and energy

Limiting the quantity of water you heat is one of the biggest savers.

Processes

'It's not what you do, it's the way that you do it.' To change a system from one state (that is, a specific pressure, temperature, volume, internal energy, etc.) to another there are many possible **different process paths** that could be followed, and each involves different amounts of work and of heat transfer.

Natural processes

As time progresses, some things tend naturally to occur. For instance:

- Water flows downhill.
- Heat flows from a hotter body to a colder one.
- Substances mix.
- Gases expand to fill the space available.
- Exothermic chemical reactions (for example, burning).
- Friction turns mechanical work into heat.

The reverse of these processes never occurs naturally – they are **irreversible**.

Try imagining what each of these processes would look like on a film or video that is run backwards – clearly unnatural.

The Second Law of Thermodynamics

There are several *equivalent* ways of stating this law:

✓ A natural process can never be reversed in its entirety.

✓ It is impossible to completely change heat into work.

Links For more on efficiency, see page 127.

✓ Heat will not flow from a colder body to a hotter one without an input of work.

✓ The **entropy** (that is, statistical randomness) of the universe is always increasing.

✓ You cannot reverse the direction of time.

You can use this law to predict the **direction of a natural change** and hence the **position of equilibrium**, for example, for chemical reactions.

Reversible processes

Reversible processes are an **ideal**, as there are no fully reversible processes. It is a **useful idea** for thinking about equilibrium and about the limits of efficiency. There are two types of reversible process: **adiabatic** and **isothermal**.

Adiabatic (thermally isolated) processes

There is no heat transfer, $Q = 0$, so:

- work input = internal energy gained
  $-W = \Delta U$, that is, no energy is wasted
- Compressing a gas **adiabatically** leads to a corresponding temperature rise.

Real systems, even if well-insulated, are never fully isolated thermally. Changing things **too quickly** for heat to be lost can be near adiabatic, but friction and viscous drag will mean it is still not fully reversible.

Isothermal (same temperature) processes

- Two bodies in thermal equilibrium (that is, at the same T).
- Heat can flow back and forth in either direction – reversibly.
- But with zero temperature difference there would be no **net** heat transfer.

Real systems with very small temperature difference, for example, **heat exchangers**, can be **almost** reversible. They are **large and slow**.

Now try this

1 The cup of coffee shown in the box above first steams and then has milk added to it. Name two natural processes that occur and explain why reversing them requires effort or may even be impossible.

2 Explain why a bicycle pump gets warm when you pump tyres up quickly.

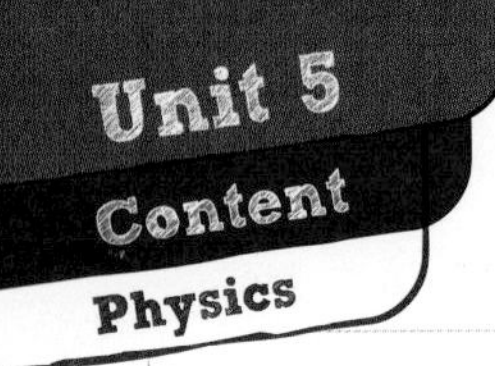

Had a look ☐ Nearly there ☐ Nailed it! ☐

Cycles

Engines, refrigerators and living organisms all use cycles of processes to cause changes and energy transfers to/from their surroundings while the system itself remains unchanged overall.

Ideal reversible (Carnot) cycle

Comprises:

- two **isothermal** processes at different temperatures, T_H (hot) and T_C (cold), with heat transfer amounts Q_{in} and Q_{out}
- two **adiabatic** processes – where $Q = 0$ by definition.

On a pressure–volume diagram, the **area** under the p–V curve for each process is the work done. Thus, the net work done in the cycle, W, is the area enclosed – positive when the direction of the cycle is clockwise, that is, for a **heat engine**.

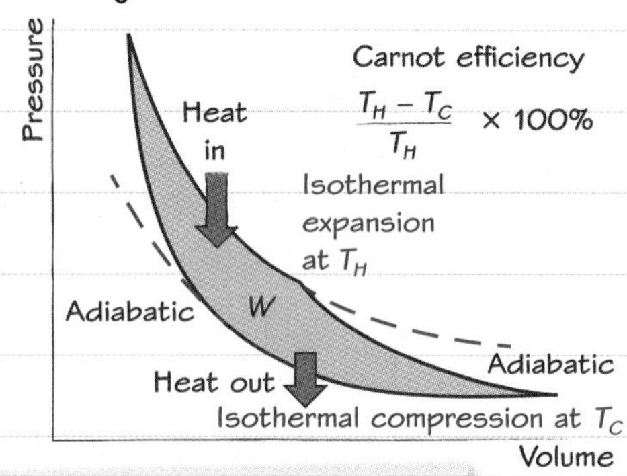

Links See page 127 for 'efficiency'.

Heat engines – steam turbines

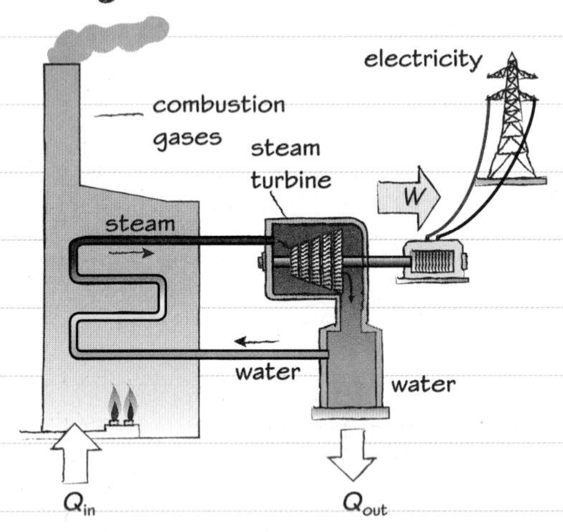

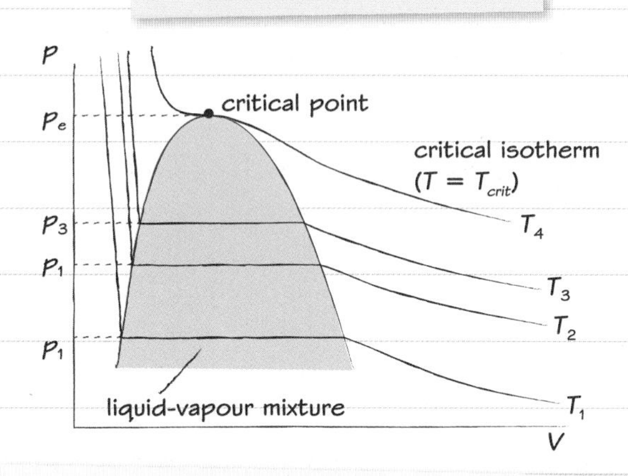

On the p–V diagram for water/steam a large part of the isothermal lines are also at a constant pressure, while the change of state is occurring.

Heat engines – internal combustion

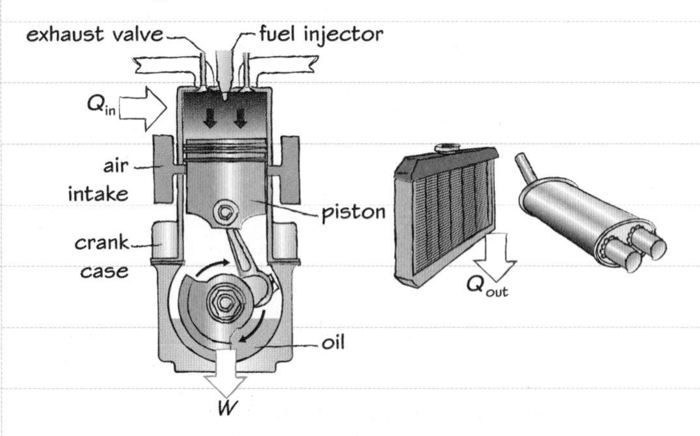

Power station steam turbines:

✓ come quite close to the ideal cycle by using very large, heavy heat exchangers.

Internal combustion engines:

✓ save the huge weight of a 'boiler' heat exchanger by burning the fuel inside the cylinder where it then expands, doing work

✓ benefit from a higher **thermal efficiency** (see page 127) due to the very high gas temperature produced by fuel/air compression followed by combustion.

Now try this

1 Both of the adiabatic processes in a Carnot ideal reversible cycle operate between the same two temperatures. Describe what is different between them, and explain why one involves more work than the other.

2 (a) Superheated steam (that is, heated above its boiling point) has a high specific heat capacity. Explain what that means, and why it is important in the steam engine cycle.

(b) Latent heat is also important in the cycle. Describe in which components of the steam engine cycle it is transferred, and in which direction.

Efficiency

How much can you usefully get out for the energy that you put in? When the input energy is heat then there are limits to what you can do with it: efficiency = $\frac{\text{'useful' energy output}}{\text{total energy input}}$

Some equations will be given to you in the exam. Check the latest SAMS on the Pearson website to see which equations appear on the formula sheet.

'Useful' and 'wasted' energy

With **engines**, heat counts as 'wasted' if it cannot be converted into 'useful' work. But for **heating buildings**, the 'waste' heat from an engine can become 'useful' heating – for example, in a combined heat and power scheme. Then only heat that escapes the building is 'wasted'.

Heat engine thermal efficiency

efficiency, $\eta = \frac{W}{Q_{in}} = \frac{Q_{in} - Q_{out}}{Q_{in}} = 1 - \frac{Q_{out}}{Q_{in}}$

- Heat input, Q_{in}, is supplied at a high temperature, T_H.
- The 'waste' output heat, Q_{out}, is at the low temperature, T_C, of the coolant and so is no longer useful for driving the engine.

Maximum efficiency (Carnot) engines

No engine can be more thermally efficient than an ideal reversible Carnot cycle engine, for which: $Q_{out}/Q_{in} = T_C/T_H$. Hence:

$$\eta_{rev} = 1 - (T_C / T_H)$$

These formulae give you efficiency as a fraction: to express it as a percentage, just multiply by 100%

A human body can be thought of as a heat engine 'burning' food energy at a temperature of 37 °C and rejecting heat from its surface, say at about 27 °C.

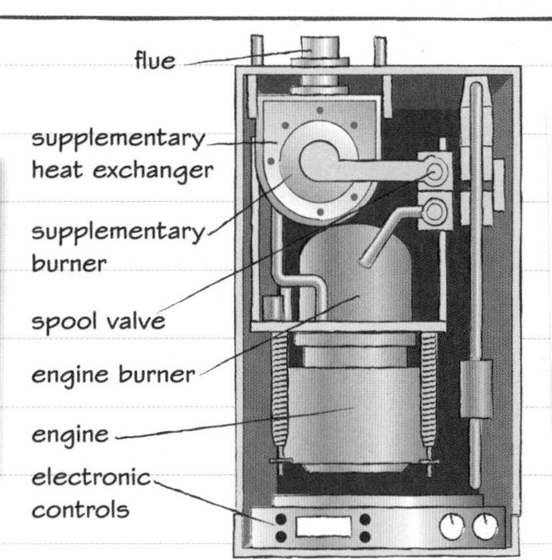

Micro combined heat and power (CHP) can generate electricity while providing heating and hot water.

Now try this

1 Calculate the maximum thermal efficiency of a reversible engine operating between the temperatures given in the box above for the human body.

2 Experiments on actual muscle fibres indicate thermal efficiency values of up to only about 0.4%.

(a) Calculate: (i) the work done by a 75 kg gymnast who performs 20 pull-ups, lifting the body each time by 30 cm, and (ii) the food energy input needed to drive that work output.

(b) Explain why exercise of this kind often makes a person perspire.

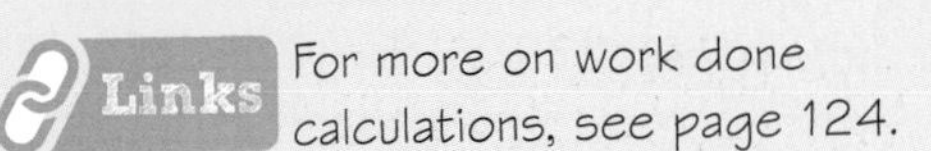

Links For more on work done calculations, see page 124.

Why is there excess heat to lose, and why must the skin be kept cool?

3 A micro combined heat and power plant, like the one shown in the box above, can generate 1 kW of electrical power while producing a heat output of 6 kW.

Estimate (a) a minimum value for the heat input that must be supplied by the gas burner, and (b) the corresponding thermal efficiency for the unit. (c) Explain why, despite its low efficiency value, micro CHP can nevertheless be considered an energy saving system, compared with a remote power station with 30% efficiency.

Had a look ☐ Nearly there ☐ Nailed it! ☐

Moving heat

Cooling or heating spaces to the desired temperature is becoming the biggest area of energy management. Reverse cycle vapour compression does this extremely cost effectively.

Refrigerators and heat pumps

Both use the same basic equipment, but:

- Refrigerators cool an enclosed space and reject heat outside.
- Heat pumps draw in heat from the ground or outside air and move that heat into a room or hot water tank at a higher temperature.

So they are distinguished by:

- the positioning of the heat exchangers, and
- which space is being temperature controlled.

Basic refrigeration

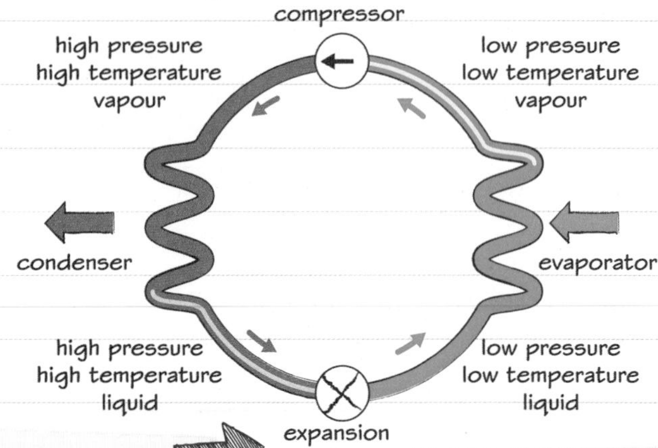

A simple expansion valve gives good temperature control, but does not recover the small amount of work done during expansion.

Notice the symmetry of the cycle diagram above

Coefficient of performance

CoP is the inverse of thermal efficiency → numbers bigger than 1.

- Heat moved, Q, is what is 'useful'.
- Work input, W, is the cost.

So, maximum theoretical (reversible Carnot cycle) CoP values are:

For refrigeration: $Q_{in}/W = T_C/(T_H - T_C)$

For heat pumps: $Q_{out}/W = T_H/(T_H - T_C)$

Heat extracted from food cabinets can be put to use.

Common refrigerants

- chlorofluorocarbons (CFCs) – but they damage the ozone layer
- alkanes – still some air pollution
- ammonia – needs large equipment
- carbon dioxide – no latent heat above critical point at 31°C.

To improve actual CoP value

You should try to:

✓ minimise temperature gap $(T_H - T_C)$ so:

- Use oversize heat exchangers.
- Tap large steady heat sources like underground water.

✓ Expand a liquid because the volume change is small, so not much work is lost.

✓ Cut friction or viscous flow losses.

✓ Keep compression work low – gas heat capacities affect this.

Choice of working fluid

High latent heat of vaporisation/condensation helps by keeping heat transfer processes isothermal and near reversible. This is why water/steam is a good fluid for heat engines. For refrigerators and heat pumps, using denser vapours keeps machinery more compact. Boiling and freezing points of refrigerants need to match with system temperatures.

Now try this

1 Calculate the maximum theoretical CoP for a heat pump taking heat from a ground coil at 8 °C and discharging heat at 30 °C.

2 Describe how you could adapt the supermarket cooling systems shown above so that the 'waste' heat is used for heating the shop.

Mention positioning of heat exchangers and control systems.

Elasticity

Elastic behaviour is the ability of a material to spring back to its original shape and size after being stretched, squashed or otherwise distorted.

Hooke's law

Force, F, is proportional to extension, Δx

$$F = k\Delta x$$

where k is a constant.

This is:

- about stretching a material (tension) or squeezing it (compression)
- not a general law
- obeyed by metals and some other materials for a limited range – that is, up to the **limit of proportionality**.

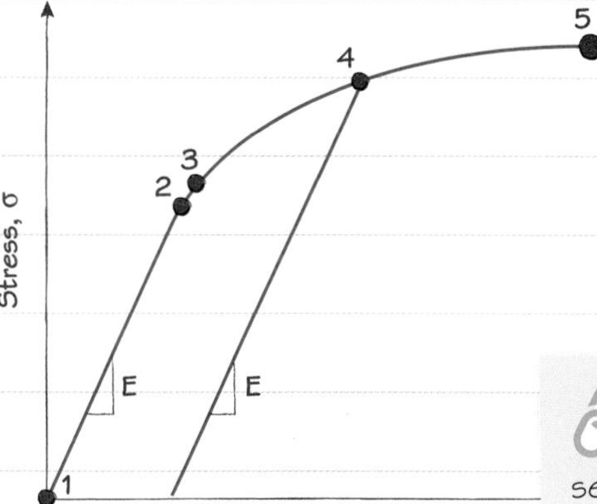

Tensile stress–strain curve for a non-ferrous metal:
1. Original state
2. Limit of proportionality
3. Elastic limit (yield point)
4. Increased yield strength after plastic deformation
5. Failure (UTS)

Links For plastic deformation, see page 130.

Tensile/compressive stress and strain

These enable you to take into account the dimensions of the original sample:

$$\text{stress} = \frac{\text{force}}{\text{cross-sectional area}}$$

$\sigma = F/A$ (Units: N m^{-2} or Pa)

$$\text{strain} = \frac{\text{extension}}{\text{original length}}$$

$\varepsilon = \Delta x/L$ (Dimensionless ratio – no units)

Some equations will be given to you in the exam. Check the latest SAMs on the Pearson website to see which equations appear on the formula sheet.

Young's modulus (elastic modulus), E

$$E = \frac{\text{stress}}{\text{strain}} \text{ N m}^{-2} \text{ or Pa}$$

Rearranging this gives another way of expressing Hooke's law: $\sigma = E\varepsilon$

Elastic limit (Yield strength):

- Highest tensile **stress** with full elastic recovery
- Beyond this the material **yields** to give a permanent **plastic** deformation
- Hooke's law and Young's modulus fail to apply from shortly before this point.

Springs and elastic energy

The work done (energy stored) is equal to the area under the F vs. Δx graph:

$$W = \frac{1}{2}F\Delta x$$
$$= \frac{1}{2}k\Delta x^2$$

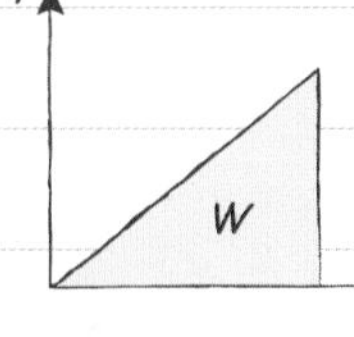

Elastic hysteresis

Elastic hysteresis occurs in materials like rubber, due to internal friction:

- Both stretching and recovery take time.
- The force is higher during loading.
- Work turns to heat.

Because they absorb energy these are useful materials for cushioning shock and for damping of oscillations

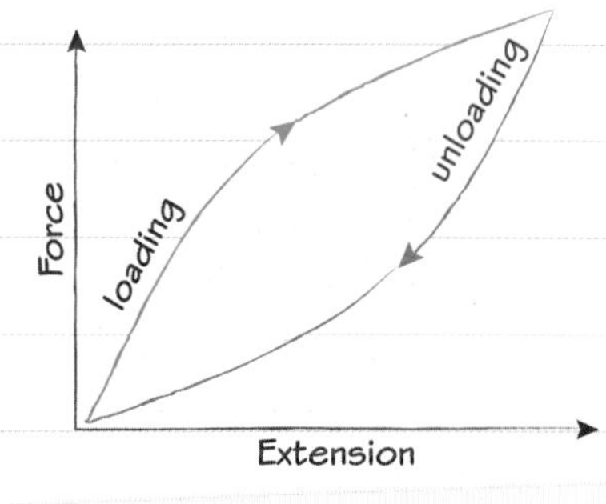

Energy lost due to internal friction equals the area enclosed by the **hysteresis loop**.

Now try this

1 A structural steel rod is 3.6 m long and 32 mm in diameter. Its Young's modulus is 200 GPa and its yield strength is 250 MPa.

Calculate: (a) the maximum tensile load it could carry without any permanent distortion, (b) the strain value at the yield point, and (c) the increase in its length when under that maximum load.

2 A spring that obeys Hooke's law is stretched by 12 mm above its original length by hanging a 300 g mass on it.

Calculate (a) the spring constant, k, and (b) the amount of elastic energy stored in the spring. (Assume the gravitational field strength, $g = 10$ N kg^{-1}.)

3 Give two examples of rubber being used to absorb energy.

Shape change

'Plastic' means 'able to be formed'. It describes materials experiencing a permanent change in shape without completely breaking up, and happens after they pass the elastic limit.

Plastic deformation

When the stress exceeds the **yield strength**, the crystalline structure of **metals** allows layers of atoms to glide over one another. When stress is removed the material stays deformed – has a **permanent set**. The amount of plasticity depends on the microcrystalline structure of the metal – so varies between materials. Though most non-metals show little or no plasticity (that is, they are **brittle**), some 'plastics' polymer materials, for example, nylon, can behave rather similarly to metals.

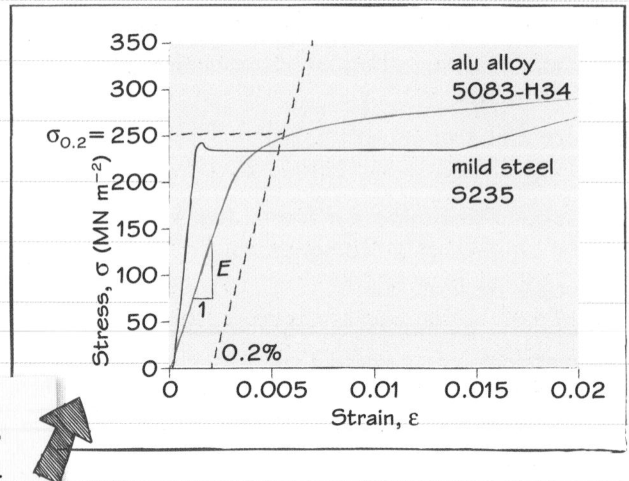

Two alloys with a similar yield stress. Steel shows a drop in stress before it begins to show 'work hardening'. Aluminium does not.

Ductility

Ductility is ability to be shaped by **plastic flow** under **tension**:

- drawing into rods, wires or tubes
- deep drawing of sheet material into bowls, cans, etc.

Ductility depends on temperature. If cooled sufficiently even metals become brittle.

Malleability

Malleability is ability to be shaped by **plastic flow** under **compression**:

- rolling or hammering into sheets
- cold forming by stamping or pressing.

Some metals – for example, lead – are malleable but not ductile. Many plastics materials and amorphous solids like clay are malleable.

How are metal products produced?

Stainless steel tubes are produced by **drawing** through a die to form the thin walls – a **tensile** process that requires **ductility**.

Cap heads of steel screws are produced by **drop forging** – a **compressive** process with a hammer and die on either hot or cold metal, which requires **malleability**.

Lead sheet is produced by **hammering** and **rolling** – **compressive** processes that depend on malleability.

Now try this

Look at the stress–strain curves shown in the box at the top this page. Explain (a) which of the materials is stiffer (that is, has a larger Young's modulus value) and (b) state what you can say from the graph about their ultimate tensile strengths (UTS).

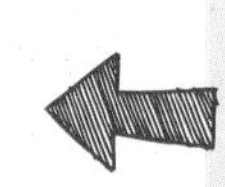

For Young's modulus, see page 129; for UTS, see page 131.

Failure

Understanding how and why materials fail is fundamental to securing safety and reliability.

Strength

Ultimate tensile strength (UTS) is the highest **stress** a material can sustain, just before it breaks.

For ductile materials, there is a lower **yield strength** – the stress at which plastic deformation starts.

Below that, a material can spring back fully to its original shape and size.

Brittle materials like concrete or glass only have a UTS – no yield point, no plastic deformation.

For more on material shape change, see page 130.

Brittle failure

- Happens when stress becomes concentrated at the tips of tiny imperfections and cracks that exist in the material.
- Cracks grow rapidly (visible under X-ray) and break right across the item.
- Most ductile materials harden as they are worked, until stress levels become high enough to cause a brittle failure.

Stress concentration due to a crack

Creep

Creep is a slow version of plastic deformation. It increases with temperature, and could eventually lead to failure if components no longer fit or if they neck in.

Fatigue

Fatigue is caused by repeated cycles of loading and unloading, which causes gradual hardening and crack growth. Eventual fatigue failure occurs at stress levels well below the normal UTS value.

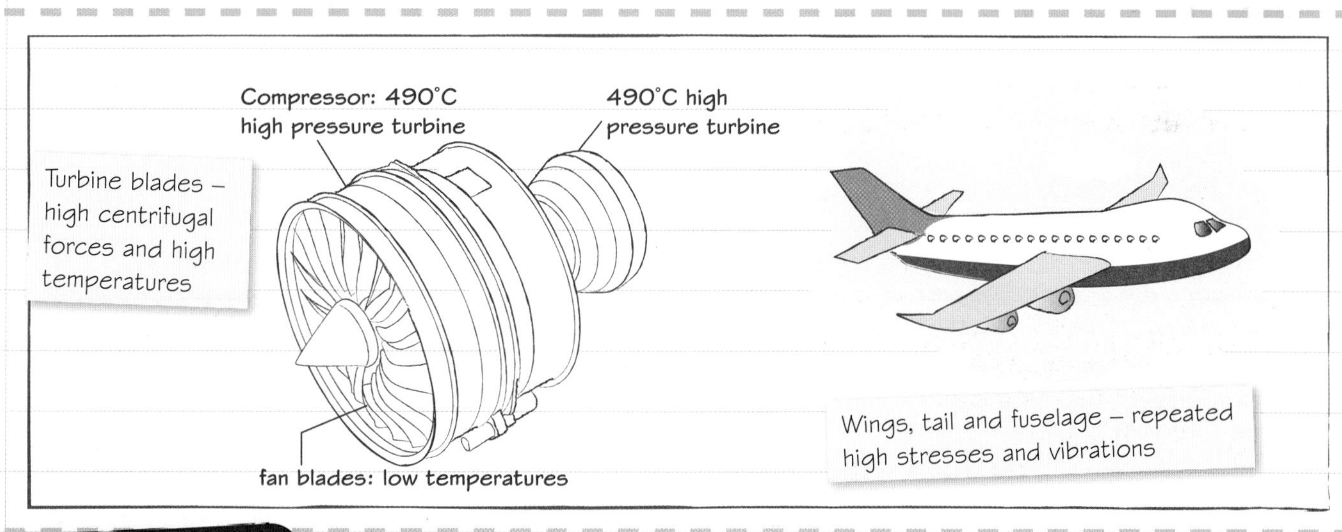

Now try this

1. Glass fibres can be very strong in tension, but only if they are plastic coated to prevent scratches developing on their surface.

   Explain how such scratches could lead to a brittle failure.

2. Concrete and stone have a high compressive strength, but are very brittle in tension.

   Suggest what features of their composition are responsible for this weakness.

3. Look at the box above. Avoiding failure in flight is an all-important concern for aircraft engineers.

   (a) Explain why creep could be expected in some turbine blades, and suggest what problems it might cause.

   (b) Explain why structural parts of an aeroplane may be subject to fatigue failure, and suggest what symptoms aircraft inspectors might search for prior to failure.

Fluid flow

Moving of liquids or gases differs from moving solids – because fluids do not have a fixed shape – but it is still about driving forces, resistance (viscous drag force) and energy used.

Streamline flow

Also known as **laminar flow**:

- It occurs at lower values of flow rate and pressure difference.
- Drift velocities of particles are all parallel and in the same sense.
- Fluid in contact with a solid surface has virtually the same velocity as that of the surface (that is, zero for a static solid surface).
- Velocity changes across the flow of the stream.
- It is the most energy efficient type of flow – so it is what engineers usually aim for unless mixing is needed.

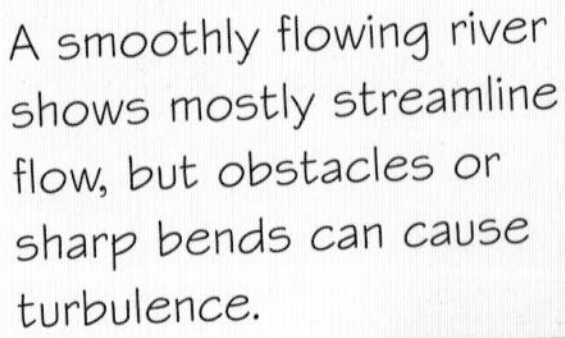
A smoothly flowing river shows mostly streamline flow, but obstacles or sharp bends can cause turbulence.

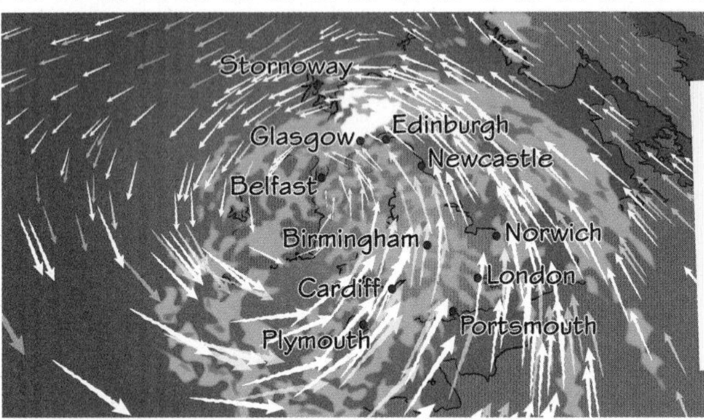

Much weather forecasting is about modelling turbulent flow, which is why there is so much uncertainty in it.

Turbulent flow:

- occurs at higher flow rates
- Includes rotational flows
- absorbs much more energy – generates more resistance to flow
- is chaotic – much more complex to model mathematically.

Viscosity

Fluids vary widely in their resistance to flow – different viscosities:

- layers of fluid moving at different speeds cause a **velocity gradient**, $\Delta u/\Delta y$ (rate of change of velocity with distance), where u is velocity in the x direction and y is perpendicular distance across the streams of flow – see diagram opposite.
- **(viscous drag force, F)/(area of layers, A)** = Shear stress, τ (measured in N m^{-2} or Pa)
- **dynamic viscosity**, η, is defined by **Newton's law of viscosity** using a simple streamline flow model, which gives the following equation:

$$F/A = \tau = \eta\,(\Delta u/\Delta y)$$

Units for η are Pa s

The onset of turbulent flow happens sooner for low-viscosity fluids.

Fluid sheared between two flat plates has a uniform velocity gradient.

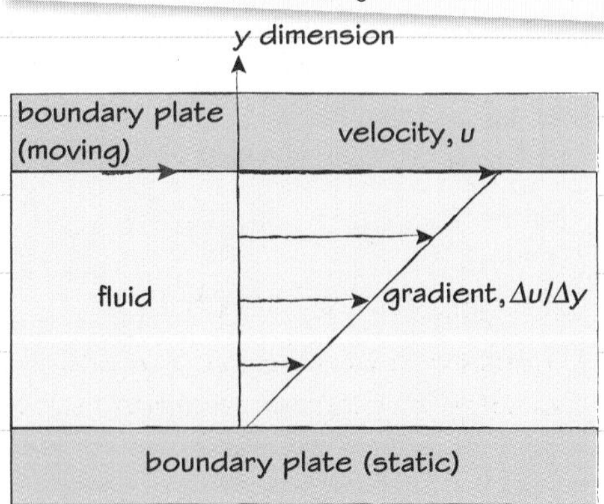

In a river or in a pipe, the velocity profile is curved (i.e. it has a changing gradient) with a maximum in the centre.

Now try this

1. Crankshaft bearings and pistons in a motor car engine are lubricated by a thin film of oil between the moving surfaces.

   Explain why, on a cold morning, a large force is needed to turn the engine over, but it gets easier once the engine is warm.

2. 'Streamline' designs for cars or aircraft are extensively tested in wind tunnels to check them.

   Explain why in these applications it is important to reduce turbulence to a minimum.

3. Suggest why turbulence in warm air convection currents could be helpful in achieving more comfortable heating of a room.

Non-Newtonian fluids

Most fluids are Newtonian, that is, their viscosity, η, is a constant value at any given temperature and pressure (liquids tend to thin with temperature rise, while gases become more viscous).
But 'non-Newtonian' liquids, whose viscosities change in other ways, have useful applications.

Shear thinning (pseudoplastic) and shear thickening (dilatant) fluids

These fluids change viscosity as soon as there is a shear stress (a velocity gradient).

- Mostly these are colloidal suspensions of solid particles or droplets in a liquid.
- Brushing, sliding or stirring a liquid provides shear stress.

Modern **pseudoplastic** paints thin when being brushed, but then thicken and don't run or drip.

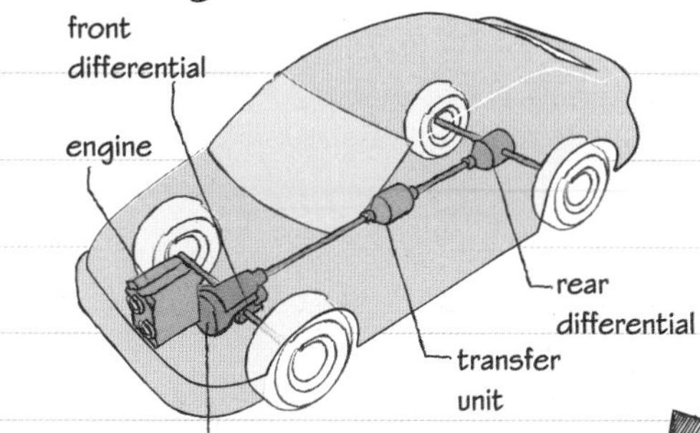

All Wheel Drive: a **dilatant** fluid in the transfer unit automatically detects slippage in the front wheels, solidifies and engages drive to the rear wheels as well.

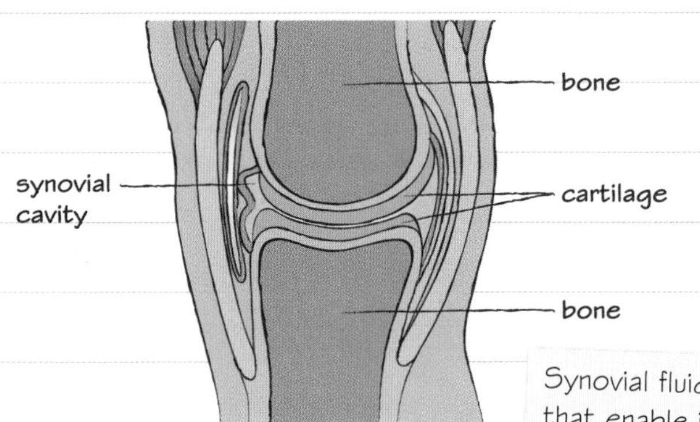

Time-dependent behaviours

- **Thixotropic** fluids thin **gradually** on stirring then slowly reset, for example, yoghurts, jellies and many other colloidal suspensions.
- **Rheopectic** behaviour is rare – a **time-dependent** thickening with shear stress, thinning again when it stops.

Synovial fluid has **rheopectic** properties that enable it to lubricate joints but also provide shock protection.

Bingham plastics

Bingham plastics behave like solids at low shear stresses, but above a yield stress they begin to flow. Examples of Bingham plastics include toothpaste, mud and slurries, mayonnaise, butter and margarine.

Now try this

Name the type of non-Newtonian fluid behaviour exhibited by each of the following:

(a) Ketchup (with xanthan gum as an ingredient) is very thick, but thins and flows when shaken.

(b) Damp sand on a beach feels soft when you are still but it stiffens to feel hard when you walk on it, making dry-looking footprints, which then disappear almost immediately.

(c) Quicksand seems firm, but gives way if you step on it and it thins even more you struggle, so you then sink even more quickly.

(d) Wet concrete or plaster mix will thin when it is stirred vigorously for a period. It then remains mobile for a while before gradually setting.

(e) Silk-screen printing ink requires some force to squeeze it from its tube and appears semi-solid, but, with firm pressure, a squeegee makes it flow freely through the fine mesh of the screen to create a print on the paper or other surface below.

Had a look ☐ Nearly there ☐ Nailed it! ☐

Fluid dynamics

By applying a few basic principles you can make useful calculations and predictions about fluid flow, despite the detail of the flow processes being too complex to tackle.

Mass flow continuity

Once a steady state flow has been established in a system of pipes, **mass flow rate, $\Delta m/\Delta t$**, must be the same:

- entering the system
- leaving from its outlet
- crossing every boundary along its length.

Mass flow continuity means that you can measure mass flow at any point that you choose along the flow.

Volume flow and pressure

Liquids in pipes: liquids are nearly **incompressible**, so mass flow rate continuity also implies continuity of **volume flow rate, Φ**, ($m^3 s^{-1}$).

The result for **streamline flow** is that flow rate, Φ, varies in proportion to the **pressure gradient**, $\Delta p/L$

where: Δp is pressure difference end to end
L is the length of the pipe.

Pressure drop in pipes

- Pressure drop increases if the viscosity, η, increases and if there is turbulence.
- Pipe radius has a dramatic effect – pressure drop varies inversely with the 4th power of r.

Bernoulli's principle

Applies the principle of energy conservation to steady flow systems:

$$\frac{1}{2}mv^2 + mgh + pV = \text{a constant value}$$

Dividing through by mass, m, the result is:

$$\frac{1}{2}v^2 + gh + p/\rho = \textbf{a constant value}$$

where ρ is density.

So when velocity, v, increases, pressure, p, will automatically drop, provided that the height, h, has not changed.

(g is the gravitational field strength)

Air has to travel further over the top of an aerofoil than underneath it. The pressure difference creates 'lift'.

lift
fast air
low pressure
slow air
high pressure

Pitot tubes, used to measure aircraft velocity, work by converting air speed into a pressure difference.

pressure transducer
full normal air pressure
reduced pressure due to speed of passing air

Now try this

1 The pressure generated by a central heating pump is sufficient to circulate water all around a certain house in 10 minutes. The pipes are currently 15 mm outside diameter with a 1.0 mm wall thickness.

Estimate what effect on the circulation time you might expect if the piping were changed to 10 mm OD 'minibore' pipe with a 0.7 mm wall.

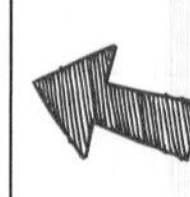

Start by calculating the pipe internal radius. Don't forget to allow for the reduced volume of the water inside the pipes, as well as the reduction in flow rate.

2 Explain why wind blowing over a curved sail causes a boat to move forward.

Your Unit 5 exam

Your Unit 5 exam will be set by Pearson and could cover any of the essential content in the unit. You can revise the unit content in this Revision Guide. In this Revision Guide, pages covering Organs and systems have the heading "Biology", those covering Properties and uses of substances are headed "Chemistry" and those covering Thermal Physics, materials and fluids are headed "Physics".This skills section is designed to **revise skills** that might be needed in your exam. The section uses selected content and outcomes to provide examples of ways of applying your skills.

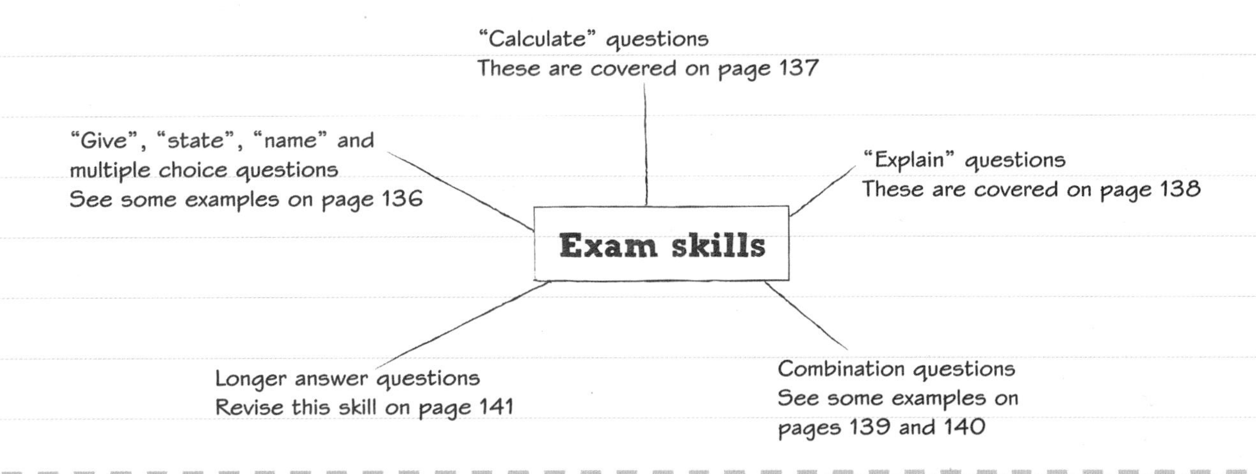

Exam checklist

Before your exam, make sure you

- ✓ Have a black pen you find easy to use, and at least one spare as well as a calculator and a ruler
- ✓ Have double-checked the time and date of your exam
- ✓ Get a good night's sleep

In your exam, you should

- ✓ Answer every question
- ✓ Always show your working
- ✓ Remember to use the formulae sheet and any other useful information at the end of the paper.
- ✓ Use any extra time to check your answers.

Check the Pearson website

The questions and sample response extracts in this section are provided to help you to revise content and skills. Ask your tutor or check the Pearson website for the latest **Sample Assessment Material** and **Mark Scheme** to get an indication of the structure of the actual paper and what this requires of you. Details of the actual exam may change so always make sure you are up to date.

Now try this

Visit the Pearson website and find the page containing the course materials for BTEC National Applied Science Look at the latest Unit 5 Sample Assessment Material (SAM) to get an indication of:

- the number of papers you have to take
- whether a paper is in parts
- how much time is allowed and how many marks are allocated
- what types of questions appear on the paper
- what additional information, such as a periodic table or formula sheet, may be given to you in the exam.

Your tutor or instructor may already have provided you with a copy of the Sample Assessment Material. You can use these as a 'mock' exam to practise before taking your actual exam.

Had a look ☐ Nearly there ☐ Nailed it! ☐

'Give', 'state', 'name' and multiple choice questions

'**Give**', '**state**' and '**name**' questions require you to recall one or more pieces of information. You usually get 1 mark for each correct part of the answer, but you might get 2 or more marks if you have to give a reason for something. Remember to think of a mark a minute being available so don't spend a lot of time on these questions. If you can't remember the answer, move on and come back if you have time at the end.

Worked example

The type of blood you have is related to the proteins found on the surface of your blood cells. State the eight main blood groups. **2 marks**

Sample response extract

A rhesus negative, A rhesus positive, B rhesus negative, B rhesus positive, AB rhesus positive, AB rhesus negative, O rhesus positive, O rhesus negative.

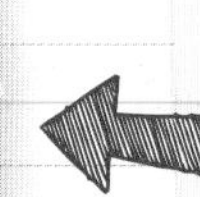

'State' here means you just have to write down the 8 answers. You get 1 mark for the ABO type and 1 mark for the rhesus negative or positive part.

For more on blood type systems, see page 91.

Worked example

Which blood group is the universal donor? **1 mark**

☐ **A** AB rhesus positive
☒ **B** O rhesus negative
☐ **C** O rhesus positive
☐ **D** AB rhesus negative

Try to read the question without looking at the responses, think of the answer and see if your answer is in the choices or there is a better answer. This way you will not be distracted by the choice of responses.

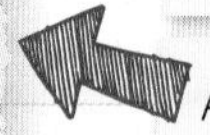

AB rhesus positive is the universal recipient.

Worked example

CO_2 is often used as a refrigerant. When it evaporates in the refrigerator's heat exchanger (evaporator), it absorbs heat at a rate of 286 kJ kg^{-1} of refrigerant changed from liquid to gas. Name that quantity. **1 mark**

Sample response extract

Specific latent heat of vaporisation.

'Name' indicates that a short, straightforward answer is wanted.

It is important to include the word 'specific', which tells you it is the quantity **per kilogram.**

Now try this

If a person has a blood transfusion, ideally they should be given a matched blood type or blood type O rhesus negative. However, if they are given only blood plasma it does not matter what the blood type is. Give a reason why this is.

'Calculate' questions

'**Calculate**' questions require you to obtain a numerical answer, showing relevant working. You must include the unit, if the answer needs one.

Worked example

CO_2 is increasingly being used as a refrigerant because most other choices of gas have a much more damaging climate change impact. CO_2 has a density of 1.977 kg m^{-3} at 0 °C and atmospheric pressure. Use the ideal gas equation to calculate an estimate of the volume of air that would be displaced by a spill of 100 g of CO_2 into a room at 18 °C. **3 marks**

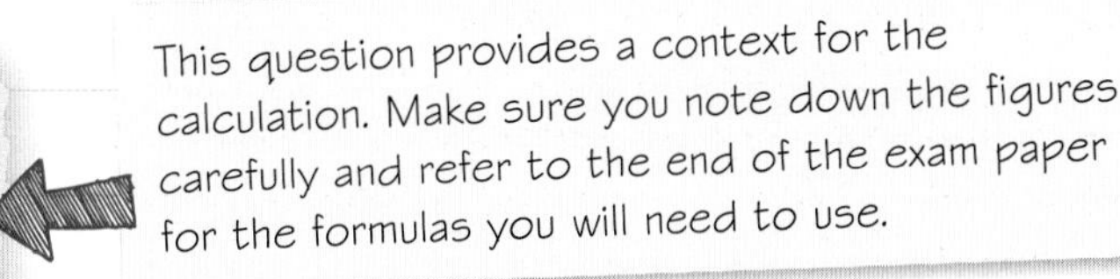

Sample response extract

Density, $\rho = m/V$

So, volume at 0°C, that is 273 K, $V_{273} = m/\rho$

$= 0.100\,\text{kg}/1.977\,\text{kg m}^{-3} = 0.0506\,\text{m}^3$

Using the ideal gas equation: $pV = nkT$, for the same pressure, p, and number of molecules, n, V will be proportional to T. So volume at 18°C, that is, 291 K

$V_{291} = V_{273} \times (291/273) = 0.054\,\text{m}^3$ or 54 litres.

The answer quotes the formula for density and the ideal gas equation, and then rearranges them as required.

The result is rounded to two significant figures, reflecting the uncertainty in the least precise data items (mass 100 g, temp to nearest 1 K), and includes the correct unit (either m^3 or litres would be acceptable).

Worked example

CO_2 liquid at −20 °C has a density of 1256 kg m^{-3} while the saturated vapour has a density of 62.3 kg m^{-3} and exerts a pressure of 2.0 MPa. Calculate the work done by 1 kg of CO_2 in expanding to its new volume. **3 marks**

This calculation has two steps, so make sure you show the formulas and all your working.

Sample response extract

$W = p\Delta V$ and $V = m/\rho$

For 1 kg of CO_2, $\Delta V = V_{vapour} - V_{liquid}$

$= (1/62.3) - (1/1256)$

$= 0.0153\,\text{m}^3$

So $W = 2.0 \times 10^6 \times 0.0153$

$= 3.06 \times 10^4\,\text{J}$

$= 30.6\,\text{kJ}$.

First calculate the volume change by using the two density values.

Be careful with the powers of 10 and don't forget to write in the units with your answer.

Now try this

Using the data about CO_2 given on page 136, and in the question above, calculate the total internal energy change, ΔU, for 1 kg of CO_2 in passing through the evaporator.

Revise the First Law of Thermodynamics on page 124.

You can use the value for the work done calculated in the second worked example above.

Had a look ☐ Nearly there ☐ Nailed it! ☐

'Explain' questions

'**Explain**' is probably the most common type of question. They require you to justify a point, using a clear process of reasoning. For maths questions, you must show your working. Some questions ask just for two things to be linked while others look for a longer series of connections and reasons.

Worked example

Cars made from mainly aluminium tend to be more expensive than those made from steel. One reason is that aluminium costs about four times more than steel. Use your knowledge of the Hall-Hèroult process to explain the costs of making aluminium. **4 marks**

Questions like this allow you to show your extending writing skills as well as the quality of your written communication.

Sample response extract

The process starts with the mineral bauxite. The costs of mining it and the equipment and the energy to transport it long distances add to costs. It then has to be purified into alumina, which uses energy and other reactants. Cryolite, used to lower the melting temperature, also has to be mined, processed and transported. However, the overall price of the cryolite is offset by the reduction in energy costs, due to the lowering of the melting point of the alumina.

The process of electrolysing alumina is continuous so makes savings on automation of the process and so labour costs. Electrical energy has to be supplied to create the high current needed to electrolyse the alumina. Some aluminium plants are sited near hydroelectric power stations to try to reduce these energy costs.

The answer tries to justify the costs in a logical order, from the start to the end of the process they have described.

Look how the answer tries to relate statements about 'energy costs' to specific examples from the process where energy is used, for example, to melt the aluminium ore.

Worked example

Rubber exhibits elastic hysteresis. Explain what this means, and why rubber is used as a shock absorber. **4 marks**

This question is worth 4 marks. Notice that the question has two parts. So your answer must cover both parts and have several points.

Sample response extract

When a stretching force is applied to rubber, the stress–strain curve follows a different path from what happens when the force is removed. A large amount of work is done stretching the rubber, but because of internal friction, much less stored energy can be recovered when it springs back.

Mechanical energy is converted into heat. This provides useful damping of vibrations, which makes rubber a good shock absorber.

This answer starts by describing the stress-strain curve. (You could also include a sketch graph to help your explanation.) Then it mentions work done and stored energy. It uses 'internal friction' to explain the difference.

Links For more on elasticity, see page 129.

The answer ends by linking the energy change to rubber's use as a shock absorber.

Now try this

Explain how the use of carbon anodes adds to the costs of the Hall-Hèroult process.

Combination questions (1)

Many questions in the exam will have several parts, which use different command words to encourage you to display different types of skills and knowledge. They often build on a topic, such as in this biology example, and in the chemistry example on page 140.

Worked example

Thermoregulation and osmoregulation can cause problems for distance runners and are linked. The body regulates temperature partly by sweating, as the evaporation of water from the skin needs heat energy. Two marathon runners, A and B, had their core temperatures recorded during a race. The graph shows the results.

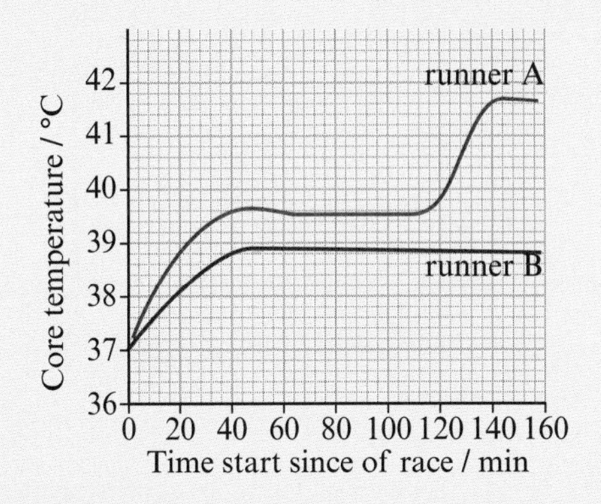

The graph shows two main things of interest:

- periods of change (both runners in the first 40 minutes and runner A after 120 minutes)
- constant periods (both runners between 40 and 110 minutes).

(a) Name the process, by which water molecules move from the tubule into the medulla. **1 mark**

The command word 'Name' ('State' might also be used') just requires a short answer and not a long explanation.

Sample response extract

Osmosis.

Links: Revise the kidney and osmoregulation on page 99–101.

(b) Runner A lost over 3 kg of water during the race.

(i) Deduce reasons for the change in core temperature after 120 minutes. **2 marks**

'Deduce' means you need to work something out – here it is the connection between water loss and inability to sweat.

Sample response extract

Not enough water available to produce sweat. Heat production will exceed heat loss.

Use the information you are given in the question.

(ii) Explain the effect of this water loss on the production of urine in the kidneys. **3 marks**

In part (ii) you need to be selective once more, keeping the answer relevant to urine production in the context of lack of water.

Sample response extract

Lower water potential of plasma stimulates ADH release from the pituitary. Collecting ducts are made permeable to water and water is reabsorbed into the blood. Urine is concentrated and low in volume.

(c) At about 2.5 hours into the race, runner A started to feel dizzy. Give a reason to explain this. **3 marks**

The command **'give a reason'** means that you should explain information given to you.

Sample response extract

The data indicates that runner A had become very dehydrated. His blood volume would have decreased and his blood pressure dropped. This could make him feel dizzy.

Now try this

Explain why water leaves the filtrate and returns to the plasma.

You can revise osmoregulation on page 100.

Combination questions (2)

Some questions in the exam will have several parts, which use different command words to encourage you to display different types of skills and knowledge. They often build on a topic, such as in this chemistry example, and in the biology example on page 139.

Worked example

A camping gas stove, containing butane, was used to heat 2.25 kg of water, in an experiment to estimate the enthalpy of combustion of butane. The temperature of the water rose from 21.0 °C to 91.0 °C.

(a) Write a balanced equation for the complete combustion of butane, C_4H_{10}.

Sample response extract

$C_4H_{10} + 6.5O_2 \rightarrow 4CO_2 + 5H_2O$

Halves are acceptable when balancing diatomic molecules such as O_2. An alternative correct answer is:

$2C_4H_{10} + 13O_2 \rightarrow 8CO_2 + 10H_2O$

(b) Calculate the energy absorbed by the water. The specific heat capacity of water = 4.18 kJ kg^{-1} K^{-1}

Sample response extract

Change in temperature = (91.0 – 21.0) = 70.0 °C

Energy = mass of water × specific heat capacity × temperature change
= 2.25 × 4.18 × 70.0 = 658 kJ

Remember a change of 1 K in temperature is the same as a change of 1 °C. Check units in your answer are consistent with data given in question.

Notice there is no sign on this answer, as it is an energy change, not an enthalpy change. The answer is given to three significant figures as the data in the question is to three significant figures.

(c) The mass of butane used to heat the water was 16.2 g. Use this information and your answer to (b) to calculate an enthalpy of combustion for butane.

Sample response extract

Amount of butane in moles = 16.2 ÷ 58.0 = 0.279 mol

Energy released by 0.279 mol = 658 kJ,

so enthalpy change = −(658 ÷ 0.279) = −2360 kJ mol^{-1}

Notice the value for the enthalpy of combustion has a negative sign, as the reaction is exothermic.

(d) The true value for the standard enthalpy of combustion of butane is −2878 kJ mol^{-1}. Suggest reasons that explain the difference between this value and your answer calculated in (c).

Sample response extract

The value from (c) is less negative, or smaller in magnitude, than the true value. This could be because the experiment with the camping gas was carried out in non-standard conditions. However, as the experimental value is less exothermic than the true value, this suggests not all the energy in the fuel has been used to heat the water. This could be due to incomplete combustion of the fuel, or to heat warming up the pan or the air surrounding the stove, instead of the water or energy absorbed by some of the water molecules changing state.

Take care when comparing negative values. Terms like 'smaller' or 'greater' can be confusing.

Now try this

Calculate the standard enthalpy change for the formation of propane, C_3H_8, using the $\Delta_c H^\circ$ data in the table.

Substance	$\Delta_c H^\circ$ (kJ mol^{-1})
C(s)	−394
H_2(g)	−286
C_3H_8(g)	−2220

Longer answer questions

There will be one or more longer answer questions. You may be given scientific information to analyse, and/or you may be asked to apply your knowledge of a topic, making links and comparisons. You need to present clear arguments and draw conclusions, emphasising the evidence.

Worked example

Explain why pregnant women are given a blood group test. **6 marks**

Sample response extract

Pregnant women are given a blood group test to see if they are rhesus positive or negative. If they are rhesus negative and their baby has inherited rhesus positive blood type from its father, the mother may be exposed to the baby's rhesus positive blood during labour. This causes the mother to develop antibodies. In later pregnancies, if the baby is rhesus positive the antibodies may attack the baby's blood cells and cause issues from anaemia to still birth.

Your answer will contain a number of scientific ideas, which need to be linked together in a logical order. You need to read the question carefully to work out what is required.

This answer clearly and logically states the reasons for the blood group test.

Revise blood types on page 91.

Worked example

Heat exchanger materials	Aluminium alloy	Copper alloy	Stainless steel
strength (UTS) (MPa)	124	380	586
ductility (% elongation)	25	65	60
relative cost	2	5	4
thermal conductivity ($W m^{-1} K^{-1}$)	180	77	16

Use the table to comment on which metal would be most suitable for making the thin-walled tubes and fins in a liquid-to-air heat exchanger. **6 marks**

Sample response extract

The requirement is to produce strong, thin-walled tubes and fins that can resist the high pressure of liquid inside, while also allowing rapid heat transfer that minimises the temperature difference between liquid and air. Stainless steel is less suitable than copper alloy because, despite its high strength and good ductility, its poor thermal conductivity would result in temperature drops between 4 and 5 times larger. Aluminium alloy has a lower ductility and lower strength than copper alloy, so the tubes and fins would need to be made about 2.5 times thicker. But its thermal conductivity is 2.4 times higher, meaning that the effect of the extra thickness is more or less cancelled out and temperature drops are similar.

Therefore, both copper alloy and aluminium alloy could be effective, but the aluminium has a cost advantage.

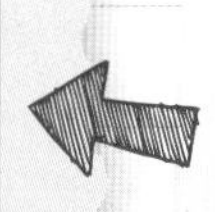

Use both the data given and your knowledge of the background to the topic. Carry out calculations and use the results to justify your answers.

Carry out calculations before you decide how to make your argument, for example, strength ratios – strength determines the minimum thicknesses that can be used. Then use the results as evidence to support your statements.

First, the answer makes clear the key requirements on which the materials are being compared.

Using copper alloy as the reference point, and using ratios to compare each of the others to it, has helped to make the arguments clear.

The answer has considered all the lines of data given.

Now try this

Find a longer answer question (often worth 6 marks) in the SAM. Answer it, using the advice above and then check your answer against the marking guide, which will give you an idea of how answers are expected to be structured.

Had a look ☐ Nearly there ☐ Nailed it! ☐

Issues and impacts

Science is moving forward fast in the 21st century and you will often find this new science has an impact on your life or the lives of people around you. This impact may be environmental, economic, social or ethical. Often these impacts will be interlinked.

Environmental impacts

The environmental impact of a scientific issue may be local or global:

Global impact: a new way of generating electricity that releases no carbon dioxide into the atmosphere will have a very positive impact. It could reduce global warming and protect habitats globally.

Local impact: if your local park is destroyed to build one of the new generating stations, the impact on your local environment will be negative, as the number of species living in the area falls.

Economic impacts

Many scientific discoveries have an economic impact, which you might not expect. For example:

- Putting raw sewage into rivers or the sea pollutes the water and reduces biodiversity.
- Many diseases are spread in polluted water.

Treating sewage costs a lot of money. But cleaning up has a positive environmental impact, which in turn has economic benefits. There is less disease and more tourists will visit – saving more money.

Lancelot Encore

When the owners of Lancelot, the dog in the photo, heard that a dog had been cloned in 2005, they stored some of Lancelot's DNA.

When Lancelot died of cancer, they paid over $150 000 to have him cloned. In 2009, the puppy Lancelot Encore was born ('encore' means 'again'). He is a clone of the original Lancelot. He fathered eight puppies in 2012 and continues to be fit and healthy.

However, cloning dead individuals has environmental, economic, social and ethical impacts.

Social impacts

Science has an enormous impact on society.

- You are probably vaccinated against many deadly diseases such as polio and diphtheria.
- You have probably had enough to eat all your life.
- You travel in cars, trains and aeroplanes.
- If you are ill, you will be treated with medicines to make you feel better or cure you.

As a result of developments like these, the human population has grown exponentially, changing social structures enormously.

Ethical impacts

Ethics relates to the morality of our actions – are they right or wrong?

As science makes progress, people voice concerns about the rights and wrongs of what is being done. Different individuals, and different countries, take different views of the ethical impact of science.

For example, the Large Hadron Collider hopes to help clarify events at the beginning of the universe. Some people think this is unethical as it disregards the actions of their God. Science is not good or bad – we have to decide what is ethical and how science is used.

Now try this

Look at the case history of Lancelot Encore.

(a) Suggest a possible environmental impact of cloning dead animals.

(b) What might be an economic impact of using cloning technology to replicate pets?

(c) Outline potential social and ethical impacts of cloning dead animals.

In each case, you need to think of how the technology might be used and the impact it will have.

Energy sources

We use **energy transfers** everywhere, from heating and lighting our homes and businesses to powering our cars and aeroplanes. The main source of the energy we use is **fossil fuels** such as coal, oil and natural gas. We need to find alternative fuels for two reasons. Fossil fuels are a **finite** resource. Eventually we will use them all up and they cannot be replaced. Also, burning fossil fuels increases the levels of **greenhouse gases** in the air. More and more evidence points to this as a cause of global warming and climate change.

Fossil fuels and global warming

When fossil fuels are burned in air, they produce **carbon dioxide**, a greenhouse gas. These gases help to keep the surface of the Earth warm enough to sustain life. Levels of greenhouse gases have been increasing steadily. The temperature of the Earth's surface is also increasing, causing global warming and affecting the climate, melting the Arctic and Antarctic ice and causing extreme weather events. Scientists think burning fossil fuels is the cause. They are working to find alternative energy sources that do not produce greenhouse gases.

Scientists are are also working on **carbon capture**. Waste carbon dioxide from electricity generation or industry is captured and moved to huge storage sites, often underground, where it cannot enter the atmosphere.

ITER

ITER is a huge international project trying to find a way to use nuclear fusion to provide the energy we all need. Nuclear fusion occurs when hydrogen nuclei collide and fuse to form heavier helium atoms. Huge amounts of energy are transferred to the surroundings. This is the process that powers the Sun.

Thirty-five nations are working together to build the world's biggest tokamak, an experimental magnetic fusion device. The scientists are trying to prove it is possible to use nuclear fusion to provide a carbon-free energy source on a really large scale. They hope one day to use this to generate carbon-free electricity.

Solar cookers and lights

In many areas of Africa and India, people do not have electricity. They have to forage to find fuel for cooking or use chemicals such as kerosene.

Kerosene can be expensive, and burning kerosene produces carbon dioxide and carbon. This affects the eyes and lungs of the people breathing it in. Kerosene lamps give poor light and are easily knocked over, causing fires.

Solar cookers and solar lights use energy from the Sun to produce heat for cooking or light for studying, working indoors or just being sociable. The energy used is free and renewable.

Now try this

Investigate one alternative energy source (either in use or potential), for example, the ITER project, solar, wind or hydroelectric power, biofuels. Find two sources of information about your chosen energy source and write notes about the environmental and economic impacts of your chosen alternative energy source.

Things to think about in your answer:

- how easily available is the energy source and who would be most likely to use it?
- what type of equipment is needed to harness the energy?
- how much useful energy does it produce and what is the payback time for the equipment?

Medical treatments

Medical and surgical treatments and techniques are changing fast. We may soon look back at the range of treatments available in the early years of the 21st century with the same mixture of horror and fascination that we now view removing a limb without anaesthetic!

Stem cell therapies

Stem cells can differentiate and become specialised cells needed in the body, for example, blood cells, muscle cells or nerve cells. Embryonic stem cells come from early embryos. Adult stem cells are extracted from adult tissues. Scientists want to use stem cells to treat lots of medical problems, for example:

- people paralysed because their spinal cord is damaged or because they had a stroke
- brain diseases such as Parkinson's, Alzheimers and other forms of dementia
- conditions where new, working tissues are needed, for example, diabetes, multiple sclerosis.

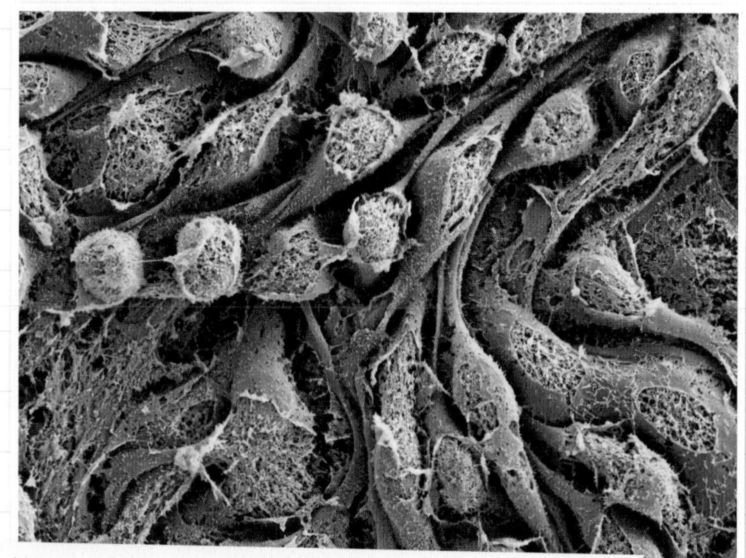

Human embryonic stem cells magnified around × 1000

Layla

When Layla was only 11 months old, she became the first person to have her blood cells gene edited. This allowed them to recognise and destroy the leukemia cells that were killing her. 18 months after the treatment, she was still doing well.

Gene editing

New techniques such as genetic engineering are opening up ways for scientists to change the information in our cells. They are looking for ways to use these techniques to cure many different diseases, for example:

- treating genetic diseases such as cystic fibrosis and sickle cell anaemia
- treating cancer by editing the immune system so it destroys cancer cells.

Surgical developments

Surgery is developing all the time. Doctors can:

- transplant organs, organ systems and even faces
- operate through tiny holes using keyhole techniques
- fit prosthetics – artificial limbs are improving all the time, for example, scientists and doctors are working to connect artificial limbs into the nervous system so they can be controlled like the original limb.

Issues arising

New medical developments often cost a lot of money per patient treated. Some of them raise ethical issues – for example the use of human embryos donated from infertility treatments as a source of stem cells. New treatments have to be tried out – would you volunteer to have a new treatment tried on you?

Now try this

Find out about a new medical treatment. List the potential advantages and disadvantages of this treatment. Include any economic, environmental, social or ethical issues raised by the treatment in your list.

This gives you practice at picking out information from one or more articles.

This may be one mentioned in the text or something you have heard about in the media. Ideally use more than one source to look up the treatment you have chosen. List your sources of information.

Pharmaceuticals

There is a constant search for new medicines to both make us feel better and to cure diseases. New technologies such as computer modelling and genome analysis are making this process faster and more efficient all the time.

The antibiotic timebomb

Penicillin was the first antibiotic and it became widely used in the 1940s. Since then, infections caused by bacteria have been cured using antibiotics. Now, bacterial pathogens are becoming **resistant** to the antibiotics that used to destroy them.

Scientists are racing to develop new antibiotics. It takes 10–12 years and over a billion pounds to discover and develop a successful new drug. So far, bacteria are developing antibiotic resistance faster than new antibiotics are being discovered. Fast and relatively cheap genome analysis of bacteria may help scientists get ahead.

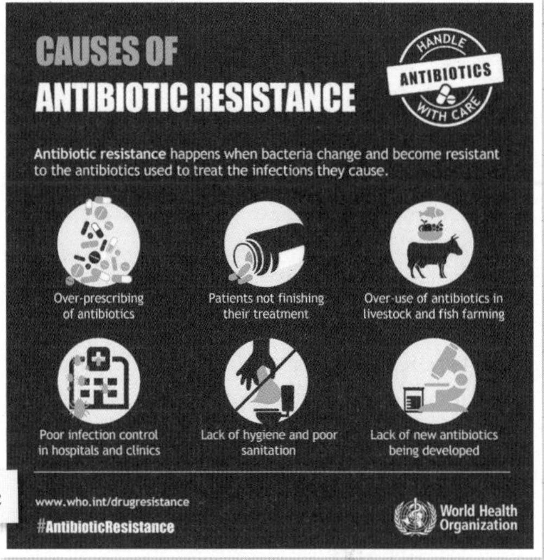

The development of antibiotic resistance

Performance-enhancing drugs

In almost every sport, there are some athletes who use drugs illegally to enhance their performance. These include anabolic steroids (to build muscles), erythropoietin (which increases the numbers of oxygen-carrying red blood cells), stimulants (to improve focus and alertness) and sedatives or beta blockers (when steady hands are needed). Professional athletes are tested regularly for drugs. It is a constant battle to develop accurate tests, to identify those who break the rules.

What is genome analysis?

Genome analysis involves identifying every single base pair and the order of the bases on every strand of DNA in the cells of an organism. Improvements in technology mean genome analysis is now highly automated. It is becoming faster and costs less all the time. Rapid, cheap genome analysis means we can identify inherited diseases and inherited risk factors and analyse the genomes of pathogens.

Pharmacogenomics and personalised medicine

The problem: Many medicines work well for some people but cause side effects or do not work in others.

1. **Pharmacogenomics as a solution**: Pharmacogenomics is the study of how genes affect the way a person responds to a drug. Using the results of big human genome analysis projects, scientists can see how different people respond to medicines. You can use current medicines more effectively and target the development of new drugs at people with specific genotypes.
2. **Personalised medicine as part of the solution**: In personalised medicine, you consider the genetic makeup of a particular patient, or even a specific cancer within a patient. You use the information to choose the most effective treatment for that individual.

Now try this

Either: produce a 3-minute presentation for young parents explaining the impact of antibiotic resistance and the issues around the development of new antibiotics,

Or: produce a similar 3-minute presentation explaining pharmacogenetics and/or personalised medicine including potential social and ethical impacts of genomic information being available.

In your assessment, you have to explain a scientific issue to a specific audience – this is good practice!

Think about the audience you are presenting to – what is their language level, how much will they know about the topic, how can you make it interesting?

Remember to be scientifically accurate. Think about the issues and make it suitable for the audience of young parents who will have small children who may quite often get ill.

Had a look ☐ Nearly there ☐ Nailed it! ☐

Chemical developments

Chemicals are developed to carry out specific functions and new chemicals often produce immediate benefits for people. Unfortunately, sometimes those same chemicals can cause unexpected problems in the environment.

Pesticides – good or bad?

- Growing enough food for people to eat is very important. Insect pests can destroy a crop. Farmers use **pesticides** to protect their crops and grow as much food as possible.
- Many crop plants such as fruit trees and bushes, sunflowers, etc. rely on insects to **pollinate** them. Without pollination they do not produce fruit and seeds and so without pollinators there is no crop.
- Bees (honeybees and solitary bees) are important pollinators.
- Some pesticides seem to have a damaging effect on bees, reducing their ability to find their way back to their hives and reducing the fertility of queen bees. For example, there is some evidence that neonicotinoids, a relatively new type of pesticide introduced to control crop pests such as aphids and root-feeding grubs, have affected bee populations in the UK and Europe.
- Science is not simple – the varroa mite is a bee parasite, which can destroy a whole hive of bees. It is spreading across the world rapidly.
- Pollinator numbers are falling fast, threatening crops worldwide. Is this the result of chemicals designed to protect crops, a new parasitic disease – or both? This is the sort of issue you have to work through as a scientist.

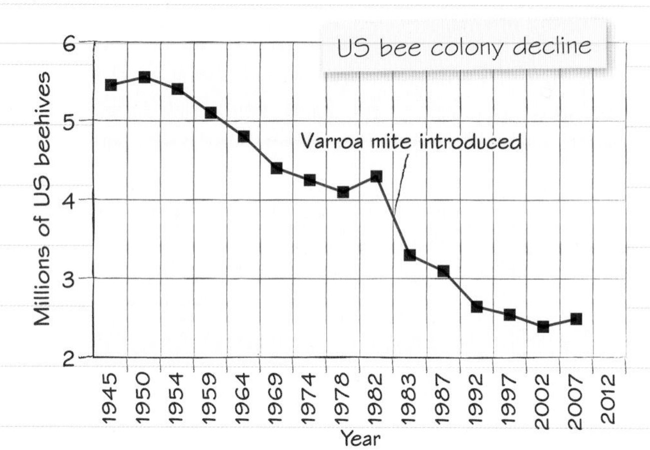

Plastics and fish

Think of how many things you use that are made at least partly of plastic. Plastics are extremely useful chemicals – but most plastics do not break down easily. Over 8 million tonnes of plastic waste gets into the oceans every year. Plastic pollution has been found in the Arctic and the Antarctic, carried by ocean currents. Plastic in the oceans can break down into tiny particles called microplastics. These are eaten by fish. Scientists are only just discovering the problems they cause.

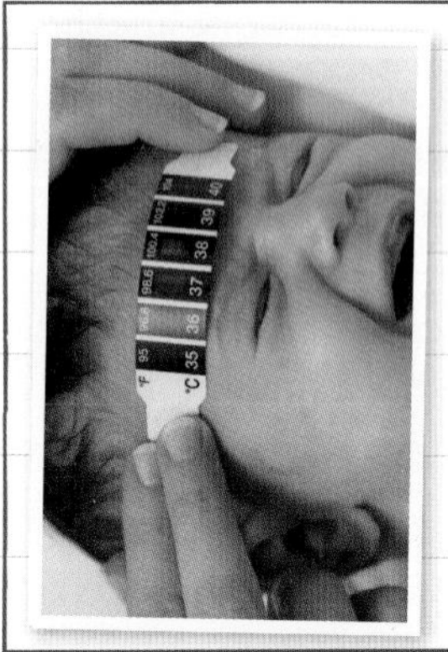

Smart materials

A smart material has one or more properties that changes in a controlled way in response to changes in the environment such as pH, temperature, stress, electricity, etc. The change is reversible and can be repeated many times. Examples include:

- shape-memory materials, which always return to their original shape
- materials that change colour at different temperatures, for example, for thermometers, especially for babies and small children.

Now try this

Plan or make an A3 poster to present to your fellow students on how we use plastics, including smart plastics. Give some of the economic, environmental, social and ethical aspects of plastics in our lives.

This question gives you practice at reading articles, selecting information and presenting it for a particular audience.

Posters are an important way of presenting information, both in industry and in scientific and medical research.

Nanotechnology

Nanotechnology is technology on a very small scale. It involves particles that are around 1–100 nanometres in diameter, in some cases, single atoms or molecules. Nanotechnology is only possible using powerful microscopes, which allow you to see and manipulate these very tiny units.

Electronics

Nanotechnology is making electronics faster and safer:

- replacing solder
- instant-on computers
- wearable electronics to record vital signs for fitness training and medical use with 100 nm thick silicon patches.

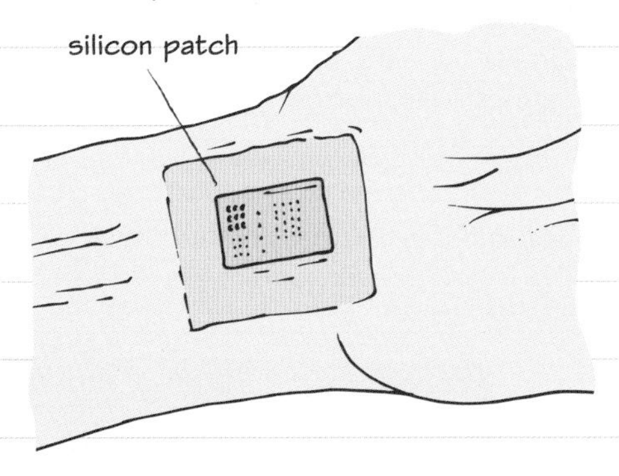

Space travel

NASA hope that carbon nanotubes and other nanotechnology will help them produce tiny space probes and rovers. Thousands or millions of these nanobots can be launched into space to explore other planets and give us the information we need to plan human space flights.

Energy

Energy needs are increasing all the time. Nanotechnology can be used to:

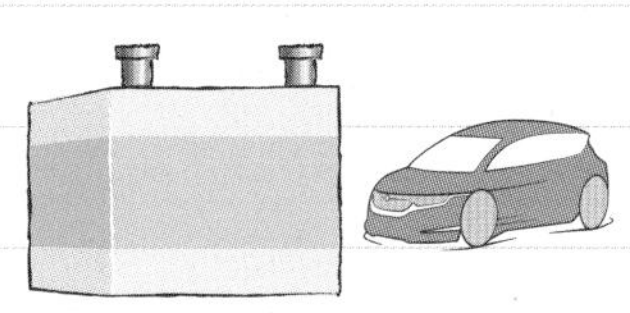

make batteries much lighter and more efficient

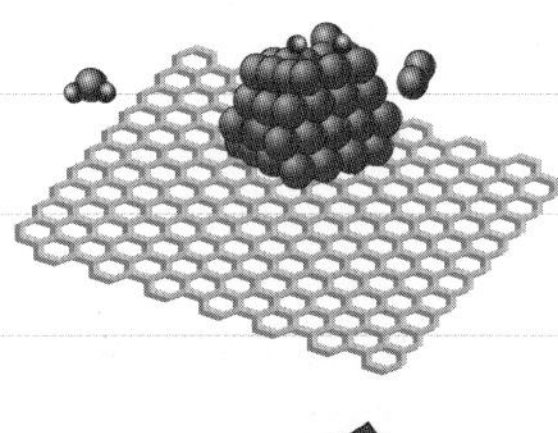

produce efficient catalysts for vehicles and manufacturing to remove greenhouse gases such as methane

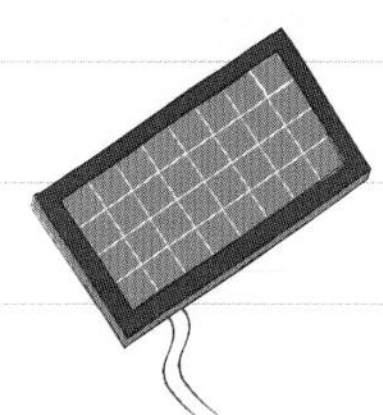

make solar panels more efficient

make special coatings to reduce water/ice build up on wind turbines.

Health

Nanosensors that can detect poisons and pathogens will help diagnose many diseases accurately and almost immediately. Nanoparticles can be engineered to recognise cancers for diagnosis and carry drugs to specific sites in the body, including the heart, the brain or tumours.

Now try this

1 Explain why the development of nanotechnology only happened after the development of powerful electron microscopes.

2 Make a table to show current uses of nanotechnology and planned uses of nanotechnology for the future.

A table can be a useful way of making notes on a topic.

Had a look ☐ Nearly there ☐ Nailed it! ☐

Food technology

Food technology applies the latest developments in science and technology to increase the amount of food available for everyone and to produce different types of processed food.

Genetically modified (GM) food

Genetic modification of food is the most powerful tool we have to help us solve global malnutrition. We can add desirable traits to our food plants and animals, by engineering the very genetic make-up of organisms. The development of new technologies such as Crispr Cas9 (a new, faster, and very precise way of inserting new genes into a genome) will make this genetic modification easier and more accurate. In many parts of the world GM crops are now widely used. In Europe there is still opposition. Most scientists see GM technology as key to producing enough food for everyone in years to come.

Some benefits of GM crops

- Can give increased yield.
- Can give increased nutritional, for example, added vitamins, more protein.
- Can survive adverse conditions, for example, drought resistant, storm/wind resistant, survive flooding.
- Can be disease resistant.
- Fruit can have longer shelf life, so less wastage.

Some concerns about GM crops

- GM genes may pass into wild populations of plants.
- GM crops will be too expensive for farmers in developing countries, which need them most.
- The inserted genes may affect people when they eat GM food (although we digest the DNA of every organism we eat anyway).

Golden rice

In many parts of the world, children go blind through lack of vitamin A. In 1999, GM Golden Rice was developed. It is very rich in vitamin A. Environmental protesters and political issues have prevented it being used for many years. Finally it won a US Patents for Humanity Award 2015 and is being trialled and developed in the countries where it is most needed. The golden colour comes from beta-carotene, the chemical that our bodies use to make vitamin A.

Novel foods

The food in this photo is made using protein from tiny fungi (mycoprotein) grown in huge bioreactors. Food technologists are searching for ways to make other new protein-rich foods from microbes.

Enzymes in the food industry

Enzymes extracted from microorganisms and other cells are used in the food industry. Uses include:

- food preservation
- producing glucose from starch and converting glucose to fructose, widely used as a sweetener
- making lactose-free products
- producing partly pre-digested baby foods.

Now try this

Look at the benefits and concerns about GM crops given here. Look online and find an example illustrating each of the points made. Make a note of the type of source you use in each case, for example, this article, Wikipedia, *New Scientist*, science journals such as *Nature*, the Mirror Online.

Some sources will be more reliable than others – and some will be biased for or against GM crops.

Links For more on validity and reliability of data, see page 157.

Links For more on supporting evidence, see page 160.

Government and global organisations

Science and scientific issues do not exist in a vacuum. You need to know some of the factors that influence the direction of scientific research. These include large **global** and **government organisations**.

Big organisations have big money behind them. They may make or at least influence laws, which affects the scope of scientific research. They have access to data that helps highlight areas where research is needed in terms of health, agriculture etc. They have influence with national governments and other international organisations.

Global organisations

- The United Nations (UN) involves 193 countries working together. It funds research into international issues such as food security and climate change.
- The International Atomic Energy Agency (IAEA) works within the UN looking at peaceful uses of nuclear technologies.
- The World Health Organization (WHO) collects data on disease and disease management around the world. It influences the focus of research into diseases and treatments.

Government organisations

Government organisations advise and influence scientific research in individual countries. They are part of the government and so can be affected by political changes. Examples include the UK Animal and Plant Health Agency (APHA), the US National Aeronautics and Space Administration (NASA) and the UK Food Standards Agency (FSA).

- They direct the work of scientists by allocating funding.
- They use research data to highlight areas for future projects.
- They may also use the results of scientific work to raise public awareness of science issues.

For example, the Food Standards Agency (FSA) produced posters, infographics and adverts to raise awareness of the dangers of *Campylobacter* in chickens for groups ranging from farmers who raise the chickens to everyone who cooks and eats chicken meat at home.

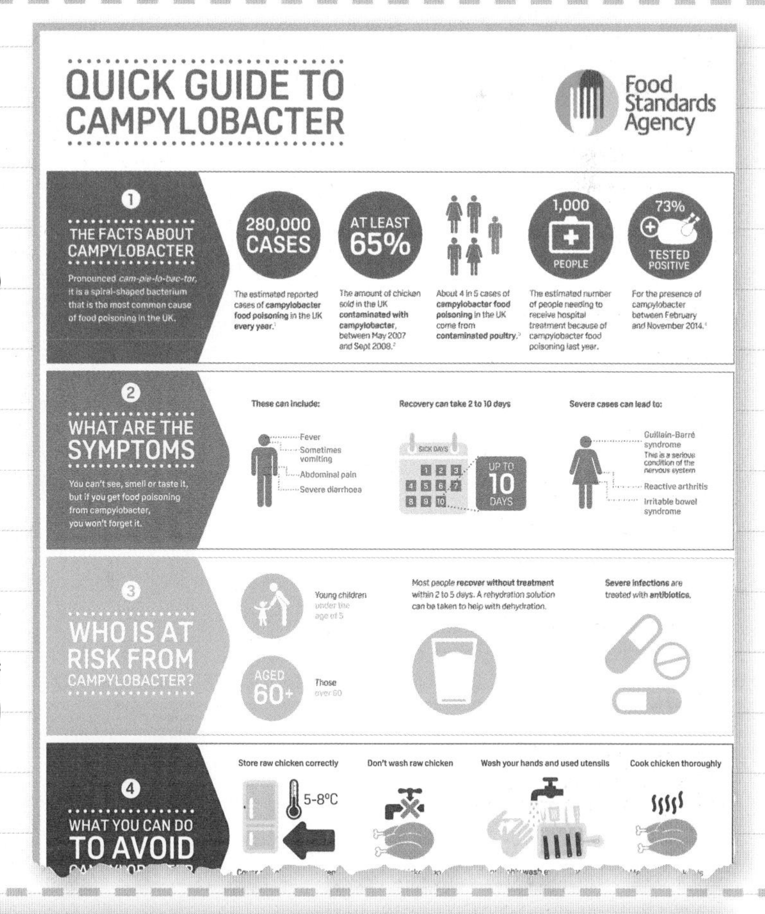

Now try this

1 State **two** ways in which global organisations can influence scientific issues.

2 State **two** ways in which government organisations can influence scientific issues.

NGOs

There are many different types of NGOs including national and international organisations, charities, learned societies and others, as you will see. Here you are going to focus on Research Councils and professional bodies.

- A **non-governmental organisation (NGO)** is a not-for-profit organisation, which is independent of the state or government. NGOs may be organised on a local, national or international level.
- Professional bodies and associations are NGOs that look after the interests of particular groups, for example, physicists, doctors. They have a major impact on science issues in the UK.

These organisations may:

consider economic, environmental, social and ethical aspects of scientific research
advise government and other bodies
allocate research funding.

UK examples

- The Royal Society of Chemistry, Royal Society of Biology and Institute of Physics are learned societies and professional associations for the three main sciences. Their roles include advising government on priority areas for research in each science. The General Medical Council (GMC) sets standards for doctors, decides if they are fit to practice and protects patients.
- Research Councils UK (RCUK) are responsible for investing public money in research in the UK – they invest around £3 billion each year. The individual science Research Councils include:

Science Research Council	Examples of current research priorities
Biotechnology and Biological Sciences Research Council (BBSRC)	agriculture and food security, bioscience for health, exploiting new ways of manipulating data, industrial biotechnology and bioenergy
Engineering and Physical Sciences Research Council (EPSRC)	energy, global uncertainties, digital technologies, healthcare technologies, quantum technologies
Medical Research Council (MRC)	infections, immunity, and antibiotic resistance, neurosciences and mental health, population medicine, global health
Natural Environment Research Council (NERC)	soil health, algal bioenergy, Arctic research, ecosystem services sustainability, human-modified tropical forests, climate change
Science and Technology Facilities Council (STFC)	big telescopes, the Large Hadron Collider (LHC), supercomputers and laser technology

- The Human Fertilisation and Embryology Authority (HFEA) is an independent regulatory body that licenses research into human fertility and the use of embryos in research. This body makes decisions about science issues based on UK and European law.
- The National Physics Laboratory maintains research into measurement in areas of science from acoustics through to electromagnets to radioactivity and time.

Now try this

1 What is a professional body? Give **two** examples.

2 Give **one** example of how professional bodies can influence science issues.

Universities and research groups

Universities and the research groups within them have an enormous influence on the direction taken by science today. Top universities and big name scientists draw funding and so they influence the direction of the research that takes place.

Funding is key

You might think that the research carried out will be on the most interesting science or the science most likely to make life better for people. In fact, it can be very hard to get money to support a research project. The university involved and the person heading the research team can make all the difference.

University departments

Top universities attract top scientists and top students. They also usually attract more funding from both governments and private investors.

As a result top universities often have a big influence on the direction of science research and the science issues considered important.

If a really important discovery is made at a university then a world famous laboratory specialising in that area of science may develop, for example, DNA fingerprinting at Leicester, Buckminsterfullerene at Surrey University (UK) and Rice University (USA).

National and international tables showing the performance of different universities are published each year. This table shows one set of listings for UK universities in 2016.

Top researchers

- Some scientists become widely known and respected by the scientific community.
- Well known scientists are more likely to be successful in grant applications for research.
- Students and other researchers want to work with top scientists.

As a result, a relatively small number of top scientists can influence the areas of science that are seen as important. In turn, these are the areas of science most likely to attract research funding.

Examples include Tim Berners-Lee, who developed the World Wide Web and Jane Goodall, whose work on chimpanzees changed the way society views and values primates.

Nobel prizes

Every year a **Nobel Prize** is awarded to the scientists responsible for the top discoveries/developments in Physics, Chemistry and Physiology or Medicine.

Nobel prize winners become very influential. They attract funding and they can influence the direction of future research.

In 2016 Jean-Pierre Sauvage, Sir J. Fraser Stoddart and Bernard L. Feringa won the Nobel Prize for Chemistry. For over 40 years, their vision has driven research into nanotechnology. Now, they will be more influential than ever.

Now try this

The ranking of a university and the fame and influence of individual scientists can influence the areas of science that attract funding. List **two** possible advantages and **two** possible disadvantages of this system.

Had a look ☐ Nearly there ☐ Nailed it! ☐

Private and multinational organisations

Science research is expensive. As you have seen, it depends on funding to take place. As a result, a lot of scientific research is funded by large private companies and multinational organisations, such as the big oil companies, the pharmaceutical industries, the Bill and Melinda Gates Foundation and some Fair Trade organisations. Each type of investor has their own agenda, which drives the science issues that are tackled.

Energy demands

Everyone wants electricity and transport. The big oil companies such as Shell and BP know that the supplies of fossil fuels such as oil and gas are running out. They are funding research into:

- new ways of extracting fossil fuels from the Earth
- more efficient ways to use fossil fuels
- cleaner, more environmentally friendly ways to use fossil fuels
- alternative renewable fuel sources to replace fossil fuels.

Big pharma

Pharmaceutical companies are constantly carrying out research into new medicines. It costs over £1 billion to produce one new medicine. They are businesses which have to make a profit, so they will tend to research:

- diseases that affect people in the richer countries of the world, which can afford to pay for medicines
- diseases that affect a lot of people – it is hard to make a profit if only a few people need the medicine.

Many pharmaceutical companies give medicines to developing countries for nothing or a very low cost – but they have to make money as well and this directs their research.

Philanthropy

Sometimes very wealthy people decide to invest some of their money in research or projects that interest them. If those projects are for the good of others, this investment is known as philanthropy.

- Bill Gates founded Microsoft. He is one of the wealthiest men in the world.
- In 2000, Bill and his wife Melinda established the Bill and Melinda Gates Foundation. This organisation provides funding for practical help in the developing world such as polio vaccines, and mosquito nets to combat malaria.
- It also funds research into malaria vaccines and drugs, new ways to combat global health challenges, and new crops to provide food and sustainable living.

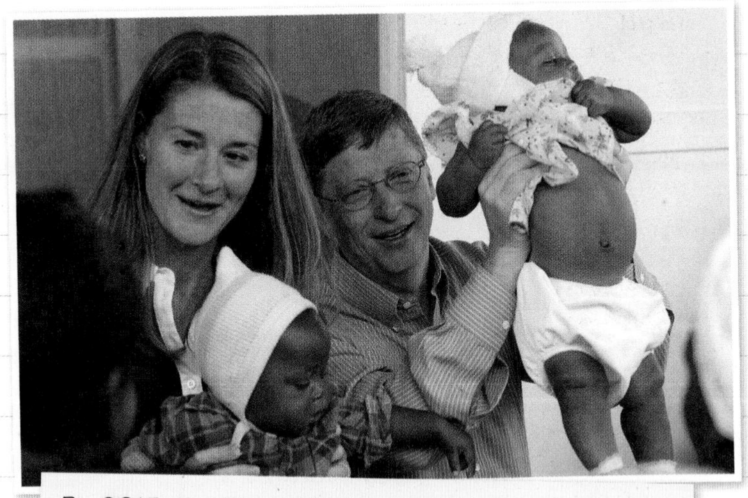

By 2015, the Bill and Melinda Gates Foundation, with their friend Warren Buffett, had given $39.6 billion to various projects. A lot of that money is funding scientific research into health care and agriculture.

Now try this

1 Give **one** similarity and **one** difference between research funding from big multinational companies and organisations like the Bill and Melinda Gates Foundation.

This will help you demonstrate that you understand the influence of different groups on scientific issues.

2 Suggest **two** problems that might arise from rich individuals funding research.

Think about the motivation of the individual and the ethics of the research.

Voluntary pressure groups

In the London Marathon, many people run to raise money for charities. In 2015, £54.1 million was raised, and, since 1981, runners have raised more than £770 million pounds for charities such as Cancer Research UK or the World Wildlife Fund (WWF). Some charities use the money raised to staff hospices or help the homeless. Many others fund research into cures for diseases, ways to prevent starvation or ways of protecting endangered habitats. Many charities support medical or environmental causes. There are relatively few charities raising funds for research in the physical sciences.

Charities

- A charity is an organisation set up to provide help and raise money for people who are in need.
- A charity is not a government organisation.
- Charities are largely run by volunteers.
- Some charities raise money to support people with a specific condition, for example, breast cancer or arthritis. They will often fund scientific research into that disease.
- Some charities raise money to help groups such as homeless people or those affected by emergencies. They are less likely to fund scientific research.

Pressure groups

Not all voluntary organisations are charities. Pressure groups such as Greenpeace and Friends of the Earth try to influence public opinion and business or government policy in the interests of their particular cause.

They may aim to change the direction of scientific research.

Pressure groups always have a particular view, which may or may not be based on scientific evidence.

Case study: The Wellcome Trust

The Wellcome Trust is an independent charity. It is the second largest provider of funding for scientific research in the UK – only the government gives more.

Globally, the Wellcome Trust funds more medical research than any other NGO apart from the Bill and Melinda Gates Foundation. Examples of research funded by the Wellcome Trust include:

- vaccine development, for example, against Ebola
- public health issues, for example, insecticide-treated bed nets against malaria
- behavioural projects, for example, the effect of infection control training for front-line health workers
- social and ethical issues in science, for example, the ethics of medical trials involving pregnant women, particularly important because of Zika virus
- the use of genome sequencing to help predict and treat disease through the work of the Wellcome Trust Genome Campus.

Now try this

Choose a charity and a pressure group. Find out what you can about them. Make a table to summarise what they do and how they use the money they raise.

You need to be able to summarise content and make tables when you make notes to take into your assessment task.

Either choose a charity and a pressure group mentioned in the text **or** choose examples you know about or are interested in personally **or** choose examples you have seen in the media. Use the internet to find out what you can about them. When making notes, you don't need to write in full sentences.

Qualitative evidence?

When you are judging the science content of a paper, article or programme, the **qualitative evidence** presented is important. Qualitative evidence includes reference to established **sources** of information such as respected scientific journals or academic websites.

Citations and references

A **citation** is a specific source mentioned or quoted in the body of a scientific paper or article. It appears within the text of the paper you are reading. Citations matter because they:

- give credit to the authors whose ideas are used
- provide a trail so you can look up the same resources
- show how the author has done background research through other papers in the same field of science
- help prevent accusations of plagiarism.

The references are a list of all the sources that have been cited in a paper or article. You can look at the references and find your way to the original cited source.

Aurora

The Aurora Borealis is an amazing sight. Tourists and scientists alike want to know when the aurora will appear, with scientists looking for ways to predict it more reliably. Papers written about their discoveries will always include citations and might read like this:

> … Using real-time estimates of the Kp index [Wing et al., 2005] or the Dst index [Sugiura, 1964], we can specify a stronger disturbance in the terrestrial field. This, in turn, correlates with much stronger auroral activity. These estimates can be used to give us the potential location and strength of an aurora [Carbary, 2005], but they have limitations…..

This is a citation.

Earth and Space Science

This is a reference.

At the end of any paper, there will be references. In an article on predicting the aurora, these might include:

Bartels, J., N. H. Heck, and H. F. Johnston (1939), The three-hour-range index measuring geomagnetic activity, Terr. Magn. Atmos. Electr., 44(4), 411–454, doi:10.1029/TE044i004p00411.Wiley Online Library

Case, N. A., E. A. MacDonald, M. Heavner, A. H. Tapia, and N. Lalone (2015a), Mapping auroral activity with Twitter,Geophys. Res. Lett., 42, 3668–3676, doi:10.1002/2015GL063709. Wiley Online Library | Web of Science®

Books, articles, newspapers and online resources

A source may be unreliable if it:

- makes statements unsupported by evidence
- uses evidence from very small or very short studies
- uses only evidence from obscure universities/scientists
- doesn't reference its sources.

The quality of the evidence given in an article, book, programme, or online resource isn't always easy to discover. Look for citations and references, acknowledgements to the scientists, or universities involved and diagrams taken from original articles. For example, in this small section from an online butterfly forum, where people are discussing butterfly conservation, you can see that the sources are not very specific and therefore difficult to verify.

> ... I read a study in the journal Nature Climate Change which suggested Ringlet, Speckled Wood, Large Skipper, Large White, Small White and Green-veined White could suffer badly if we keep getting droughts. What do you think? I'm going to check out the long-term population data collected through the UK Butterfly Monitoring Scheme for some of my local species. Can you do the same? ...

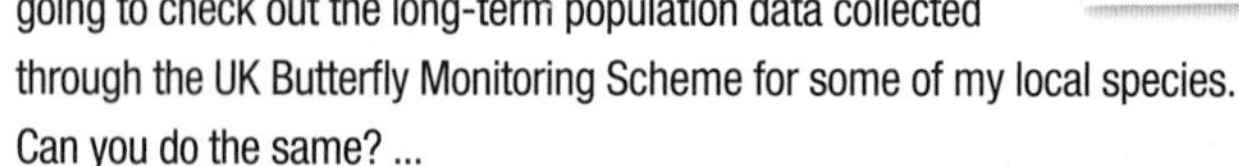

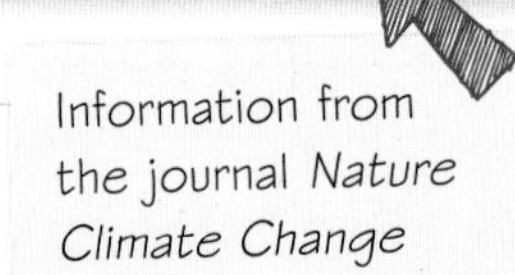

Information from the journal *Nature Climate Change*

Now try this

Look at the dates on the citations and the details of the references given for the paper on the aurora. What qualitative evidence does this give you about the paper?

Quantitative evidence?

In most areas of science, **quantitative evidence** is key. Quantitative evidence can include numerical data and calculations. It may be displayed in many different ways including graphs, tables, and charts. For each set of data, you need to decide how to display the patterns most clearly.

Pie charts

A pie chart is a type of graph where a circle is divided into sections. Each section represents a portion of the whole amount, for example, qualitative data on energy use in the US.

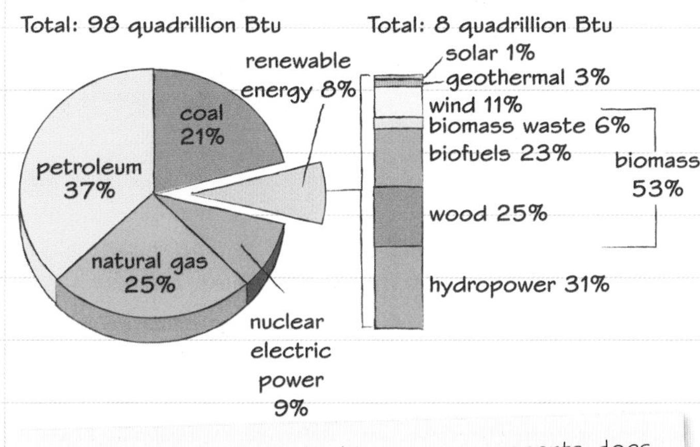

Note that the sum of the biomass components does not equal 53% due to independent rounding.

Line graphs

Line graphs give you the relationship between variables, for example, the amount of carbon dioxide in the atmosphere over time.

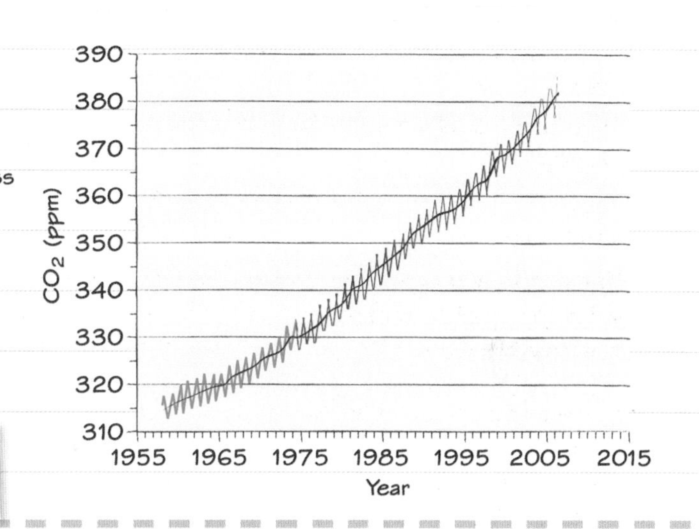

Bar charts

A chart that represents the frequency of a category by bars. The height of the bar represents the frequency, for example, data on global causes of death from the WHO.

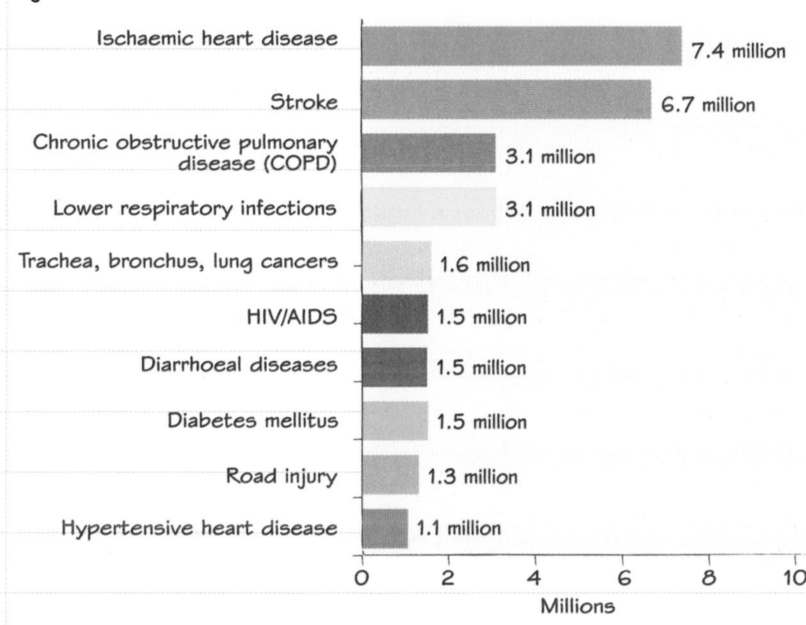

Tables

Tables can be used in many ways to record data from experiments, investigations or surveys.

One type of table is a frequency table. For example the table below shows the results of an investigation into the resting pulse rates of 100 people.

Resting pulse rate (beats/ minute)	Number of people (out of 100)
60–64	3
65–69	11
70–74	40
75–79	27
80–84	16
85–89	3

Now try this

Look at the data on resting pulse rates in the table above. Display this same data as (a) a pie chart and (b) a bar chart. (c) Which type of chart do you think shows the patterns in this data most clearly? Explain why.

Decide which way of displaying the data shows the patterns most clearly.

Had a look ☐ Nearly there ☐ Nailed it! ☐

The importance of statistics

Statistics is the science of collecting and analysing **large amounts of data**. For any scientific work to be valid, it needs to have been carried out on many different subjects or repeated many times. Statistical tests are applied to show the significance of the data collected. Are the findings the result of chance or a relevant observation? Before you accept any science you read about, ask if the results are **statistically relevant**.

Range, mean, median and mode

All of these terms are helpful – you need to know what they mean to help you make sense of quantitative data.

- **Range**: range between the minimum and maximum values
- **Mean**: the average – add all the values up and divide by the number of readings
- **Median**: the middle number of a dataset
- **Mode**: the most frequently occurring value in a dataset.

Error bars and error bands

These are lines or shaded areas shown on graph, bar charts and histograms. They represent the variation in the results and so give you an idea of how reliable the data is.

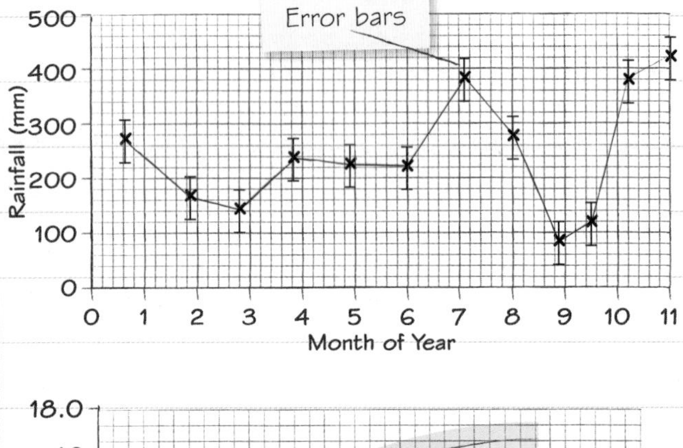

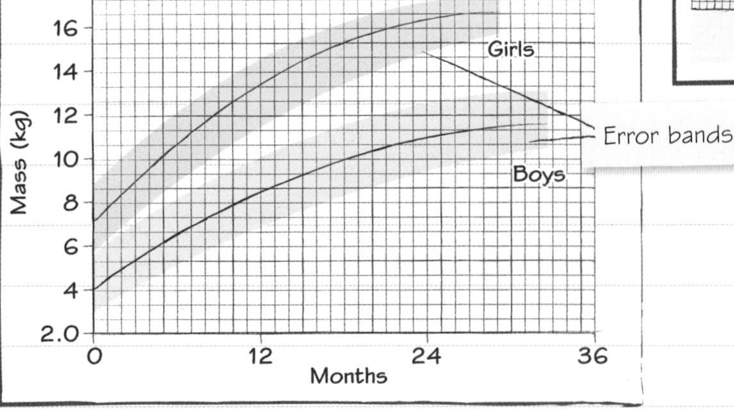

Normal distribution curves and standard deviation

A normal distribution curve results when you plot continuous variation data on a graph.

Links For more on processing data, see page 56.

- Most values will be close to the mean value – the numbers at the extremes are low.
- The standard deviation of the data is a measure of how spread out it is. The greater the standard deviation, the greater the spread of the data.
- Standard deviation gives you the probability that a particular result falls within the normal distribution.

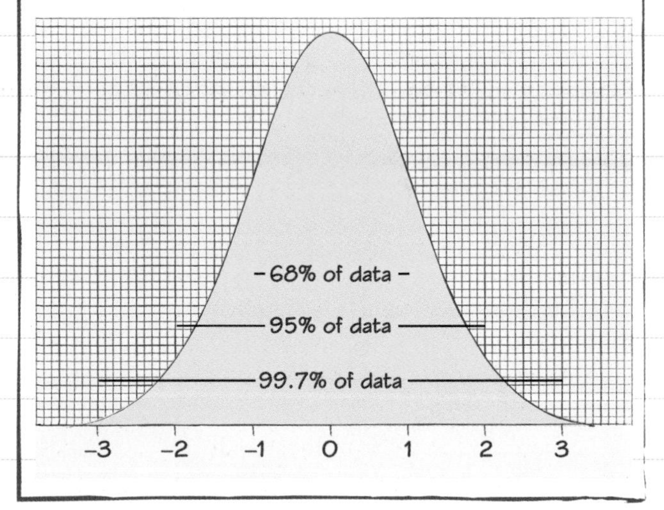

Now try this

One set of data used in a newspaper article has small error bars. The study was only on 10 individuals. Another article quotes a different study. The graphs have much bigger error bars but the study involved 30 000 people. Explain which data you are most likely to trust and why.

The validity and reliability of data

Data can look very official and intimidating, whether it is in a scientific paper, a newspaper or in a blog. **But remember: data is only useful if it is both valid and reliable.**

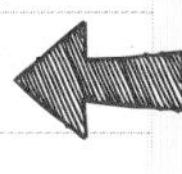

If results are **valid**, they measure what they are supposed to measure. If results are **reliable**, the investigation produces stable, consistent results, which other people can replicate.

References and citations

A long list of references at the end of a paper or article suggests it has a lot of citations in it and has been well researched.

Lots of references also allow you to check on the **reliability** and **validity** of the data collected and presented.

If you read a paper or article with only a small number of references it may mean:

- The paper is not as rigorous as it might be or was not well researched.
- The paper is on a new or unresearched area of science, so there are not many other papers to reference.

If one paper is cited in lots of other papers, usually (but not always) means it is high quality work.

Size matters

The **size** of the sample used is very important. As a general rule:

- The bigger the dataset, the more likely it is that results will be reliable.
- Evidence based on conclusions from very small datasets can raise a potential issue but must be considered unreliable until further work is published.

1. In 1998, a study linking the MMR vaccine with autism was published. It was subsequently shown to be fraudulent and completely wrong. It was based on only 12 children.

Links For more on analysis of data, see page 57.

2. The Whitehall II study into a range of diseases studies 10 308 men and women. They have been monitored since 1984 and the results are used and respected globally.

Authenticity

How can you be sure what you are reading about is real? Look for sources of data and information in an article and when they were published – they may be hopelessly out of date. For example, in July 2016, the **Mirror** reported online about a strange, rare 'ghost fish':

The 'ghost fish' had never been seen alive before.

> An incredibly rare 'ghost fish' has been seen alive for the first time ever. The ethereal deep ocean dweller, measuring just 10 cm long, has never before been spotted by researchers, let alone caught on camera. It lives in the murky depths of the sea – 2 km below the surface. The odd-looking fish is pale, with almost translucent skin and bulbous, glowing eyes.

Signs of authenticity come later in the article:

> The elusive fish was filmed by Okeanus Explorer during an expedition of the Northern Mariana Islands, in the Pacific Ocean. Dr Bruce Mundy, from the NOAA National Marine Fisheries Service, said: 'This is just remarkable. I am sure that this is an Aphyonid and I am sure that this is the first time a fish of this family has ever been seen alive.'

Check this out online – it is an exploratory vessel for the US National Oceanic and Atmospheric Administration (NOAA)

Dr Bruce Mundy is a fisheries biologist working for NOAA – a highly respected science organisation.

You must be prepared to look up information to see if the science reported in an article or programme is authentic.

Now try this

Look up the studies called Whitehall I and Whitehall II using Wikipedia. Write a series of bullet points to explain why these studies are so highly regarded by scientists and doctors.

Look at the size of the studies, the length of time they ran for, the scientists who carried them out and the percentage of the target groups who took part.

Had a look ☐ Nearly there ☐ Nailed it! ☐

Use and misuse of data

Scientists develop a **hypothesis** and look for **supporting evidence**. Most scientists are honest. Most of the data you will study will be reliable and valid. However, sometimes data is deliberately misused or altered by scientists to back up their ideas, even when they are wrong.

When the misuse of data is discovered it causes a scandal. Here are some examples:

Making SCNT human stem cells

In 2004 and 2005, a South Korean scientist Woo-Suk Hwang and his team published two papers in the journal *Nature*. They claimed to have successfully produced embryonic stem cells from adult human cells by a process known as SCNT. This caused great excitement because of the potential for treating human diseases.

A short time later the story unravelled.

- The eggs used had been obtained unethically from researchers.
- The photos had been forged.
- DNA testing showed the embryonic stem cells were **not** from an adult donor cell at all. They came from IVF embryos.

MMR vaccines

- In 1998 Dr Andrew Wakefield published a paper in the *Lancet* suggesting a strong link between the MMR vaccine and autism.
- Almost immediately, epidemiological studies on thousands of children showed no link.
- Ten of the original authors withdrew the paper.
- The *Lancet* admitted Dr Wakefield had a financial interest in the MMR vaccine being found responsible for autism.
- Dr Wakefield had taken blood from children unethically without permission.
- It was found that some of the evidence had been changed and some was completely false.
- Dr Wakefield was completely discredited and struck off the medical register.

Gravitational waves

- In 1915, Albert Einstein predicted the existence of gravity waves.
- In the 1970s, Joseph Weber at the University of Maryland was detecting them almost daily.
- Other scientists were convinced by Weber's results, until it became clear that no-one else could reproduce his findings.
- Finally, in 2016, using a completely different experimental set up known as LIGO, gravitational waves were detected for real.
- Scientists using LIGO regularly inserted false data to make it look as if gravity waves had appeared to test the equipment. When real gravity waves were detected they were amazed but knew it must be genuine.

Cholesterol, vitamin B and heart disease

In 1969 Kilmer McCully was a young doctor. He published a paper on his findings that high homocysteine levels in children are linked to atherosclerosis and heart disease. He went on to suggest that adult heart disease is often a result of vitamin B deficiency. Vitamin B is needed to reduce homocysteine in the body. The graph is based on data from 18000 people in Norway.

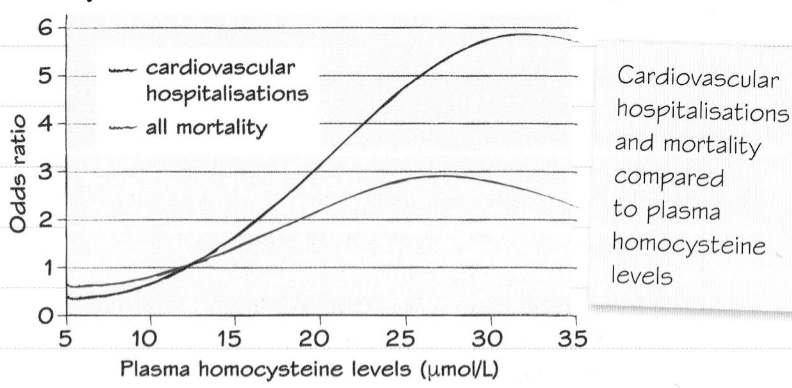

McCully's work undermined the idea at the time that heart disease was all about cholesterol and fat in the diet. As a result it took 30 years for McCully's ideas to be fully accepted. Many people may have died as a result.

Now try this

Find out more about one of these cases. Produce a summary of the results that were published and what was wrong with them. Describe how the misuse of data was discovered.

Information on all of these cases is easily available online. Take care to use a reliable source.

Potential areas for R&D

There are many areas of science that have huge potential for research and development (R&D). R&D is more than simple scientific research. It also involves commercial development of an idea or a technology so it can be used to make money.

Every large company and government has funds for R&D. Many different factors affect whether a particular piece of scientific research attracts funding for further R&D within a commercial company.

R&D Funding

1 Moving technology forward, for example, Very Large Hadron Collider, xenotransplantation:
much R&D is at the cutting edge of science and medicine.

2 Fills a need:
R&D often targeted at a particular need, for example, smaller phone batteries, new antibiotics.

3 Has impact and status:
often R&D only targeted at projects that will get noticed – small projects, or projects for developing world where there is little money, may often be ignored.

4 Potential for wide use:
for example, the potential to make money from new discoveries often drives industrial R&D and so it may focus on areas that affect large populations of people. It's important as also makes money for more R&D.

5 Commissioned research:
for example, work on malaria vaccines sponsored by the Bill and Melinda Gates Foundation. Individual philanthropists or big organisations can commission and pay for particular R&D projects.

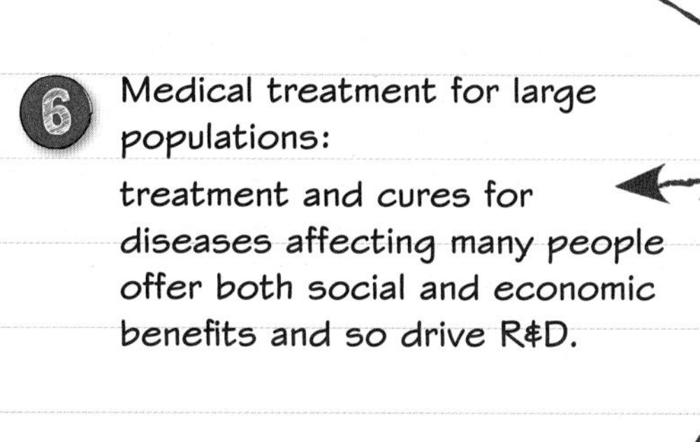

6 Medical treatment for large populations:
treatment and cures for diseases affecting many people offer both social and economic benefits and so drive R&D.

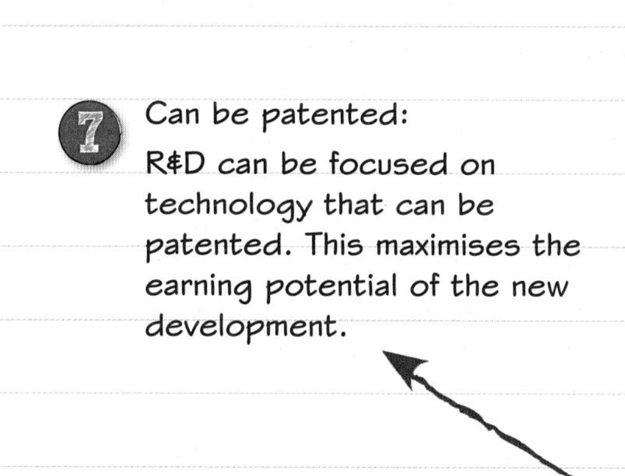

7 Can be patented:
R&D can be focused on technology that can be patented. This maximises the earning potential of the new development.

Now try this

Blue skies research is a term used to describe research that is carried out simply because it is an interesting theory. Suggest why it can be harder to get funding for blue skies research than for research into a potential new antibiotic, an environmentally friendly fuel or a new computer system.

Look through all of the main drivers for R&D. Think how each one does – or does not – apply to blue skies research.

Had a look ☐ Nearly there ☐ Nailed it! ☐

Supporting evidence

When you read a scientific paper, report or article, always look for the **supporting evidence** and evaluate how much weight you should give it. Some evidence is much more useful than others.

Personal and anecdotal experience

Personal experience is based on things that have happened in your own life. **Anecdotal experience** is based on second-hand knowledge – stories about people you know.

It is almost always based on very small samples.

Both personal and anecdotal experiences are biased by your own beliefs and world view.

This type of evidence is not a good way to support a theory. However, it can be useful in triggering ideas about areas of science that might be investigated

Scientific evidence

Scientific evidence should be **unbiased**. The best supporting evidence for any hypothesis or claim involves lots of data collected from many subjects or experiments. It is repeated over time and by different people. The evidence must be analysed and be statistically significant.

Personal, anecdotal supporting evidence	Scientific (statistical) supporting evidence
very small samples which may be biased	large, representative samples involved
factors are observed in uncontrolled conditions	precise measurements are made in controlled conditions
any other factors are ignored	as many relevant factors as possible that might affect the end results are controlled
causal relationships are assumed very easily	causal relationships are only suggested when large bodies of evidence support the idea

Now try this

An advert recommends a dietary supplement to help you get fitter.

(a) Why might you be suspicious of this claim?

(b) What could you do to help you decide whether to try the supplement?

Think about whether the advert tells you how to access research about the supplement and who will benefit if the supplement does not work as advertised.

Target audiences

When you read or write a science report, you need to know the **target audience**. The way the science is presented will vary depending on the audience the authors are aiming at.

Identifying the target audience helps you make a realistic evaluation of what you are reading. You also need to be very aware of your target audience when you do a piece of science writing yourself!

Target audiences

There are a number of clues that can help you decide who has written a particular piece of scientific writing, and who their target audience is.

The scientific community

This includes school students, technicians, undergraduates, postgraduates, research scientists, doctors, vets, etc. Writing by the scientific community aimed AT the scientific community usually has:

- a high level of literacy
- a lot of technical language
- citations and/or references.

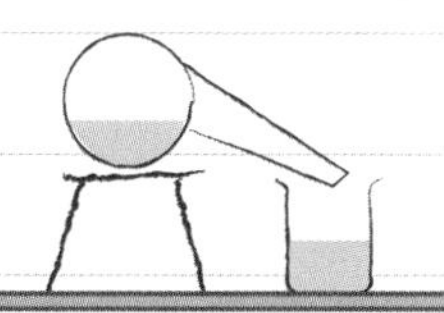

The general public

Scientific writing aimed at the general public has to be accessible by as many people as possible. The average reading age in the UK is only 9 years old, so writing aimed at the general public usually has:

- a relatively low reading age
- limited technical language
- lots of photos/diagrams.

Pressure groups

Writing by pressure groups (for example parliamentary lobbyists, environmental campaigners, etc.) is trying to win support for a particular point of view. It may be aimed at the general public or at people in power. It usually has:

- one strong, simple message
- emotional appeal
- selective use of data.

Political representatives

Political representatives include MPs and local councillors. When a piece of scientific writing is aimed at political representatives you will see:

- very clear, simple explanations, as many politicians have little science background
- clear, strong evidence
- the economic implications.

When you are writing

You must remember your target audience at all times! Use language at the right level from start to finish. Make your explanations complex or simple, depending on who you are trying to reach.

Now try this

Here are three quotes about scientific developments. Suggest the target audience for each quote and explain your choice.

Think about the level of the language, the amount of information and even the size of the font.

Look on page 162 if you need more clues.

(a) **28 June 2016** Today the LHCb collaboration has submitted for publication two papers .... reporting the observation of four 'exotic' particles decaying into a J/ψ and a φ meson, only one of which was well established before. … Each of the four particles is observed with a significance exceeding five standard deviations.

(b) **Love doves dying out!** Turtle dove numbers are falling fast. The birds are often used as a symbol of love but they are failing to breed ..!

(c) Scientists from California Polytechnic State University found genes for antibiotic resistance in the gut bacteria found inside 1000 year old Inca mummies. They reported their surprising finding at the Annual Meeting for the American Society for Microbiology.

Presentation of science

Presentation styles affect how easily you can understand the content of a science report. This will always depend on the target audience. This also depends on the skill of the author in getting ideas across. The following ideas will help you work out the target audience of a report. If you are writing a report, remember to take these factors into account.

Detail and accuracy

- Technical details make it harder to read.
- Are the details accurate?
- **Citations** and **references** help you check up on the accuracy of the details.

Language level and readability

- Short sentences are easier to read and understand.
- Lots of clauses make a sentence harder to read.
- Difficult words make it harder.
- **Keywords** in bold can help.

Writing style and terminology

- First person (for example, 'I did this) or active (for example, 'As you can see ...') writing is easier to read than passive, commonly used in scientific journals, (for example, 'The two chemicals were mixed ...').
- Lots of technical words make it harder to read.
- Sub-headings break up the text.

Use of visuals

- Photos can be just pretty or really useful.
- Diagrams need to be clear and well labelled.
- If visuals have clear captions, it is easier to work out what is going on.
- Graphs and other ways of presenting data can make things easier or more difficult to understand.

Example 1: Super sniffers!

Popular newspaper report: fun headlines, short sentences, little detail about science, pretty image.

Top boffins are breeding mutant mice with a super-sense of smell. The GM rodents may soon be used to detect explosives or drug hauls at airports around the country.

Example 2: The basal ganglia

Medical article on Parkinson's disease: complex language, lots of technical terminology, citations, photo shows relevant details, well labelled.

The area known as the substantia nigra pars compacta (SNc) projects dopaminergic neurones to the striatum where they exhibit both excitatory and inhibitory effects .... Some scientists, however, now believe that D1 and D2 receptors colocalise neurones in the striatum (Aizman O et al, 2000), which would mean that the actual pathway is something completely different, or that not all of the neurones from the SNc are dopaminergic.

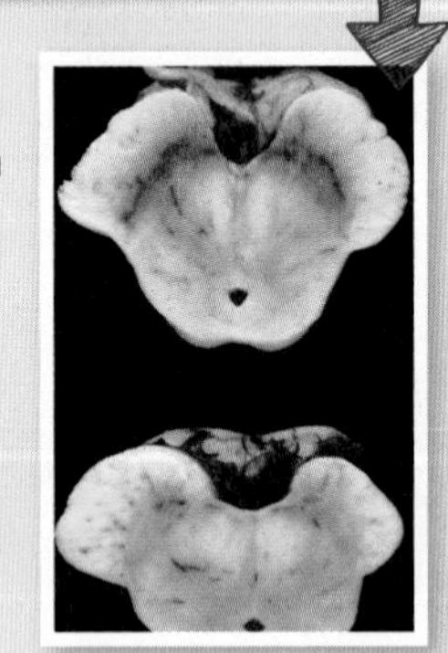

Now try this

Scientists have made a new adhesive. It is gentle enough to be used on plasters for the skin of very old people and premature babies. Shining a light on the adhesive stops it being sticky so the plasters can be removed very easily.

(a) Write a headline and a few sentences for a report on this exciting new science for the general public, for example, for a popular newspaper.

You need an eye-catching headline and very simple sentences for a reading age of 9 years old.

(b) List at least three more pieces of information you would need before you could report on this for an audience of scientists or people who have a special interest in science.

Remember all the factors you have learned that make a piece of writing useful to scientists.

Quantity, quality and bias

The **quantity** and **quality** of information in a science report make an enormous difference to how it can be used. **Bias** is another important fact – a biased article is flawed even if it contains some good science.

Quantity and quality

Both the quantity and quality of scientific information are important in science reporting. The amount written can reflect the value of a report. In general more quantity means more scientific information can be given. For example:

- Broadsheet newspapers usually publish longer science articles than tabloids.
- Scientific papers are usually longer than an article in *New Scientist*.

Quantity is less important than quality. Quality reflects the amount of accurate information, references to sources, citations and data found in a piece of science reporting.

In general, scientific articles are more likely to be of a higher quality than a tabloid extract about the same thing.

But remember: scientific papers can be fraudulent, and tabloid extracts can be very scientific.

For example, look at this tabloid article:

The Mirror, July 2016

Very strong painkillers are often ineffective for lower back pain

People with chromic lower back pain are sometimes prescribed morphine derivatives such as codeine and fentanyl for pain relief, but a review of studies has found they are generally ineffective.

The analysis, published in JAMA Internal Medicine pooled data from 20 high-quality randomised trials that included 7,295 participants.

The studies tested various drugs and most studies were funded by pharmaceutical companies. Seventeen of the studies compared the opioid, a morphine derivative, with a placebo, and three compared two opioids with each other.

The drugs relieved pain slightly, but the effects were not clinically significant and the medicines did little to improve disability. There was some evidence that larger doses worked better, but most trials had high dropout rates, some up to 75%, because of the adverse side effects or because the drugs did no good.

Details of the journal, the trial details and the number of people involved

Funding given (bias?)

Clinical significance given

Dropout rates given

What is bias?

Bias is when someone supports, or is prejudiced against a person, group or scientific idea. You need to know if the person writing a science report is biased as this will change the way they present the evidence. If you write about science you should be unbiased. If you **are** biased, you should say so. Lobbyists are usually biased because they support one particular point of view, for example, groups who are against nuclear energy or who support conserving rain forests.

Now try this

Both of the people in the figure above write an article about fossil fuels and global warming. They both use the same data.

How might their scientific reports differ?

Had a look ☐ Nearly there ☐ Nailed it! ☐

Print media

The way science is reported often depends on the medium in which the reports appear. Science papers and reports have appeared in print since 1665 in print media ranging from scientific journals to newspapers.

Specialist journals

- Published by scientists for scientists.
- **Peer-reviewed**.
- On paper or now also online.
- Highly technical, with citations and references.

For example: *Nature*, *The British Medical Journal (BMJ)*, *Science*, *Cell*, *The Astrophysical Journal*

Science magazines

- Published for the scientifically literate/ interested public.
- Some articles will have citations and references.

For example: *New Scientist*, *Scientific American*

National newspapers (and newspaper websites)

- Aimed at general public.
- Different papers aim at different sections of the public.
- Some national papers are print heavy, others have lots of pictures. Both types can be very influential.
- Some national newspapers cover more science news than others.
- Some refer to their sources.
- Look for headline science – detail not always explained well or accurately.

Examples of national newspapers are: *The Daily Telegraph*, the *Daily Mirror*, *The Sunday Times*, the *Daily Mail*, *The Sun*.

Local papers

- Aimed at local communities.
- Science reported if it affects local issues
- Science often only reported if it is about the local area, for example, Nature notes.

For example: *The Ringwood and Fordingbridge News*, the *Lincolnshire Standard*.

As the spring days lengthen, you will notice the buds on the trees unfurling. Out in the Forest the heather is still brown and dead-looking, and last year's bracken fronds remain as auburn sentinels – but in sheltered glades, the early spring flowers are opening their faces to the sun.

Notice that the language used here is poetic and about the experience, rather than a scientific explanation of the processes going on

Now try this

Look through the pages of two different national newspapers, for example, *The Daily Telegraph* and the *Daily Mirror*. Count the number of science-related articles. Find articles about the same science topic in the two different papers and make a table to compare the way they are reported.

TV and digital media

Science is no longer confined to journals, textbooks, magazines and newspapers. You can also read about science, watch programmes about science and even download academic science journals using your TV and the internet.

The presentation of science on:

TV

- May be for general audienceor scientifically literate (for example the Open University, or OU) May be for education, e.g. *Trust Me, I'm a Doctor*
- May be mainly for entertainment, e.g. *Life on Earth*
- Presenters range from real scientists to popular actors.

Journals

- Internet journals are often online versions of paper journals. Some only exist online.
- Online journals involve scientists writing for scientists.
- They have citations, references etc.
- Subscription journals only available if you pay.
- Open access journals make scientific available to everyone.

Websites

Websites often claim to be scientific. Some are and more are not:

- audience is general public
- unregulated
- often have adverts
- many written by reputable or good science writers
- some poor – inaccurate or misleading science, wrong science, biased opinions
- animations, videos etc. can make the science more accessible.

Alkaline treatment cures cancer

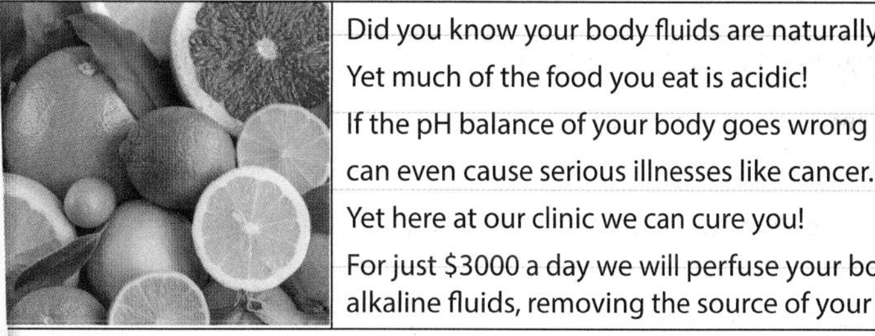

Did you know your body fluids are naturally alkaline?
Yet much of the food you eat is acidic!
If the pH balance of your body goes wrong it
can even cause serious illnesses like cancer.
Yet here at our clinic we can cure you!
For just $3000 a day we will perfuse your body with alkaline fluids, removing the source of your

Social media

Social media includes blogs, tweets etc:

- may get information straight from the scientist
- may get comments, information straight from the scientists
- handle with care as a source of information; rarely any authentication.

A naturalist's blog

This tiny plant made my day! Bog orchids (*Hammarbya paludosa*) are the smallest orchids in Britain – see my finger for scale! After three weekends of searching in likely spots, my husband and I finally discovered a small colony of them half an hour from home....

Now try this

Look at the science content of a reputable website using the criteria outlined above. Explain what specific features tell you that this is reliable science reporting.

Remember to look at factors such as the scientific institutions, scientists and funding involved to help you judge if a source is reliable.

Your Unit 7 set task

Unit 7 will be assessed through a task, which will be set by Pearson. In this assessed task you will need to read and analyse articles relating to contemporary scientific issues, as well as writing an article yourself based on information given.

Set task skills

Your assessed task could cover any of the essential content in the unit. You can revise the unit content in this Revision Guide. This skills section is designed to **revise skills** that might be needed in your assessed task. The section uses selected content and outcomes to provide examples of ways of applying your skills.

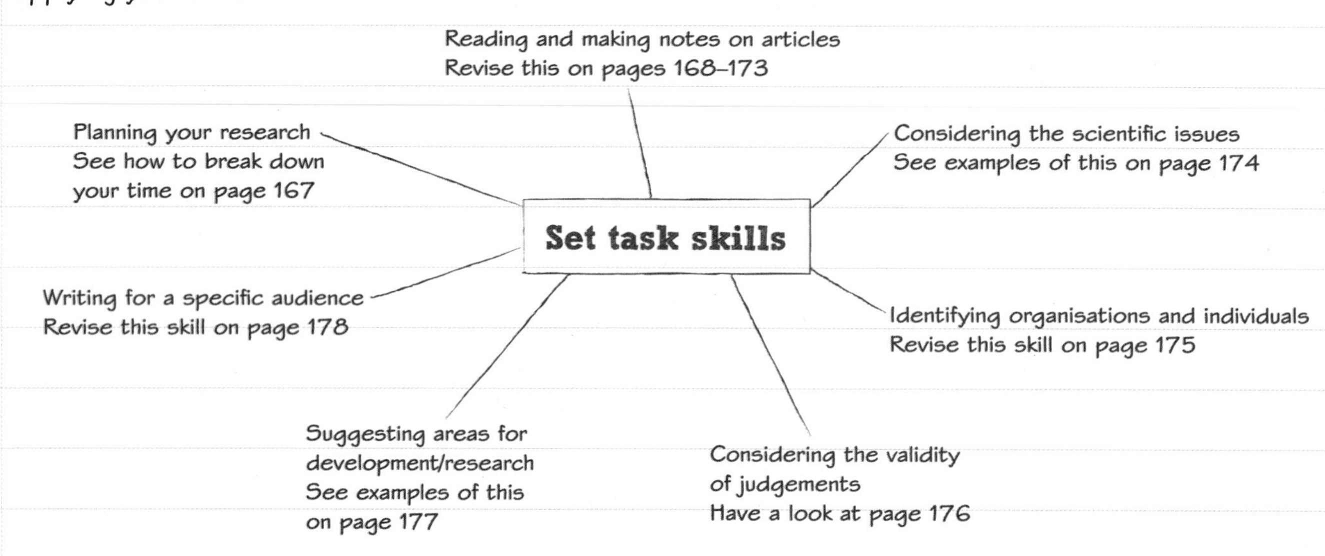

Preparatory notes

You may be allowed to take some of your preparatory notes into your supervised assessment time. If so, there may be restrictions on the length and type of notes that are allowed. Ask your tutor or check the latest Sample Assessment Material on the Pearson website for more information. Details of the assessment may change so always make sure you are up to date.

Check the Pearson website

The questions and sample response extracts in this section are provided to help you to revise content and skills. Ask your tutor or check the Pearson website for the most up-to-date **Sample Assessment Material** and **Mark Scheme** to get an indication of the structure of your actual assessed task and what this requires of you. The details of the actual assessed task may change so always make sure you are up to date.

Now try this

Visit the Pearson website and find the page containing the course materials for BTEC National Applied Science. Look at the latest Unit 7 Sample Assessment Material for an indication of:

- The structure of your set task, and whether it is divided into parts
- How much time you are allowed for the task, or different parts of the task
- What briefing or stimulus material might be provided to you
- Any notes you might have to make and whether you are allowed to take selected notes into your supervised assessment
- The questions you are required to answer and how to format your responses

Plan a flow chart for yourself of what you need to do to approach your set task.

Planning your research

You will need to plan the **time** that you are given to read, analyse, carry out further research and make notes on the articles provided to you. You can see an example of a plan below. You can use this to create your own plan of how to approach your research.

This plan breaks down the main activities into a logical structure, but you can adapt it to work best for your revision.

Task for Part A
Read through all three articles.
Annotate the articles, looking for information to answer the questions.
Carry out further research into any areas identified, such as: • key scientists or institutions • checking source data referenced in articles • looking for further articles or data to support or challenge the issues raised, making sure to keep details of the sources.
Make notes on the articles, focused on the areas that will be covered in the questions: 1 What is the scientific issue being discussed in the article? 2 Do the articles have different approaches to the issue? If so, how are they different? 3 Who are the individuals and organisations identified in the articles? 4 How might those individuals/organisation have an influence on the scientific issue? 5 For each article, what evidence is there that it has made valid judgements? 6 What potential areas for further development and/or research of the scientific issue could be suggested based on each of the articles?
Organise and make sure they're clear and not too long to refer to in future. Be sure to note the details of any sources other than those provided.

Ensure you allocate most time to the research and note making. Check how much preparation time you will have in total and allocate times to each part of your plan to help ensure you stay on track.

In this Revision Guide, the articles provided are used as examples to help you practice the skills you need. The content of the articles given to you in your assessment will be different each year and the format may be different.

Now try this

Find a sample task (for example, from the Sample Assessment Materials) and write a section of notes based on this task.

Had a look ☐ Nearly there ☐ Nailed it! ☐

Reading Article 1

You will need to read and make notes on different articles for your assessment. Below is an extract from an exemplar article from *The Telegraph* about ageing, which is similar to one you may get in your assessment. After reading it, look at the sample student notes on page 169, which relate to the numbers in the text.

Want to live to 120? Here's how...

A new study suggests that slowing the ageing process is more achievable than people realise. Britons could live to 120 if they just exercised more, ate healthily and took beneficial drugs such as statins ❶, an influential panel of health experts and scientists has concluded. Simple lifestyle changes such as walking regularly, cutting down on sugar, salt and fat, and taking advantage of drugs that already exist could extend life ❶. Members of the Longevity Science Panel ❷, set up to advise policymakers, predicted that if all the population followed the advice of health experts, the average life span could rise from 80 to 84 ...

In their report "What is ageing? And how do we delay it?" ❸, the panel of experts delved into the science behind growing older and looked at how it might be prevented through 'wonder drugs' or behavioural changes. "The ageing process is a biologically complex thing," said Dame Karen Dunnell ❹, chairman of the panel. "What we were trying to do is look into the biology of ageing and see what does really work. We found that having a long and healthy life is largely related to lifestyle and diet "...

... The ageing that we experience happens because of problems in cell division ... . Many billions of cell divisions occur in a lifetime and errors creep into the process. As well as the random errors, lifestyle factors such as smoking and drinking increase mutations. Over-eating is thought to increase mutations because it speeds up cell division. ❺ The build-up of senescent cells in an organ prevents the body's ability to repair damage. It is what causes wrinkles and age spots.

... Experiments have shown that a calorie-restricted diet, such as fasting on alternative days, can increase the lifespan of animals by up to 65 per cent. A Mediterranean diet lowers the incidence of age-related disease. Drugs can also help. Statins are known to reduce cholesterol and so prevent heart disease, but they also have anti-inflammatory powers which lower mortality risk. ❻

Richard Faragher, professor of Biogerontology at the University of Brighton ❼, said living healthier for longer was more important than increasing lifespan. "Ageing occurs because the mechanisms which keep us in good health fail over time. .... in effect, all age-related diseases are being driven by a few mechanisms which if you could control them, could be the difference between somebody hobbling down the street, or jogging past you."

However trying to persuade people to do what is good for them has proved tricky, as Cardiff University ❽ found. In 1979, 2500 men were asked to follow five simple rules – eat well, work out, drink less, keep their weight down and never smoke. Four decades on, just 25 pensioners managed to stick to the plan. But they are all far fitter and healthier than those who gave up. ❾

Source: *The Telegraph* online, Want to live to 120? Here's how..., last accessed February 2017 

Now try this

Read through the article and summarise the scientific issue being addressed. Then list the main individuals and organisations given in the article.

Analysing Article 1

Read the extract from the sample article on the previous page. This page shows you some examples of the type of notes you could make on this article.

Sample notes extract

1 Generalised statements about ways of increasing life expectancy. No original data or graphs given with article.

> Consider how useful the article is – hard evidence and quantitative data might be more reliable than generalisations.

2 Organisation mentioned: Longevity Science Panel. From online research, established in 2009.

> General statements about health experts and scientists doesn't tell you much about the reliability of the science. The name of the panel means you can research more details into how and when it was established, how it is funded, its aims and its members.

3 Name of source article: **"What is ageing? And how do we delay it?"**

> Information about the sources used to write the article means that you are able to research the validity of what is written.

4 Chair Of Longevity Panel: Dame Karen Dunnell.

> The name of the person who chaired the panel is given. You can look her up and find out if she is well-known and respected in her field. In this case, she is!

5 Scientific issue: Ageing. Ageing process is the result of problems with cell division. Lifestyle factors like drinking, smoking and over-eating are thought to increase mutations during the cell division process, and may accelerate ageing.

> In this section, you are given some insight into the science behind the ageing process. You can do further research and improve the quality of your notes by checking if the claims made in the article are supported by other scientific studies.

6 Interventions believed to reduce mortality or rate of age-related diseases: calorie-restricted diets, fasting, Mediterranean diet and statins.

> Try and find out more about the evidence mentioned – look it up and see if you can discover what the data actually shows.

7 Name of scientist involved: Professor Richard Faragher.

> When you look up the scientists involved, you should consider how influential they are, what is their institution like, where does the research funding come from?

8 Name of institution: Cardiff University.

> You will need to be able to identify the different organisations in the article. By looking them up and finding out more about them, you will be able to comment on how they may have an influence on the scientific issue.

9 From further research: study started 1979. Cohort of 2500 men asked to eat well, work out, drink less, keep weight down and not to smoke. Only 25 followed the plan for 40 years. Study leader was Professor Peter Elwood of Cardiff University. All those who followed it were fitter and healthier than those who gave up, but results as reported in this article are anecdotal.

> Full details of the study are not given in the article, so you need to do some detective work to find the original details. This will help you see if the evidence supports the general theory given here.

10 Source: Telegraph online. This is a broadsheet newspaper, so the article is written for non-scientists with a relatively high reading ability.

> You have considered the target audience of the article.

Now try this

Find out a bit more about the individuals and organisations mentioned in the article. Consider how they may have an influence on the scientific issues.

Reading Article 2

Below is a set of extracts from the type of scientific study you might get given as one of the articles to read in your assessment. It focuses on the topic of ageing in Japan. After reading it carefully, look at the sample student notes on page 171, which link to the numbers in the text.

Okinawa Centenarian Study ❶

Evidence-based gerontology

One of the most important things about the Okinawa Centenarian Study is the fact that it is based on solid evidence. The most important evidence needed for any centenarian study is reliable age-verification data. Throughout Japan (including Okinawa prefecture), every city, town, and village records birth, marriage and death data (among other data) in a koseki (family register). This system was instituted throughout Japan in the 1870s. ❷

FOCUS AREAS AND FINDINGS

After examining over 900 Okinawan centenarians and numerous other elderly in their seventies, eighties, and nineties, some fascinating findings have emerged. … Below are some of the key findings and what they mean in terms of healthy ageing – for the Okinawans, and the rest of us.

Cardiovascular health and ageing

Elderly Okinawans were found to have impressively young, clean arteries, low cholesterol, and low homocysteine levels when compared to Westerners. These factors help reduce their risk for coronary heart disease by up to 80% and keep stroke levels low. ❸

The graph indicates that the higher the plasma homocysteine (a new risk factor) level is, the more people suffer from cardiovascular disease. Homocysteine is an amino acid that causes damage to arterial walls. It is higher in people who don't get enough folate (for example, green leafy vegetables) and vitamins B6, B12 – but low in Okinawans. ❹

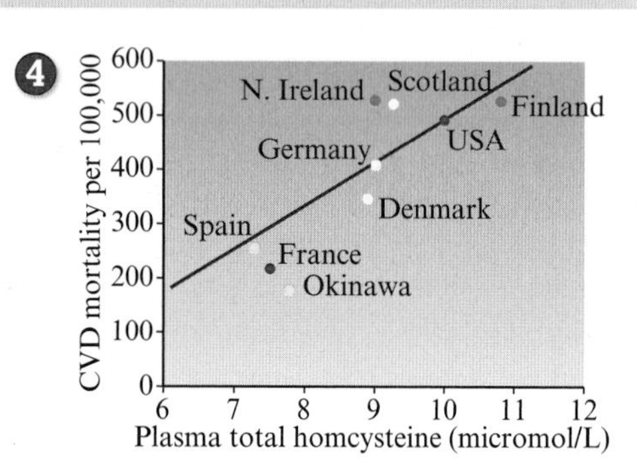

Hormone-Dependent Cancer Risk ❺

Yearly Cancer Deaths (per 100,000 people)

Location	Life Expectancy	Breast	Ovarian	Prostate	Colon
Okinawa	81.2	6	3	4	8
Japan	79.9	11	3	8	16
Hong Kong	79.1	11	3	4	11
Sweden	79.0	34	10	52	19
Italy	78.3	37	4	23	17
Greece	78.1	29	3	20	13
USA	76.8	33	7	28	19

Adapted from World Health Organization 1998; Japan Ministry of Health and Welfare 1996

Cancer and ageing

Okinawans are at extremely low risk for hormone-dependent cancers, including cancers of the breast, prostate, ovaries, and colon. Compared to North Americans, they have 80% less breast cancer and prostate cancer, and less than half the ovarian and colon cancers. Some of the most important factors that may protect against those cancers include low caloric intake, high vegetables/fruits consumption, higher intake of good fats (omega-3, mono-unsaturated fat), high fiber diet, high flavonoid intake, low body fat level, and high level of physical activity. ❺

Women's health and ageing

Women in Okinawa tend to experience menopause naturally and non-pharmacologically with fewer complications such as hot flashes, hip fractures, or coronary heart disease. Lifestyle determinants include diet, avoidance of smoking and exercise in the form of dance, soft martial arts, walking and gardening. ❻ Okinawan women also have a very high intake of natural oestrogens through their diet, mainly from the large quantities of soy they consume. Soy contains phytoestrogens, or plant oestrogens called flavonoids (isoflavones). Recent double-blind placebo controlled studies support the ability of soy isoflavones to slow the bone loss ❼ (Alekel D, et al. Am J Clin Nutr 2001;72:844–52) and hot flashes (Albertazzi P, et al. Obstet Gynecol 1998;91:6–11) ❽ that occur with menopause.

	Isoflavone intake per day (mg)
Okinawans	32
Japanese Canadian First generation	3

Based on Wilcox, B., et al. Am J Clin Nutr 1995; 61(4):901.

Isoflavone intake per day (mg) for Okinawan women and Japanese Canadian, first generation women.

Okinawans (n=18)

Japanese Canadian first generation (n=8)

Now try this

Read through the article and list all the evidence such as articles cited, graphs and sources that can be used to consider the validity of the claims being made.

Analysing Article 2

Read the extract from the exemplar article on page 170. This page shows you some examples of the type of notes you could make on this article.

Sample notes extract

1. The article is a selection of extracts from the website of the Okinawa Centenarian Study, the longest running study on people living healthily to over 100 years.

The content is written for scientists and interested non-scientists. There are many references and the biographies of the scientists involved in the study. This suggests the information in this article will be scientifically reliable.

2. Births, deaths and marriages in Japan have been registered since the 1870s so the age-verification data collected is very reliable.

You will need to consider the validity of claims made in the articles so you should look for evidence of how reliable (or otherwise) they are.

3. Okinawans have reduced cardiovascular problems evidenced by clean arteries, low cholesterol levels and low homocysteine levels.

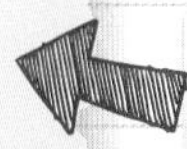

Consider whether these statements are supported by evidence.

4. Homocysteine levels may be linked to longevity of Okinawans (low levels = lower mortality from cardiovascular disease). Evidence given in graph plotting plasma homocysteine levels against mortality in different countries.

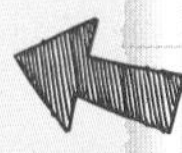

The graph gives hard data and is based on one published in the *Lancet*, a very prestigious peer-reviewed journal, in 1997. This provides good evidence for the validity of the claims in the article.

5. Data in the table gives evidence to back up statements in the text that Okinawans have much lower risk of hormone-dependent cancers. Sources of data are WHO and Japanese Ministry of Health and Welfare.

The sources of data are reliable, but from 1996. More recent data could confirm the findings.

6. Generalised statements about experience of menopause in Okinawan women and about what could cause this. No evidence or data given.

Evidence that Okinawan women do not use hormone therapies to help with the menopause would give more weight to these statements.

7. Reference to a double-blind placebo-controlled trial, so data collected should be of high quality.

It would be helpful to also know the size of the trial.

8. Citations: Alekel D, et al. Am J Clin Nutr 2001;72:844–52 and Albertazzi P, et al. Obstet Gynecol 1998;91:6-11. Both peer-reviewed journals.

Peer-reviewed journals are generally a reliable source of data.

9. Evidence of the different isoflavone intake of women in Okinawa compared to Japanese Canadian women. Relatively small sample size.

The comparison between women in Okinawa and Japanese Canadian women makes the data more valid because the two samples are similar. However, the small sample size may mean it is difficult to make assumptions about what this means in the wider population.

Now try this

Using your notes for the question in 'Now try this' on the previous page, write a short paragraph describing three pieces of evidence that the claims made in the article are valid.

Make sure you support your description by giving examples.

Had a look ☐ Nearly there ☐ Nailed it! ☐

Reading Article 3

Below are the conclusions of the long report produced by the Longevity Panel (you will have already come across this panel in Article 1). This might be the kind of content you get to read as your third article in your assessment. Next, look at the sample student notes on page 173.

What is ageing? Can we delay it?

Conclusions

It is difficult to draw firm conclusions in an area about which so much is unknown and in which there is some controversy, but there is a consensus that ageing is a complex and multifaceted process ...

We consider three theories of why ageing occurs. They are not mutually exclusive and in due course an overall theory may incorporate elements of each.

We examine a range of cellular factors which are associated with ageing, but in the interviews with experts there was little consensus about which of these were important in causing ageing in humans. (1)

The science of ageing is clearly in its infancy and much more research is needed to elucidate the processes involved. But it is a difficult area of science and whilst there are measures of increasing frailty, there is not yet a reliable measure of biological age. (2)

Model systems are valuable in studying particular aspects of the problem, but caution must be exercised in transferring findings from one species to another and especially to humans.

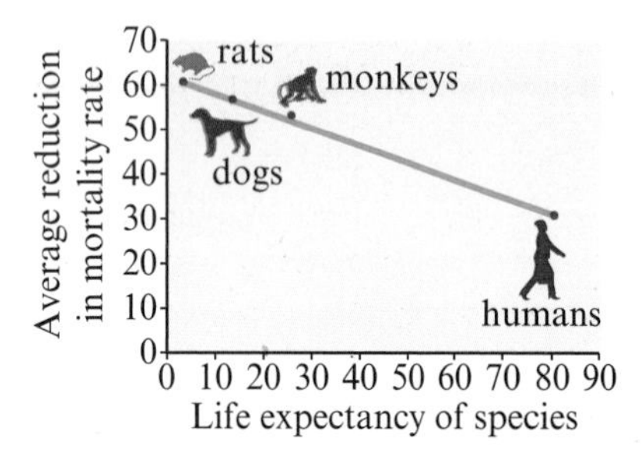

There was fair consensus among the experts that improving health span was a main driver to research on ageing and that elucidating the causes of this process would have beneficial effects for the prevention and management of the many human diseases for which age is a key risk factor.

The opposite is almost certainly also the case, in that reduction in the incidence of the major causes of mortality will almost inevitably increase lifespan to some extent. (4)

It is possible to delay ageing by a number of means in model systems, and the paper reviews briefly the possible existing and future approaches for doing so in people. Lifestyle factors, such as increasing exercise, Mediterranean diet and calorie restriction are effective and potentially achievable.

But there are problems with many anti-ageing interventions, be it unacceptable side effects (rapamycin), poor evidence (resveratrol), and lack of uptake and/or compliance (statins; behaviour changes). (5)

There are also many who desire to avoid excessive medicalisation of a normal physiological process in the general population ...

... a dramatic increase in lifespan would only occur with an intervention that was very effective in slowing the rate of ageing and was applied relatively early in life and sustained.

Nothing of that ilk is waiting in the wings at present nor seems likely to appear in the next 10 years. Given the multiple mechanisms involved in ageing, it is very unlikely that there will be a single drug that will significantly reduce the rate of ageing, and that preventive strategies are likely to have the most impact. (6)

The most common view was that longevity will continue to increase but not at the same rate as has been seen in recent decades. ...

References and bibliography (7)

1 National Institutes of Health. Single gene change increases mouse lifespan by 20 percent [Internet]. 2013. Available from: http://www.nih.gov/news/health/aug2013/nhlbi-29.htm ...

8 Kirkwood TBL. Understanding the Odd Science of Aging. Cell. 2005 Feb;120(4):437–47...

19. Miller RA, Harrison DE, Astle CM, Fernandez E, Flurkey K, Han M, et al. Rapamycinmediated lifespan increase in mice is dose and sex dependent and metabolically distinct from dietary restriction. Aging Cell. 2014 Jun;13(3):468–77.

Source: *Longevity Science Panel* (online), last accessed October 2016. 

Now try this

Identify **three** areas for development and research of scientific issues from the article.

Analysing Article 3

Read the extract from the article on page 172. This page shows you some examples of the type of notes you could make on this article.

Sample notes extract

1 The first three paragraphs show that there is no agreement between experts on ageing and how it occurs.

You could be asked to discuss potential areas for development and research of the scientific issues. The lack of agreement between scientists may show that there are plenty of areas for research.

2 Chronological age is the number of birthdays you have had, but no agreement on how to measure biological age means different scientists may define it differently.

This could affect how easy it is to compare the results of different studies. This is something you could mention when discussing validity.

3 Graph showing one example of comparing the same intervention in different species. Shows that there are problems comparing animal studies with effects in humans. Source is given.

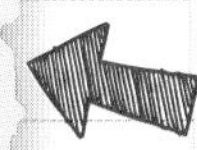

Consider what the impact of this could be on other studies mentioned in the articles you are considering.

4 Scientists agree that increasing health and reducing mortality rates will have the effect of increasing overall mortality, and studies have been done to model this.

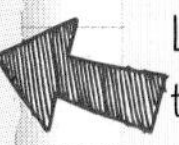

Look for examples of studies that support this suggestion.

5 Some scientists don't believe we should turn ageing into a disease and treat it with drugs.

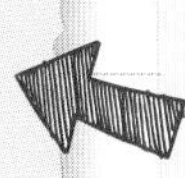

These comments show how public opinion and the opinion of scientists affect scientific issues. You will need to consider this when you think about the implications of the scientific issue being discussed.

6 A general statement suggesting that major interventions to prevent ageing seem unlikely at the moment.

Consider whether all scientists involved in research in this area agree and what the implication is of this for the scientific issue of ageing.

7 The references and bibliography extract shows that the science reported is up to date and from top researchers into ageing in the UK.

A good set of up-to-date references will help you to decide how valid the judgements in the article are.

8 Source: conclusions of the report of the Longevity Science Panel. Based on work of many scientists in the field of ageing. Written for scientists and interested non-scientists.

Some references are in the extract, with lots more in the original report. There is extensive coverage of the work of the individual scientists in the full report. This suggests the information in this article will be scientifically reliable.

Now try this

Write down **three** implications of the scientific issues being discussed in this article.

Had a look ☐ Nearly there ☐ Nailed it! ☐

Considering the scientific issues

In your assessment, you will need to consider the scientific issues in the articles you have read.

What to consider

You will need to demonstrate a comprehensive knowledge and understanding of the scientific issues by identifying and selecting relevant implications from all articles. You will also need to draw a wide range of links to, and between, any ethical, social, economic or environmental implications. Your answer will need to provide a clear, coherent and logical discussion within a well-developed structure. Below is an extract from one learner's response, discussing the implications of scientific issues identified in the articles on the previous pages.

You should:

- **Plan** your structure carefully, so your discussion makes sense.
- Indicate clearly when you are **quoting**.
- Add information from **other sources** with clear references.
- **Check** your spelling and grammar carefully.

You can plan by jotting down bullet points, but make sure your final answer is written in clear, full sentences.

Sample response extract

The three main scientific issues identified in these articles are all linked to ageing. They are the causes of ageing and longevity, ways of preventing ageing and living healthily for longer, and the difficulties of studying ageing and longevity. These issues are tackled very differently in the three articles. For example, Article 1 makes preventing ageing sound simple, by eating less, having a Mediterranean diet and taking statins. Article 2 looks at factors that appear to promote long and healthy life in a specific community in Okinawa. These include diet, exercise and genetics. Article 3 links prolonging life with treating disease. It indicates that there are no current quick fixes, comments on a lack of clear evidence for preventing ageing, and it references research.

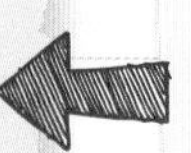

You need to show that you have read all three articles. Use them to give examples to support your answers.

The learner has identified common links between all three articles, and commented on how these are tackled in each article.

This answer is explaining the scientific issues and starts to back them up with examples from all three papers.

Now try this

Look at the extract from a learner's answer. Make notes on:

(a) where the answer brings in different social and ethical issues from the three articles, and

(b) where the answer makes links between the ethical/social/economic and environmental implications from the articles themselves and from wider reading.

The articles highlight many issues linked to getting people to change their lifestyle. Article 1 refers to a study showing a social issue – people find changing diet and doing more exercise very difficult to maintain, even though it will lengthen their lives.

Article 2 relates to a very specific population in Japan. It highlights another social issue as it shows that when people move away to different areas the benefits may be lost as their diet and lifestyle change. Article 3 refers to ethical and economic problems with anti-ageing interventions from lifestyle changes to drugs.

Living longer raises many economic and social issues. For example, what happens about jobs, pensions and housing if people live longer? Who cares for all the very old people if they are not healthy? Does everyone want to live for a very long time? These are not questions for scientists – but they need considering if everyone is going to live longer.

This is a good, direct reference to a social issue from Article 1.

Identifying organisations and individuals

You should be able to identify the different organisations and individuals mentioned in the articles and suggest how they may have an influence on the scientific issues.

Here are some examples of one student's notes about organisations and individuals, along with an extract from a full response they have written focusing on the articles on the previous pages.

Sample notes extract

Longevity Science Panel: Referenced in Article 1 line 6 and the whole of Article 3. The panel was set up to advise policy makers in government.

Made up of top statisticians and scientists specialising in ageing, for example, Tom Kirkwood.

In your notes, make sure you include in which article the person or organisation was referenced, so that you don't have to try and find it again later on. Your research should include more information about the organisations and how they might influence an issue.

Sample notes extract

World Health Organization data and Japanese Ministry of Health and Welfare: referenced in Article 2. Two large and well-respected organisations.

Make sure you explain the position and influence of each named individual.

Sample notes extract

Okinawa Centenarian Study: referenced in Article 2. Internationally renowned study group looking at the very old but healthy people in Japan.

ref http://www.okicent.org/study.html

Give links, references and use footnotes.

Sample notes extract

Richard Faragher: referenced in Article 1 final paragraph. Professor of Biogerontology at the University of Brighton. Very influential in UK and internationally, moving forward research into ageing.

https://www.brighton.ac.uk/staff/prof-richard-faragher.aspx

For individuals, make sure you find out more about any institutions or organisations they belong to.

Sample response extract

The World Health Organization is the public health arm of the United Nations. It has the support of most nations in the world and provides leadership on global health matters. One of its stated objectives is 'shaping the research agenda and stimulating the generation, translation and dissemination of valuable knowledge' (Ref: http://www.who.int/about/en/). There is huge concern in the WHO about the effect of an ageing global population and so they are driving forward research into ageing in many areas. (Ref. http://www.who.int/topics/ageing/en)

You have explained how the WHO influences the issues

Use quotes whenever you can – and reference them.

Now try this

Make a table of all the main organisations and individuals and which article they appear in Articles 1–3 on pages 168–173.

You need to list all of the main organisations and individuals highlighted in the articles you have been given. You are expected to show where they are referenced if you can. You will need to explain what they do and why they might influence the science issue – in this case, research into ageing.

Had a look ☐ Nearly there ☐ Nailed it! ☐

Considering the validity of judgements

You will need to discuss whether **some of the judgements within the articles you have read are valid or not.** Below are some things to remember when writing about this, along with a sample response from one student, focusing on the articles on the previous pages.

What to consider

- Make it clear why valid and reliable data is important.
- Explain what you mean by **valid data** – data that effectively measures what is being investigated – and **reliable data** – data that can be reproduced.
- **Only** refer to the articles you have been asked to cover, if there are instructions about this. You may not need to consider all three articles you have previously been given.
- Remember to support your answers consistently throughout with specific examples from the article(s).

You need to:

- consider how the scientific information has been interpreted and analysed to support the conclusions/judgements being made
- consider the validity and reliability of data. Looking up and using the names of some of the scientists and institutions involved helps make a good answer
- include referenes to other sources of information. Refer to other articles/papers/data you have found that support the judgments put forward in the articles you are given
- have a well-developed structure, which is clear, coherent and logical. You can use bullet points when planning your answer, but don't use them in your final answer.

Sample response extract

The judgements in this article appear to be valid, based on data that measures what is being investigated, and which has been reproduced in a number of studies. The data they have used comes from a wide number of reliable sources including the World Health Organization, which collects data from scientists and doctors all over the world. They have also used data from the Okinawa study, which has been running for many years and involves scientists from several countries.

There are 62 references, all to well-cited papers ...

This is a good answer so far. It would be even better if the learner referenced:

- scientists such as Professor Tom Kirkwood
- metadata studies, for example, Nakagawa et al
- some other resources supporting the article.

Now try this

Create your own plan to answer the question 'Discuss whether Article 3 has made valid judgements'. You will need to expand this beyond the examples given above.

If you make time to do some reading around the subject and make notes before you do your answer, it will be better practice for your real assessment!

If you have time, writing a full response would give you even more practice for your assessment.

Suggest areas for development/ research

You may be asked to suggest any potential areas for development and/or further research of a scientific issue in the articles you have been given. Below is an extract from a response which suggest possible further development or research around the issues raised in the three articles on the previous pages.

Sample response extract

Article 1 contains expert suggestions on things that can increase lifespan, including restricting calorie intake (fasting on alternate days), a Mediterranean diet and statins. Research into these areas could investigate which of the options is most effective at increasing life expectancy. Once more evidence is available, it would be possible to develop a lifestyle package, which people can follow to give increased lifespan ...

Research involves the systematic investigation of ideas or materials to establish knowledge and come to new conclusions.

Development involves using the results of scientific research and extending the ideas to produce a useful product or process on an industrial scale.

From the topics covered in Article 2 there are several that would be good subjects for further research. For example, scientists could look at the effect of homocysteine levels on cardiovascular health and ageing. They could focus on the different levels in people who live a long time compared with those who die young, the effect of an Okinawan diet on homocysteine levels or the effect of homocysteine supplements on cardiovascular health. Another possible area for more research highlighted in Article 2 is the impact of diet on different cancers. For example, scientists could try to identify which aspect of Okinawan lifestyle lowers rates of breast, prostrate, ovary and colon cancers.

Your ideas **must** be based on science from the articles themselves, not any further reading you may have done.

Many women suffer badly with the menopause. Another important area of research suggested by Article 2 is the effect of soy isoflavone supplements on menopause symptoms. If they are found to help, this could lead to development work on dietary supplements for people who eat a more Western diet to give them the benefits of Okinawan diets...

Article 3 states that the science of ageing is in its infancy and much more research is needed to discover the processes involved. This could include research into preventing and curing diseases where age is a key risk factor, and record the effect on longevity. The article also suggests research into possible drugs to increase longevity. If this is possible, there would be a lot of development work needed to develop a commercial product. Any potential drug would need to be safe to take for many years and cheap enough for everyone to use if it is to be ethical. This would involve a lot of development work.

This covers research topics clearly. More about development possibilities would make the answer better.

Highlighting the possible developments which might emerge from the research improves this answer.

Now try this

Using the sample answers on this page as your starting point, answer the question 'Suggest potential areas for further development and/or research of the scientific issue from the three articles on pages 168, 170 and 172.'

Had a look ☐ Nearly there ☐ Nailed it! ☐

Writing for a specific audience

You may be asked to **compose** a formal piece of writing for a specified audience using the information contained in the articles you are given. This might, for example, be a newspaper article, or a some copy for a website. To do well in this type of question, your answer must show a **well-developed structure** which is **clear, coherent and logical**. Read the extract from a sample student response below to understand how you might start a piece of writing to ensure it is engaging and suitable for the relevant audience.

Sample response extract

Imagine if we could all live to at least 100 years old – and be healthy! Scientists around the world are finding evidence that this may soon be a real possibility. From collecting evidence from healthy centenarians in Japan, America and Sicily, to analysing all the best research, top scientists are working to find the secret of a healthy old age!

It isn't all plain sailing. Some scientists have shown that people like you and me find it very hard to make changes to our lifestyles ...

This student is writing an article for a charity, to be published nationally, in order to raise awareness of research into ageing and attract funding for it to continue. The response needs to inform the general public – and persuade them to give generously to support the research! They've done this well by starting with an engaging opening that appeals to everyone's sense of imagination, to try and draw them into the topic.

The student explains what is known about healthy ageing, then highlights what research is needed. They will need to go on to give more detail about the potential further work required.

The persuasive and familiar language must be maintained throughout the whole article to encourage charity giving.

Remember

You must demonstrate that you have not only **read** all the articles thoroughly, but that you have also **understood** their content and can **put the ideas together so they make sense**.

Don't forget

- ✓ Choose your writing style, language and approach to suit your **target audience**. For example, you may be writing a scientific report or trying to persuade people to take a particular point of view.
- ✓ You must write in the **same style** throughout your answer.
- ✓ Choose the bits of **evidence** you use from the articles with care to support your arguments.
- ✓ Always be accurate in your use of **scientific terminology, but** choose your terminology carefully to suit your target audience.
- ✓ Make sure you use **all** the articles referred to in the question.

Now try this

A big charity is trying to attract more funding for research the science of ageing. You have been asked to raise awareness of research into potential ways of living longer healthily.

- Write a suitable article which will be published nationally in leaflets and on the charity website.

Planning your answer carefully is important as you will be given credit for a clear, well-planned piece of writing as well as for explaining the science clearly and at the right level.

Answers

Unit 1, Biology

1 Cells and microscopy

image length = 7 × 0.06 mm = 0.42 mm.

$$\text{magnification} = \frac{\text{image size}}{\text{real size}}$$

$$\text{real size} = \frac{\text{image size}}{\text{magnification}}$$

real size = 0.42/400 = 0.001 mm (correct to 1 significant figure).

2 Cells

Two membranes	One membrane	No membranes
nucleus mitochondria	lysosomes rough and smooth ER Golgi apparatus (or Golgi body)	ribosomes centrioles nucleolus

3 Prokaryotes

1 The results show that there are some gram-positive bacteria present.
2 Some bacteria stained purple because they are gram-positive. The stain is absorbed by the thick cell wall because these bacteria do not have an outer membrane.

4 Plant cells

	Plant (eukaryotic) cell	Animal (eukaryotic) cell	Bacterial (prokaryotic) cell
Cell wall	✓	✗	✓
Chloroplasts	✓	✗	✗
Nuclear membrane	✓	✓	✗
Cell membrane	✓	✓	✓
Ribosomes	✓	✓	✓
Centrioles	✗	✓	✗

5 Specialised cells: plant cells

1 A
2 Root hair cells are adapted for their function of absorbing water and dissolved minerals:
- a long cell extension into the soil increases surface area for absorption
- thin cell wall makes it easier for substances to cross into the cell
- many mitochondria supply energy for active transport of minerals from the soil into the cell.

6 Specialised cells: animal cells

Any three from:
- sperm cell has flagellum
- overall shape sperm cell is e.g. streamlined
- sperm cell has fewer mitochondria / other organelles
- sperm cell has acrosome
- egg cell has zona (pellucida) / jelly layer
- egg cell has cortical granules
- differences in food store types
- sperm cell has less cytoplasm.

7 Epithelial tissue

Ciliated columnar cells have hair-like projections called cilia that move particles or mucus, for example, in the lungs they move inhaled particles out of the lungs. In order to keep moving, the cilia need a lot of energy. This is provided by respiration, which occurs in the mitochondria.

8 Blood vessels and atherosclerosis

The endothelium is made up of a single layer of long, flat cells, orientated lengthways in the direction of blood flow. Because they are flat, the surface is smooth, which helps blood to flow more easily. Also the diffusion distance is small, so substances can diffuse easily in and out of blood vessels (especially capillaries) into the surrounding tissues.

9 Fast and slow twitch muscle

Slow twitch muscles.
They can contract slowly over long periods of exercise and do not tire easily. Energy is from aerobic respiration, so they have many capillaries to transport oxygen and glucose to the cells for respiration and many mitochondria for respiration to take place.

10 Nerve tissue

Progressive de-myelination means that the myelin sheaths gradually break down.
When the myelin sheath breaks down somehow, it causes nerve signals or impulse to be transmitted more slowly, disrupting how the nervous system normally communicates.

11 Nerve impulse

Potassium ions leave the cell because they are following down the diffusion gradient and are attracted by the negative charge.

12 Electrocardiogram (ECG)

The P wave represents the depolarisation of the atria, whilst the QRS complex represents the depolarisation of the ventricles.
The QRS complex is much larger than the P wave because the ventricles have far more muscle mass than the atria.

13 Synapses

(a) It prevents the release of acetycholine.
(b) Acetycholine is a neurotransmitter, which enables nerve impulses to stimulate muscles. If its release is prevented by botulism toxin, this would mean the muscle would be paralysed.

14 Brain chemicals

Dopamine agonists mimic dopamine. They bind to dopamine receptors at synapses.
They trigger action potentials.

Unit 1, Chemistry

15 Writing formulae and equations

$C_2H_5OH + 3O_2 \rightarrow 2CO_2 + 3H_2O$

16 Electronic structure of atoms

(a) 11 electrons so $1s^2 2s^2 2p^6 3s^1$
(b) 10 electrons so $1s^2 2s^2 2p^6$
(c) 22 electrons so $1s^2 2s^2 2p^6 3s^2 3p^6 4s^2 3d^2$
(d) 20 electrons so $1s^2 2s^2 2p^6 3s^2 3p^6 3d^2$

17 Ionic bonding

NaCl, as it contains the Na^+ ion, which is smaller than the K^+ in KCl. The charge is the same in both cases, as is the Cl^- so neither factor is relevant in this comparison.

18 Covalent bonds

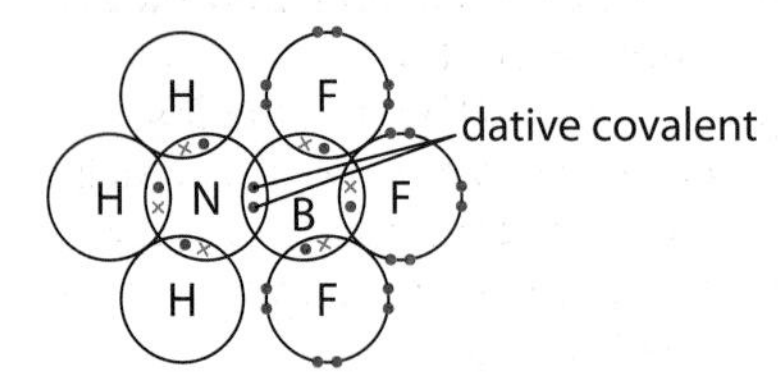

19 Metallic bonding

Structure in metals has positive ions in layers, which can slide across each other when hammered. The electrons move with the layers, so the electrostatic attraction to the nuclei stays intact preventing splitting.

20 Intermolecular forces

Both compounds form van der Waals forces. HCl has greater van der Waals forces than HF as it has more electrons. HF can also form hydrogen bonds and HCl can only form dipole–dipole interactions. Hydrogen bonds are much stronger than dipole–dipole interactions and van der Waals forces, so more energy is required to break the hydrogen bonds, so HF has a higher boiling point.

21 Relative masses

392

22 Amount of substance: The mole

$49.6/0.62 = 80.0\,g\,mol^{-1}$

23 Calculating reacting masses and gas volumes

amount of calcium nitrate = 4.20/164.1 = 0.026 mol
ratio of calcium nitrate to NO_2 formed = 2:4,
so moles of NO_2 formed = 0.026 × 2 = 0.052 mol, so mass of NO_2 = 0.052 × 46.0 = 2.39 g
ratio of calcium nitrate to O_2 formed = 2:1,
so moles of NO_2 formed = 0.026/2 = 0.013 mol, so mass of O_2 = 0.013 × 32.0 = 0.416 g
total mass of gases formed = 2.81 g

24 Calculations in aqueous solution

amount of Na_2CO_3; $n = c \times V$, so $n = 0.016 \times (500/1000)$ = 0.0080 mol
mass of Na_2CO_3 using; mass = amount × molar mass = 0.0080 × 106 = 0.85 g

25 The periodic table

$H(g) + e^- \rightarrow H^-(g)$

26 Trends in the periodic table

Carbon has a giant covalent structure, which consists of many, strong covalent bonds. In order to melt the carbon, these bonds have to be broken, which requires a lot of heat energy.

27 Reactions of periods 2 and 3 with oxygen

They should wear gloves and goggles. Calcium oxide is a metal oxide, so is likely to form alkaline calcium hydroxide if it comes into contact with moisture from the skin/eyes. This may be corrosive or irritant.

28 Reactions of metals

Aluminium is a reactive metal that will react with water and form aluminium hydroxide and hydrogen gas, which is highly flammable, so may ignite and explode. Also, a powder has a large surface area, which would mean the reaction would be extremely fast.

$2Al + 3H_2O \rightarrow Al_2O_3 + 3H_2$

29 Oxidation and reduction

Fe loses 2 electrons to form Fe^{2+}, which is oxidation. H^+ ions from the acid gain electrons to form hydrogen gas, which is reduction. Overall the reaction is a redox reaction.

Unit 1, Physics

30 Interpreting wave graphs

1 0.6 ms
2(a) 1.7×10^{-6} m
(b) $\frac{3}{4}$ of a cycle

31 Wave types

1 Molecules displaced along the direction the wave travels.
2 Antennas perpendicular to direction of travel; these waves can be polarised.

32 Wave speed

1 1.20 m
2 $0.004\,kg\,m^{-2}$
3 $2.03 \times 10^8\,m\,s^{-1}$

33 Wave interference

1 Curve that cuts axis at ~ 0.004 m, 0.029 m and 0.054 m, has a maximum at ~ 0.016 m and a minimum at ~ 0.042 m with amplitude ~ 3.0 scale divisions.
2 Reflected wave is weaker than incident wave, so amplitude of resultant never cancels completely to zero.

34 Diffraction gratings

1 The white line is the transmission beam. Either side of it the light has been scattered by the grating. That causes a path difference, which increases with the angle of scatter. The first coloured spectrum occurs where the path difference is causing a phase difference of 1 whole cycle. That happens at slightly different angles for the different wavelengths. A second spectrum occurs where the phase difference is 2 whole cycles, etc.
2 You can identify the elements present by matching line frequencies with known spectra.

35 Stationary waves

1 Stationary wave stores energy and has a fixed pattern (of nodes and antinodes). Progressive wave transmits energy and has a moving pattern of displacements.
2(a) Reflecting the wave back on itself creates a stationary wave; the position of the first antinode gives a large amplitude signal.
(b)

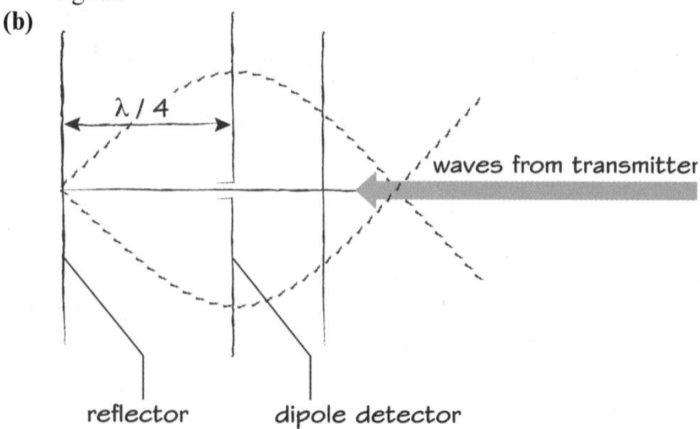

36 Musical instruments

(a) Tension and mass per unit length of the string are still the same.
(b) 1.5 times higher frequency.

37 Optical fibres

1 45.6° or 0.796 rad.
2 Copper: for local connections to, for example, TV, Hi-fi, lab equipment etc. that need easy changing; optical fibre: for permanent longer distance links, for example, between distant router cabinets.

38 Endoscopy

Both instruments require illumination: in the endoscope, light is carried by one bundle of fibres into the site of interest. Both produce focussed images using objective and eyepiece lenses, but optical fibres that carry the endoscope's image back out of the body make it narrower, much longer and flexible compared with a short, fat, rigid microscope. However, the endoscope image is pixelated – one pixel per fibre.

39 Analogue or digital?

(a) Diagram that shows thermistor still connected to electronic circuit, but the voltmeter connected in parallel with, or replaced by, the input to an analogue to digital converter (ADC). The ADC digital output shown connected to a data logger or computer.
(b) An explanation that includes two of the following points: labour-saving as measurements can be automated; remote monitoring possible; data readily processed by computer.

40 Electromagnetic waves

1 Ultraviolet (~ 400 to 10 nm); X-rays (~ 10 to 0.01 nm); γ-rays (<0.01 nm).
2 Microwaves (1 to 40 GHz, i.e. 300 to 75 mm) because they are not absorbed and high frequency allows high data transmission rates.
3 Long and medium wave radio broadcasts cannot escape into space, but reflection allows them to travel between continents – line of sight is not required.

41 Waves in communications

1 Bluetooth® is very short range, device-to-device and uses a range of frequencies. Wi-Fi is a networked system that can connect many devices, has a range up to about 100 m and uses a dedicated frequency. Their coexistence is helped by Bluetooth® frequency-hopping, which means it will only interfere with nearby Wi-Fi for very short bursts of time. Wi-Fi data packets can be re-sent if not acknowledged as received.
2(a) UHF radio and microwaves.
(b) Mobile phone operators each have specific allocated bands, but within that band they utilise different frequencies for upload, download and between different cells; satellites generally have separate bands allocated for upload and download signals. Mobiles use simple antennae creating spherical waves. Satellites use dish antennae to create parallel beams

Unit 1, Skills

43 'Give', 'state' and 'name' questions

A strong electrostatic attraction between two nuclei and the shared pair(s) of electrons between them.

44 'Calculate' and 'write' questions

$I_1 = k/(r_1)^2$; $I_2 = k/(r_2)^2$; so, 0.223 mol dm^{-3} (to 3SF).

45 'Explain' questions

During analogue to digital signal conversion, both sound and pictures are initially collected as analogue signals. Even though camera detectors are pixelated, each pixel's analogue signal has to be sampled and converted to a digital signal that represents its colour and intensity using numbers. A to D conversion is carried out electronically.

Both sampling rate and sensitivity affect the quality. Higher quality reproduction (for example, HD TV) requires more data to be transmitted and so takes up more bandwidth.
For pictures, the number of pixels multiplied by the sampling rate determines the number of data items to be transmitted each second. The picture sampling sensitivity determines the number of distinct colours that can be reproduced; more colours requires each data item to have a higher number of bits.
For sound, the sampling rate and sensitivity limit the frequency range (high or low pitch) and dynamic range (soft and loud) of the sounds that can be reproduced.
Multiplexing divides the available bandwidth into a defined number of channels, each with a different frequency. Each channel can carry several different TV stations' signals because the data speed is high enough to send data for each of them within the time defined by the sampling rate. Receivers decode the multiplexed signals and split them back into separate TV stations.
An example of compromise is the sound and picture quality versus the number of channels that can be transmitted within a given frequency band. HD channels require higher data speeds and higher processing power, so can be slower to load.

46 Answering longer questions

The second ionisation energy is greater than the first ionisation energy, as the second electron is removed from a positive ion so the attraction to the nucleus is greater.

47 'Describe' and 'discuss' questions

Path difference between the waves coming from adjacent lines on a grating depends on the angle through which the light is scattered. If the path difference is a whole number of wavelengths, then the waves will be in phase and interfere constructively giving bright lines. In between, when the phase difference is near 180° (π rad), destructive interference causes darkness.

48 'Compare' questions

The answer can be as a clear paragraph, or in table form, as below.

Red blood cells	White blood cells
do not have a nucleus	have a large nucleus.
they can move through narrow blood vessels but do not leave blood vessels	they can squeeze through capillary walls and are found in the blood and tissues
they carry oxygen in the blood	they fight pathogens
both made in bone marrow, but white blood cells also made in lymph nodes	
both found in blood	

Unit 3: Science Investigation Skills

50 Developing a hypothesis

(a) The higher the concentration of acid, the faster the metal will dissolve.
(b) The more cells in a circuit the brighter the bulbs will be.

51 Planning an investigation

An answer to include:

(a) Equipment

- hydrogen peroxide (known concentration) – this is the substrate required for the reaction. The concentration must be kept the same, as the concentration of substrate affects the rate of an enzyme-controlled reaction.
- potato – this contains the catalase enzyme being investigated.
- conical flask – to hold the reaction mixture.
- delivery tube – to transfer the oxygen produced to the gas syringe to be measured.
- gas syringe
- bung – to prevent any oxygen escaping from the conical flask.
- knife/potato cutter – to cut the potato into equal sized cubes.
- measuring cylinder – to measure out 30 cm^3 of hydrogen peroxide.
- stop clock – to time 60 seconds.
- thermostatically controlled water baths set at the temperatures being investigated – the conical flasks and reaction mixture will stand in water baths set at each temperature being investigated.

(b) Technique
Volume of gas collected in the gas syringe after one minute.

(c) Standard operating procedure
Four 1 cm^3 cubes of potato (with the skin removed) are added to 50 cm^3 of hydrogen peroxide at room temperature in a conical flask. The delivery tube connected to a bung is placed in the conical flask and is attached to a gas syringe. Start the stop clock and time for one minute. After one minute note the volume of gas given off. Repeat each temperature a further two times. Repeat the steps for solutions at a range of temperatures: 20 °C, 30 °C, 40 °C, 50 °C. Use an ice and/or water bath to maintain the solutions at the correct temperature. The volume and concentration of the hydrogen peroxide are kept the same for each experiment, so that only the temperature can have an effect on the rate. The size and number of potato cubes are kept the same for each experiment, so that the enzyme concentration is the same each time. A range of temperatures is used so that a pattern can be seen. The gas syringe gives accurate measurements of the volume of gas produced in one minute. Repeating the experiment would allow identification of anomalous results and calculation of a mean for each temperature investgated.

52 Risk assessments

For example:

Hazard	Harm that could be caused	Severity of harm	Likelihood of harm	Control measures	Procedures if harm occurs
catalase enzyme	irritant/ harmful to skin/ eyes	low	low	gloves goggles lab coat	wash off skin eye bath
hydrogen peroxide	irritant	medium	low	gloves goggles lab coat	wash off skin eye bath
oxygen gas	oxidising	low	low	no naked flames	fire blanket raise alarm and evacuate
glass ware	cuts from broken glass	medium	low	take care clean up any breakage with designated glass brush into glass bin	first aid

53 Variables in an investigation

(a) For example, type of surface, thickness of surface.
(b) Height that the ball bounces.
(c) For example, the softer/thicker the surface, the lower the height that the ball bounces.
(d) Drop ball from same height each time. Use the same ball each time. Control the drop (for example, use same person or method of releasing the ball).

54 Producing a method

Select 3 sample areas under trees and 3 not under trees. Obtain a map of each area and divide it up into smaller areas. Assign coordinates to the area. Randomly generate numbers and place 1 m^2 quadrats in the position determined by the random numbers and count number of different plant species in each quadrat (the species richness). Repeat this for 20 randomly placed quadrats. Repeat for each sample area on the same day if possible. Measure the light intensity at the time of recording using a light meter. Note: you will find it is difficult to control the variables for an investigation like this and this will need to be considered in your results. You may wish to record other variables such as soil pH and soil moisture in order to be sure of your conclusions.

55 Recording data

(a) Use a gas syringe to collect the O_2 produced in one minute.
(b) Repeat 0.50 m test as the second test seems to be anomalous.
(c) Two or more results the same can confirm results are correct. So you repeat to get concordant results.

56 Processing data

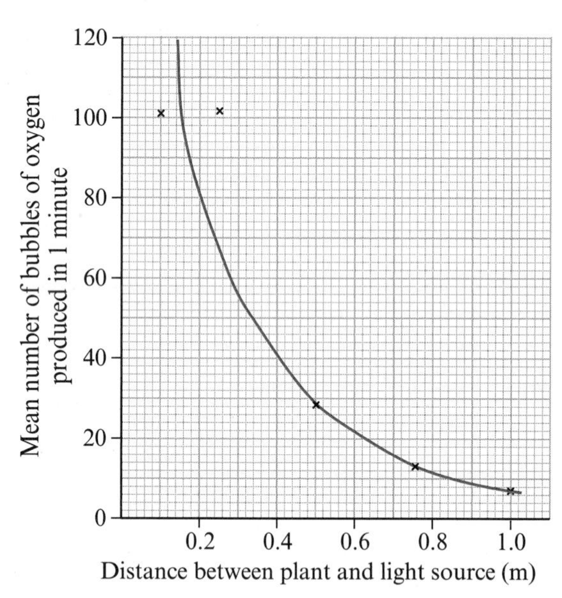

You would choose a line graph and plot distance between plant and light source (mm) on the x-axis and mean number of bubbles produced in one minute on the y-axis. Your graph should have ascending equidistant intervals on each axis and an appropriate scale. You must include axis labels with units and plot all of your points correctly. Here a smooth curve of best fit is most appropriate.

57 Interpretation and analysis of data

(a) The number of bubbles produced decreases the further away the light is from the plant. When the light is very close to the plant the number of bubbles produced does not change.
(b) The closer the light is to a plant the more it will photosynthesise. This is shown by the increase in number of oxygen bubbles produced when the light is closer. Plants use light energy to photosynthesise and this produces oxygen. So, the more light that is available, the faster the rate of photosynthesis and so the more bubbles produced in one minute. The optimum rate of photosynthesis is a limiting

factor and there is a point when it does not matter if the light is closer, the rate of oxygen being produced will not increase. This is around 0.25 m, as the rate of photosynthesis does not change after that point. In order to know the optimum distance, more tests must be carried out in smaller increments around 0.25 m distance.

58 Evaluating an investigation

Examples might include:

Possible errors	How they could be minimised in the future
low precision of balance (only recording to two decimal places)	use a more precise balance (for example, four place balance)
reading the thermometer can be subjective (based on opinion)	use a digital thermometer
low precision of thermometer (can only be read to nearest degree)	use a more precise thermometer. A; a digital thermometer may read to nearest 0.005 °C
use of measuring equipment with relatively high in-built error, for example, measuring cylinders	use measuring equipment with lower error, for example, graduated pipette
measuring equipment may give inaccurate readings due to it not being calibrated (for example, burette, mass balance, pH meter, thermometer)	calibrate equipment before using it by comparing the measurements it makes to known standards and adjusting the equipment accordingly
inaccurate observations due to subjectivity of measurements, for example, judging the incorrect colour	compare the measurements to a standard chart or image, for example, have a colour chart to compare the colours of a flame test to

59 Enzymes: Protein structure

Cysteine: $H_2N-CH(CH_2SH)-COOH$ (R group: CH_2SH)

Threonine: $H_2N-CH(CH(CH_3)OH)-COOH$ (R group: $CH(CH_3)OH$)

Condensation reaction

Cysteine + Threonine → Dipeptide + H_2O

$H_2N-CH(CH_2SH)-CO-NH-CH(CH(CH_3)OH)-COOH$ + H_2O (peptide bond: $-CO-NH-$)

Dipeptide

60 Enzymes: Active sites

The enzyme has an active site that has a specific shape. The substrate has a complementary shape that binds to the active site of the enzyme. This allows the reaction to happen. The product of the reaction will not have the same shape and so is released, allowing the active site to be available for another substrate to bind.

61 Enzymes: Biological catalysts

1 A reaction cannot happen unless the the substrate collides with the enzyme's active site.

2 The enzyme provides a reaction pathway with lower activation energy so less energy is needed for the reaction to happen. This means that more of the substrate molecules are likely to have enough energy to react, resulting in an increased rate of reaction.

3 The initial rate of reaction must be recorded, as the rate will decrease as the substrate is used up and its concentration decreases.

62 Enzymes: Factors affecting activity

Protease will show its highest activity at pH 1–2, as this is the pH of stomach acid. It will slow down when the pH is above this and may not work at all at higher pH. It will have the highest activity at about 37 °C, as this is body temperature. Above 40 °C it will start to denature and so the activity will stop. Below the optimum temperature the activity will also be lower, the lower the temperature is. Increase in concentration of substrate will increase the rate of the reaction until the enzyme's active sites are all occupied and so the rate of reaction will not increase further.

63 Diffusion of molecules

1 Molecules from the cake will diffuse from the area of high concentration to areas of low concentration. The energy from the heat in the oven will allow the molecules to move faster.

2(a) You would smell the cake near the oven first, and, (b) at the furthest part of the house from the kitchen last.

64 Kinetic theory and diffusion

1 Gas particles are spaced out and flow easily around each other and around other gas particles, such as in the air. Gas particles are energetic so move into free space down their concentration gradient.

2 The particles are packed close together and do not have enough energy to move.

65 Plant growth and distribution

Plant ten cress seeds in a pot of soil. Leave in a sunny room kept at room temperature. (make sure you measure the temperature but it should be around 20 °C).
Repeat for four more pots of identical soil. Leave each in a sunny room at the following temperatures: 10 °C, 30 °C, 40 °C, 50 °C.
At the same time each day, observe any germination. Record any measurements of cress growing above the soil. Ensure that the volume of water supplied to each pot is the same and that they are watered at the same time. Do not place pots in direct sunlight but use well-lit rooms. Control the temperature of the rooms using a thermostat/place pots in incubators.

66 Improving plant growth

1 Potassium and iron; possibly magnesium.

2 Decreased rate of photosynthesis due to not enough water being available in the soil. Plants cannot take up sufficient minerals and nutrients from the soil leading to mineral deficiencies. The vacuoles will not be filled with fluid and so the plant may droop.

67 Sampling techniques

Random sampling allows for reliable and valid data collection. For example, place a grid over a field and assign each section coordinates. Use a random number generator to decide the coordinates of where the quadrat is placed in the field.

68 Investigating fuels

(a) 1 Use a 100 mL measuring cylinder to measure out 100 cm^3 of water in a 250 cm^3 beaker to act as a calorimeter.

2 Record the temperature of the water using a thermometer.

3 Clamp the calorimeter in position about 10 cm above a heat-proof mat.
4 Find the mass of a spirit burner containing a fuel using a mass balance.
5 Place the burner under the calorimeter and light the wick.
6 Place the heat shield around the apparatus.
7 Stir the water with the thermometer.
8 Stop heating when the temperature has risen by 20°C.
9 Replace the cap on the burner. Reweigh the burner as soon as possible.
10 Repeat the steps for the same fuel in order to obtain concordant results.
11 Repeat for different fuels with longer carbon chains.

(b) 1 × 100 cm^3 measuring cylinder (to measure out the water for the colorimeter)
1 × 250 cm^3 glass beaker (to act as the calorimeter, to hold the water whilst it is being heated)
1 × thermometer (to measure the temperature)
1 × clamp and stand (to hold the beaker at a height of 10 cm above the spirit burner
1 × mass balance (to weight the mass of the spirit burner before and after burning)
1 × heat shield (to prevent heat loss to the surroundings during the experiment)
A range of spirit burners containing different fuels (to compare the energy released from different fuels).

69 Risks of investigating fuels

Carbon monoxide and soot will form. Carbon monoxide is an odourless toxic gas that combines with haemoglobin in our blood and prevents oxygen from being carried by blood cells. This can be fatal. Soot is an irritant and can cause respiratory difficulties.

70 Units of energy

(a) heat energy = 50 × 4.2 × 44 = 9240 J = 9.24 kJ
(b) molar mass of ethanol = (2 × 12) + (6 × 1) + (1 × 16) = 46
(c) energy released (heat of combustion) = $\frac{9.24}{3} \times 46$
$= -141.68\ kJ\ mol^{-1} = -140\ kJ\ mol^{-1}$ (2s.f.)

71 Symbols in electrical circuits

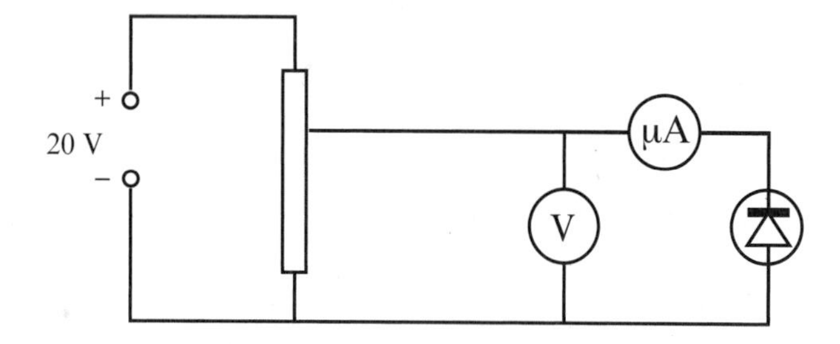

72 Power equations

1 power = current × voltage
60 = I × 120
60/120 = 0.5 A

2 power = $\frac{\text{work done}}{\text{time}}$
800 = work done/60
60 × 800 = work done = 48000 J or 48 kJ

73 Energy usage

Appliance	Current used (A)	Power (W)	Voltage used (V)	Suitable fuse rating (A)
printer	0.5	50	100	3
hairdryer	10	2200	220	10
tumble dryer	11	2500	227	13

75 Types of task

(a) The learner has stated the length of the filter paper; this will help to control the surface area for diffusion as this will also affect diffusion rate.
- The learner has stated that they will use a line to indicate the start and finish position of the dye each time. This means that the distance that the dye has to travel is controlled and does not affect the rate of diffusion between each repeat.

(b) The learner has stated that they will use different concentrations of dye, but they have not said how many different concentrations there will be or what the concentrations will be.
- The learner has only stated how long the strips should be and not how wide the strips should be. If the strip width is not the same in all repeats for all concentrations, then the surface area for diffusion will be different.
- The learner has stated that the time taken for the dye to move up 5 cm up the paper will be recorded. The learner has not said how this will be done or to what precision, for example, using a stopwatch to the nearest second.
- The learner has not included repeats for the same concentration of dye in order to produce concordant results.
- The learner has not stated the volume of dye that will be used or what piece of equipment would hold the dye solution. This would need to be standardised across repeats as it could affect the rate of diffusion.
- The learner has not said how temperature will be controlled. This is important as temperature affects the rate of diffusion.

(c) The data collected will not be reliable, as the learner has only suggested testing each concentration once. Repeat measurements of the same concentrations should be included in order to obtain at least two concordant results.

76 Recording data

Length of wire (mm)	Current (A)			Voltage (V)			Resistance (Ω)
	repeat 1	repeat 2	mean	repeat 1	repeat 2	mean	

Table should have:
- independent variable in first column
- dependent variables in subsequent columns
- headings with units
- border around the outside of the table.

77 Dealing with data

1 **(a)** 0.04942 = 0.05 (2 d.p.), 0.049 (2 s.f.)
(b) 168.472 = 168.47 (2 d.p.), 170 (2 s.f.)
(c) 0.11 (2 d.p.) 0.11 (2 s.f.)
(d) 19464.6843 = 19464.68 (2 d.p.), 19000 (2 s.f.)

2 **(a)** 4.94×10^{-2}
(b) 1.68×10^{2}
(c) 1.09×10^{-1}
(d) 1.95×10^{4}

78 Calculations using data

(a) Mean energy
1.8 is anomalous, so do not use in calculation
(3.7 + 3.8 + 3.9 + 3.7)/4 = 3.8 kJ (rounded to 1 decimal place as same accuracy as results)

(b) Standard deviation
$3.7 - 3.8 = -0.1, (-0.1)^2 = 0.01$
$3.8 - 3.8 = 0, (0)^2 = 0$
$3.9 - 3.8 = 0.1\ (0.1)^2 = 0.01$
$3.7 - 3.8 = -0.1, (-0.1)^2 = 0.01$
$0.01 + 0 + 0.01 + 0.01 = 0.03$
$0.03/3 = 0.01$
$\sqrt{0.01} = 0.1$ = standard deviation

Standard error
sample number = 3 (remember 1.8 has not been included, as it is deemed anomalous)
square root of 3, $\sqrt{3} = 1.73$
standard deviation 1.73 = 0.1/1.73 = 0.06
standard error = 0.06

79 Statistical tests: Chi-squared

There is a significant difference between the number of right-handed and left-handed individuals observed and what was expected.
The value of $\chi 2$ = 18.50. This is bigger than the critical value for these degrees of freedom (3.841). This means that there is a less than 0.05 (5%) probability that the difference between the observed and expected numbers was due to chance. (In fact, from the probability table you can see that there is a less than 0.01 (1%) probability that the differences were due to chance.)

80 Statistical tests: t-test

The probability that there is no significant difference between the mean height of German shepherds and the mean height of Siberian huskies is between 10 and 20%. (This means there is a 10–20% probability that the differences in the height was due to chance.) Scientists accept anything greater than 5% as being due to chance.

81 Displaying data (1)

1 Bar charts must have:
- Both axes labelled fully with units where appropriate.
- The y-axis (with the numbers on) must have an ascending scale with equidistant intervals.
- The x-axis should be used to display the categories.
- The bars should all be the same width.
- If the data is discrete, then there should be spaces between the bars.
- If the data is continuous, then there should not be spaces between the bars.
- If the bar chart is displaying more than one set of data for each category, then a key must be provided.

2

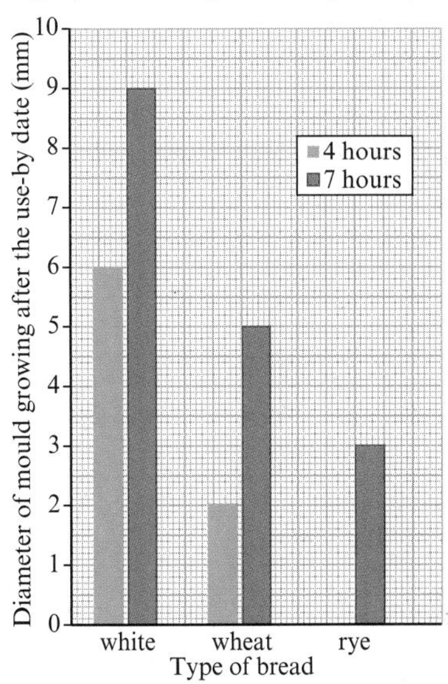

82 Displaying data (2)

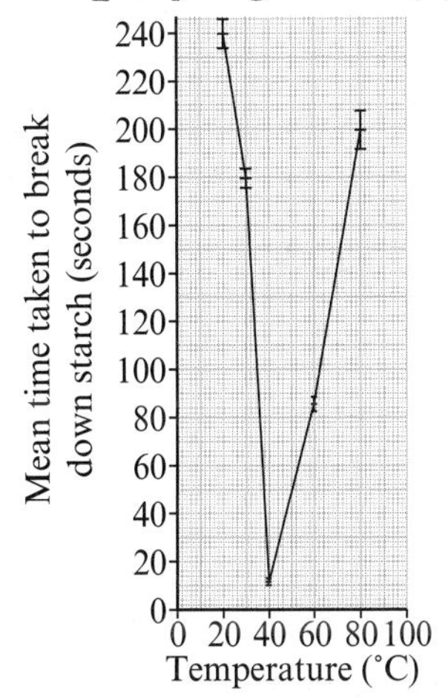

83 Interpreting graphs

1 As time increases, the volume of oxygen increases until about 90 seconds, when the graph levels off showing the increase in oxygen is slowing.

2 $\% \text{ error} = \frac{\pm 0.005 \times 100}{45.55} = 0.011\%$ (2 s.f.)

3 Human reaction time also affects the error of the stopwatch as although it can record the time to the nearest 0.01 second, a person is unable to press stop this quickly when required.

84 Correlation analysis

(a) There is a strong positive correlation between temperature and rate of reaction.

(b) There is a weak negative correlation between the length of index finger and intelligence.

85 Using secondary evidence

You will need to analyse your secondary data to see if this is correct. You should comment on how the variables were changed and what was measured. Look at how long the investigation was performed for and if repeats were performed. Did they use the most precise equipment available? Did they fully investigate the hypothesis?

86 Writing a plan

Hypothesis – the wider the wire the less resistance the wire will have. This is because there will be a larger area for the electrons to move through, so although the electrons move at the same rate, more electrons will move through at the same time, making it easier for the charge to move through the wire. Therefore, there is less resistance.

1. Set up a circuit with a bulb (to show the circuit is working), an ammeter and a voltmeter and a power supply.
2. The ammeter should be placed in series with the wire to be tested. The voltmeter should be placed in the circuit in parallel to the wire to be tested.
3. Place a 50 cm long copper wire into the circuit. Measure the current and the voltage.
4. Allow the wire to cool, as temperature affects resistance and then repeat the experiment.
5. Repeat again for five different widths of wire.
6. Repeat any anomalous results.
7. Calculate the average current and voltage for each wire.
8. Calculate the resistance using the equation $V = IR$.

Control variables: same bulb, ammeter, powerpack and voltmeter.
Same length wire each time.
Allow wire to cool between each reading.

The ammeter and voltmeter allow the current and voltage to be measured so that the resistance can be calculated.
You can plot a graph of resistance against wire width to show how wire width affects resistance.

Ensure you do not touch the hot wire or bulb as it may be hot and may cause burns.

This is a good answer as it has covered all the points in the question. The hypothesis is explained using scientific ideas. There is a method that could be followed to gain results. Control variables are given.
There are few risks but these have been discussed. If there are no risks, it is a good idea to state this and say how you know there are no risks.
Plotting a graph is the method for data collection analysis.

87 Constructing an answer

Mark your answer from page 86 using the following points as a guide. Add any information you have missed in order to improve it.
Have you included the following?

- A hypothesis that links the two variables together and is supported by scientific knowledge.
- The equipment (including sizes and numbers) you will use.
- The technique / procedure you will use.
- The reasons why you have chosen that method.
- The risks and how you would minimise them.
- Identification of the control variables and descriptions of how they would be controlled.
- Identification of the dependent variable and how will it be measured? What units will it be measured in? To what precision will the measurements be made?
- Identification of the independent variable and the range of measurements to be used.
- Description of how you will analyse the data.

88 Evaluation questions

Advantages
Using five different alcohols will show a trend and this is good as a range of alcohols will be investigated. The water is stirred which allows the heat to be transferred evenly, so a true temperature rise can be measured.
Disadvantages
It is difficult to follow this method and get the same results each time. This is because of the following:

- There is no mention of keeping the burner at the same difference from the water each time. Using different differences could cause different temperature rises and affect the results.
- There is no mention of how heat loss to the surroundings would be minimised. The learner could have used a heat shield around the flame to ensure that most of the energy went into the water.
- The learner did not say how much water to use. The volume of water heated will affect how much the temperature rises. The volume of water heated is also needed to calculate the energy released from the fuel.
- The start temperature has not been recorded. The learner would not know how much the temperature had changed by without the start temperature. Temperature change is needed to calculate the energy released from the fuel.

The learner would not get enough reliable evidence following this method to be confident in their conclusion. There are no repeats, so the learner would not be able to identify if there were any anomalous results.

The answer above is a good response to a question asking you to evaluate a given method. The answer evaluates the method, the results and the conclusion. There is an explanation of each of the comments given. The answer includes what is good about the method and what could be improved. The answer also explains how the disadvantages would affect the quality of the results that would be obtained.

89 Answering evaluation questions

Your evaluation should include:

- Advantages of the method. What points of the method will allow the learner to produce valid results and why?
- Disadvantages of the method. What points of the method need to be improved and why?
- Is there enough information for the method to be repeated exactly as it was carried out by someone else?
- Suggested improvements.
- Explanation of whether this method will fully and adequately investigate the hypothesis.

A good answer would be:
The method is good because it describes how you would detect whether the enzyme is active or not. It describes how iodine is used to detect whether the starch substrate had been broken down. This is important as you would need to know what results to collect and what these results show. The method is also good as there is a description of how it would be performed using starch agar plates and iodine. However, the method also has disadvantages as there is not enough information to follow it exactly. The learner does not state what pH values would be investigated. A range of pH between pH2 and pH9 would be appropriate. The learner does not state how other variables such as enzyme concentration, substrate concentration and temperature would be controlled. These would be controlled as they also affect enzyme activity. Temperature could be controlled by placing the plates in a thermostatically controlled incubator between adding the filter paper discs and adding the iodine.
The learner does not state how long the plates would be left for

before adding the iodine. This would need to controlled for all pH values investigated and all repeats. For example, the plates could be left for 12 hours in a 20 °C thermostatically controlled incubator. The learner does not say how many filter paper discs are placed on each plate and does not mention whether each pH investigated would be repeated. It is important to carry out repeats in order to obtain concordant results and to identify any anomalous results. The learner does not say how the yellow/brown circles are to be measured. The diameter could be measured with a ruler to the nearest mm. The method will not fully investigate the hypothesis of whether pH affects enzyme activity unless the range of pH is stated and the other variables that affect enzyme activity are controlled.

Unit 5, Biology

90 The heart

This keeps oxygenated and deoxygenated blood separate which means as much oxygen as possible being carried to the cells.
It also allows the two sides to be different in their amount of muscle, leading to a higher pressure on one side (pumping to the body) than the other (pumping blood to the lungs).

91 Blood vessels and types

Any two from:

- capillary walls are one cell thick
- no elastic tissue/collagen/muscle/multiple layers in the capillary (walls)
- no valves in capillaries
- capillaries have a very narrow lumen
- capillaries are porous/have pores.

92 The cardiac cycle and the heartbeat

It passes into the right atrium (right atrial systole), through the tricuspid valve into the right ventricle to the lungs (right ventricular systole) via the pulmonary artery, from the lungs to the left atrium via the pulmonary vein. Through the tricuspid valve into the left ventricle (left ventricular systole) and to the rest of the body via the aorta (left atrial systole).
(content in brackets for information only)

93 Cardiovascular disease (CVD): Risks and treatment

(a) Decreasing HDL always leads to an increase in CVD, decreasing LDL always leads to a decrease in CVD.
(b) Reduce levels of LDL and increase levels of HDL cholesterol.

94 *Daphnia* heart rate

The biggest problem in ensuring accuracy is that the high heart rate is difficult to count. One way to deal with this is to cool the *Daphnia* and slow the heart rate. Another way would be to film the *Daphnia* and slow down the film in order to count the heart rate more accurately. Another possibility is to use a strobe light to freeze the motion. The rate of strobing equals the heart rate when it appears to be frozen.

95 The human lungs

The walls of the alveoli and blood capillaries are one cell thick. There is a dense network of capillaries around the alveoli and many alveoli provide a large surface area. All of this maximizes the rate of diffusion of gases.

96 Lung ventilation

A

97 Measuring lung volumes

Exercise increases tidal volume, breathing rate, respiratory minute volume and oxygen consumption because muscles require more energy and oxygen is needed in the cells to supply the increased energy requirement through respiration.

98 Structure of the kidney

The renal artery brings blood with wastes (urea) to the kidney to be cleaned. The renal vein takes cleaned blood from the kidney to the heart.

99 Kidney function

Filtrate is hypertonic at the base of the loop, deepest into the medulla. This is because water has left the filtrate by osmosis as it passes through the medulla. Filtrate is isotonic to the medulla tissue.
Filtrate becomes hypotonic as it reaches the DCT as ions diffuse and are actively pumped out of the tubule as it passes out of the medulla.

100 Osmoregulation

Salt will lower the water potential of the blood plasma.
This fall will be detected by the osmoreceptors in the hypothalamus.
The pituitary will be stimulated to produce ADH.
ADH will make the walls of the collecting ducts permeable to water.
Water will leave the filtrate by osmosis leaving concentrated urine.

101 Maintaining balance

Lower blood pressure causes juxtaglomerular cells in kidneys to produce renin. This in turn causes angiotensin I and II to be produced. Angiotensin causes adrenal cortex to produce aldosterone. This causes kidneys to reabsorb more sodium ions and increase secretion of potassium and hydrogen ions. Increase in sodium ions causes more water to be reabsorbed, increasing blood volume and increases blood pressure.

102 Surface area to volume ratio

Cube with side 2 has area of 2×2 (each side) $= 4 \times 6$ (number of sides) $= 24\,\text{cm}^2$. Its volume is $2 \times 2 \times 2 = 8\,\text{cm}^3$

$$\frac{\text{SA}}{\text{V}} = \frac{24}{8} = 3$$

Cube with side 8 has area of 8×8 (each side) $= 64 \times 6$ (number of sides) $= 384\,\text{cm}^2$. Its volume is $8 \times 8 \times 8 = 512\,\text{cm}^3$

$$\frac{\text{SA}}{\text{V}} = \frac{384}{512} = 0.75$$

103 The cell surface membrane

The membrane has a phospholipid bilayer with the hydrophobic tails of the phospholipids in the centre and the hydrophilic heads on the outside, facing water. There are proteins in the membrane, some on either surface, some within the membrane and some crossing right through it. The membrane will have glycoproteins on the outside and cholesterol within it.

104 Passive transport

They should be small molecules and be soluble in lipids.

105 Active transport, endocytosis and exocytosis

Insulin is engulfed in a membrane-bound sac called a vesicle. This then moves to the cell membrane where it fuses with the membrane. This is possibly due to the fluid nature of membranes. The insulin is shed to the outside as the vesicle membrane opens.

Unit 5, Chemistry

106 Metal oxides and hydroxides

The nitric acid can be neutralised using calcium hydroxide.
$Ca(OH)_2(s) + 2HNO_3(aq) \rightarrow Ca(NO_3)_2(aq) + 2H_2O(l)$

107 Aluminium and titanium

Carbon is a non-metal, so is not ductile or malleable, so cannot be shaped or repaired easily.

108 Useful products from electrolysis of brine

Chloride ions can pass through the diaphragm but not the membrane, so sodium hydroxide from the diaphragm cell is contaminated with sodium chloride

109 Formulae in organic chemistry

110 Alkanes

Molecules in diesel must be bigger than molecules in kerosene. Hence, they have more electrons so greater London forces, so higher boiling point range. Fractions contain a mixture of molecules. Some of the molecules in diesel must be a similar size to some of the molecules in kerosene, as the boiling point ranges overlap.

111 Alkenes

There is greater electron density in the carbon-carbon double bond in alkenes so more likely to be attacked by reactants such as electrophiles. In addition, the pi bond in alkenes is weaker than the sigma bond in alkanes, so is more likely to break.

112 Naming hydrocarbons

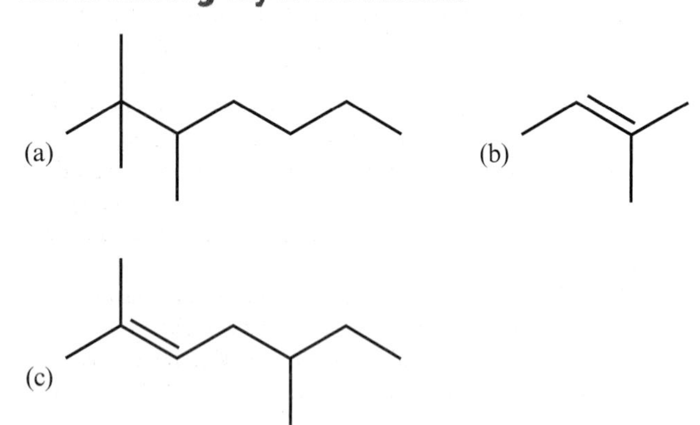

113 Reactions of alkanes

$C_{16}H_{34} + 16\frac{1}{2}O_2 \rightarrow 16CO + 17H_2O$
OR
$C_{16}H_{34} + 8\frac{1}{2}O_2 \rightarrow 16C + 17H_2O$
Other equations are possible forming a mixture of C, CO and CO_2

114 Reactions of alkenes

1-chlorobutane and 2-chlorobutane are formed. 2-chlorobutane is the major product, as the positive carbon in the intermediate has two alkyl groups attached. These push electrons towards the positive carbon and so stabilise the carbocation intermediate, so major product forms more readily.

115 Hydrocarbon reactions of commercial importance

$3CH_2{=}CH(C_6H_5) \rightarrow$
$-CH_2-CH(C_6H_5)-CHCH(C_6H_5)-CH_2-CH(C_6H_5)-$

116 Enthalpy changes in chemical reactions

Platinum is a rare metal, so expensive. Extra cost of buying better catalyst will not be offset by increased effectiveness of reaction. Platinum may be 'poisoned' by reaction with impurities in the reactants, that adsorb onto its surface and do not easily desorb.

117 Measuring enthalpy changes in chemical reactions

$Q = mc\Delta T$, so $\Delta T = Q/mc$
$\Delta T = 7400/(50 \times 4.18)$
$= 35.4\,K$ (or °C)

118 Enthalpy changes of formation and combustion

$q = mc\Delta T$, $q = 500 \times 4.18 \times 60 = 125\,400\,J = 125.4\,kJ$
amount of butane = 2.80/58 = 0.0483 mol
$\Delta_c H = 125.4/0.0483 = -2600\,kJ\,mol^{-1}$ (2 significant figures)
Calculated value likely to be lower/smaller in magnitude as heat lost to surroundings/incomplete combustion.

119 Enthalpy change of hydration

$\Delta_{sol}H = -(-646) + (-307 - 314)$
$= +25\,kJ\,mol^{-1}$

120 Calculations using enthalpy changes

$(-394) - (-238 - 286) = +130\,kJ\,mol^{-1}$
Calculated value uses standard values at 100 kPa and 298 K, value from industrial process is at 600 K and 2000 kPa, which are not standard conditions.

Unit 5, Physics

121 Energy and work

1(a) master piston area = $\pi r^2 = 78.5\,mm^2$ (= $7.85 \times 10^{-5}\,m^2$), so volume displaced = $78.5\,mm^2 \times 25\,mm = 1960\,mm^3$ (= $1.96 \times 10^{-6}\,m^3$)
(b) $p = F/A = 12\,N / 7.85 \times 10^{-5}\,m^2 = 1.53 \times 10^5$ Pa (= 153 kPa)
(c) $W = p\,\Delta V = 1.53 \times 10^5\,Pa \times 1.96 \times 10^{-6}\,m^3 = 0.3\,J$
2 slave piston area = $\pi r^2 = 1960\,mm^2$ (= $1.96 \times 10^{-3}\,m^2$) so force on brake $F = pA = 1.53 \times 10^5\,Pa \times 1.96 \times 10^{-3}\,m^2 = 300\,N$

122 Thermal energy

1. Heat is an energy transfer associated with random, uncontrollable, microscopic, thermal motions of molecules. Work is an energy transfer that changes the position or speed of a large object, i.e. large enough for us to use or control.
2. The temperature rises quickly at first, then more and more slowly as the thermometer reaches thermal equilibrium with the child's mouth.
3. The temperature falls fairly steadily until it reaches the freezing point. Then it stays at the same temperature for a long time – until all the water has frozen – before it begins to fall further.

123 Gases and thermometry

1(a) 78 K
(b) $V = NkT/p = 1.74 \times 10^{25} \times 1.38 \times 10^{-23}\,J/K \times 78\,K/1.01 \times 10^5\,Pa = 0.185\,m^3 = 185$ litre
(c) $V_1 = NkT_1/p$; $V_2 = NkT_2/p$; N, and p are the same in both cases and k is a constant.
Hence, $V_1/V_2 = T_1/T_2 = 293/78 = 3.75$ times larger.
(d) Nitrogen displaces air and so can cause asphyxiation. 3.75 × 185 litres = $7.5\,m^3$, that is, comparable with the volume of a small room.
2 For any given T and p, the ideal gas equation predicts that V depends directly on N, the number of particles. So, for example, the volume occupied by a mole of gas at NTP (293 K and 1.01×10^5 Pa) is 22.4 litre for any material. If the volumes are the same, then the gas density $\rho = m/V$ depends directly on the molecular mass.

124 Energy conservation

(a) 2 min = 120 s; so, energy used = 3.0 kW × 120 s = 360 kJ

(b) ΔT = 85 K; $m = Q/(c\Delta T)$ = 360 ~~kJ~~/(4.18 ~~kJ~~ kg^{-1} ~~K~~$^{-1}$ × 85 ~~K~~) = 1.01 kg (~~strikethrough~~ indicates units that cancel out)

(c) unit conversions: 2.26 MJ kg^{-1} = 2260 kJ kg^{-1}; 3.0 kW = 3.0 kJ s^{-1}
so t = 2260 ~~kJ kg~~$^{-1}$ × 1.01 ~~kg~~/3.0 ~~kJ~~ s^{-1} = 761 s (= 12.7 min)

125 Processes

1 The coffee gives up heat to the steam as it vaporises and expands. Energy would be needed both to reheat the coffee and to re-compress the vapour. The latent heat recoverable as the vapour condensed would not fully match the energy input.
Mixing of the milk into the coffee happens naturally (though it can be speeded up by stirring). Once mixed, separation is near impossible because the particles involved are microscopic and randomly distributed, being held in suspension by electrostatic forces; but it might be partly achieved using a centrifuge, which requires a substantial input of work.

2 Pumping compresses the air very quickly, so even without insulation the process is near adiabatic. That means all the work is transformed into increased internal energy and hence higher temperature in the gas. Frictional and viscous forces mean that extra work is being done, which also converts directly into heat and raises the temperature even more.

126 Cycles

1 The compression process occurs between higher pressures, and $W = p\Delta V$, so the work done in it is higher than the work recoverable from the expansion process. (The same is true of the work done in the isothermal processes.)

2(a) High specific heat capacity means a large quantity of heat is needed to raise the temperature of the superheated steam. This helps a large quantity of heat to be transferred into the system at the high temperature, and that increases the store of energy that can be transferred as work during the turbine expansion, thus aiding the efficiency of the cycle.

(b) Latent heat of vaporisation has to be input to the system in the boiler to form vapour (steam). It is then rejected out of the system in the condenser to reform liquid water and to pull a low pressure at the turbine exit. The condenser is itself cooled by an external water source.

127 Efficiency

1 max. theoretical efficiency, $\eta_{rev} = 1 - T_H/T_C$
= 1 – 300 K/310 K = 0.032 = 3.2%

2(a) (i) W = 75 kg × 9.81 N kg^{-1} × 0.30 m × 20 = 4.41 kJ
(ii) 4.41 kJ × 100/0.4 = 1.1 MJ

(b) To avoid the athlete overheating, ΔU must be kept to zero. So all the food energy released in biochemical reactions must be output either as work or heat. The calculations in part (a) used 0.4% efficiency, which means that 99.6% has to be lost as heat. Evaporating perspiration allows much of that heat to be transferred as latent heat of vaporisation of water and carried away with the vapour. Keeping a cool skin temperature is important for keeping up the body's thermal efficiency, which depends on maintaining the temperature difference.

3(a) Q_{in} = 1 kW + 6 kW = 7 kW

(b) efficiency, η = 1 kW/7 kW = 14%

(c) Because its heat output, Q_{out}, is not wasted, but used for water and space heating.

128 Moving heat

1 max. theoretical CoP = 303 K/(30 – 8) K = 13.8

2 If the condenser is air-cooled, the warm air from it could be ducted into the shop instead of to the outside atmosphere. If it is water-cooled, the warmed coolant water could be directed to a water-to-air heat exchanger inside the shop. To avoid overheating the shop space on a warm day, controls would be needed to redirect the warm air or water once the shop space reaches its target temperature.

129 Elasticity

1(a) cross-sectional area = πr^2 = 8.0 × 10^{-4} m^2, so max load = 8.0 × 10^{-4} m^2 × 250 MPa = 200 kN

(b) $\varepsilon = \sigma/E$ = 250 MPa/200 GPa = 0.125%

(c) $\Delta x = \varepsilon L$ = 4.5 mm

2(a) weight = 3.0 N, so k = 3.0 N/12 mm = 250 N m^{-1}

(b) energy = $\frac{1}{2}k\Delta x^2$ = 250 N m^{-1} × (0.012 m)^2 /2 = 0.018 J = 18 mJ

3 For example, gym mat, door stop, etc.

130 Shape change

(a) The steel is stiffer because the initial part of the curve is steeper.

(b) The graph does not show the fracture point (it occurs at >2% strain). But the work hardening slope for the steel indicates the line will rise above that for the aluminium alloy, making steel the stronger of the two.

131 Failure

1 Stress concentration around scratches is high because the material is very stiff (high Young's modulus). Stresses become high enough to form tiny cracks that grow and further increase stress concentration. Rapid brittle failure results from multiple cracks spreading and joining.

2 They are composite materials made of particles bonded under pressure or by chemical reaction. They typically contain many boundaries and tiny voids that can initiate cracks.

3(a) Creep can be expected because of the elevated temperatures in some blades and the sustained high centrifugal forces. Creep can cause elongation, which could allow blades to strike the casing. It can also result in crack formation leading to brittle fracture.

(b) Stresses, especially on elements like wings and tail planes, are not constant, but they are frequently loaded and then unloaded. Cycling of stress can cause microscopic defects to move or enlarge over time, leading to crack formation leading to fatigue failure. X-ray photos can be used to detect tiny cracks.

132 Fluid flow

1 Oil viscosity is high when cold, so creates a large viscous drag force. Viscosity falls on warming.

2 Turbulence increases viscous drag forces and thus wastes energy (turning it into heat).

3 Turbulence causes mixing of air streams, so will result in a more uniform temperature.

133 Non-Newtonian fluids

(a) shear thinning (pseudoplastic)
(b) shear thickening (dilatant)
(c) shear thinning (pseudoplastic)
(d) thixotropic
(e) Bingham plastic

134 Fluid dynamics

1 Internal radii are $(15 - 2)/2 = 6.5$ mm and $(10 - 1.4)/2 = 4.3$ mm. So circulation time would be about $(6.5/4.3)^4 = 5$ times longer.
2 Air travelling round the curved front side of the sail travels further, and so faster, than air passing more directly by the rear of the sail. The slower air at the rear of the sail exerts a higher pressure and pushes the boat forward.

Unit 5, Skills

136 'Give', 'state', 'name' and multiple choice questions

Blood plasma does not contain blood cells that would carry the blood type antigens, therefore the body will not make antibodies to attack the blood cells and will not cause a reaction.

137 'Calculate' questions

From previous question, $W = 30.5$ kJ.
Latent heat is absorbed during evaporation, so Q is negative.
Using data for L from page 136:
$-Q = mL = 1\,\text{kg} \times 286\,\text{kJ kg}^{-1} = 286\,\text{kJ}$
$\Delta U = Q - W = -286 - 30.5 = -316.5\,\text{kJ}$

138 'Explain' questions

The alumina itself comes from the mineral bauxite. Bauxite is quite common but the costs of mining it, including equipment and energy then transporting it long distances, add to its cost. It then has to be purified into alumina, which uses energy and other reactants. The cryolite used to lower the melting temperature also has to be mined, processed and transported. However, the overall price of the cryolite is offset by the reduction in energy costs, due to the lowering of the melting point of the alumina.
The process of electrolysing alumina is continuous so makes savings on automation of the process and so labour costs.
However, even with the cryolite, the melting point of the alumina is very high, so a lot of energy is required to melt the alumina, so that it conducts. Additional electrical energy has to be supplied to create the high current needed to electrolyse the alumina. The high levels of energy required for both these parts of the process add to the high costs, although some aluminium plants are sited near hydroelectric power stations to try to reduce these costs.
As oxygen is produced at the positive electrode, it will react with the electrodes to form carbon dioxide.

$$C_{(s)} + O_2(g) \rightarrow CO_2(g)$$

This means the carbon electrodes require replacing regularly, adding to the cost.

139 Combination questions (1)

The concentration of solutes in the plasma is more than in the filtrate, so water molecules move from filtrate into the plasma. ADH causes channels in the walls of the collecting duct, which allows water molecules to pass through.

140 Combination questions (2)

$-106\,\text{kJ mol}^{-1}$

Unit 7: Contemporary Issues in Science

142 Issues and impacts

(a) If lots of dead animals are cloned, there will be even more animals to feed because animals will continue to reproduce as well, which will put a strain on the environment
(b) It would make a lot of money for the firms that can clone pets. This would allow them to do more research into cloning.
(c) Poor people would not be able to clone their pets but rich people would; this would be unfair. A lot of money might be spent on trying to clone rare animals but the success rate is poor – it could waste money which would be spent better in other conservation projects. Some people might try to clone extinct animals, which could cause major ecological issues – any sensible, thoughtful points like this.

143 Energy sources

For example: Solar energy
Solar energy is the energy produced by the Sun, which we can capture and use.
Environmental impacts
Positive:
- Renewable.
- Doesn't need mining or bring up from ground so this reduces environmental impact.
- Doesn't produce greenhouse gases, so removes any global warming.
- Silent (no noise pollution).

Negative:
- Takes up a lot of room/ground cover.
- Pollutants can be caused by the process of making solar panels, etc.

Economic impacts
Positive:
- Can be made available to everyone – especially as poorer countries are often in very sunny regions.
- Can make electricity cheaply in the long term.

Negative:
- Materials to make solar panels can be expensive in the short term (but long term probably cheaper than conventional methods).
- Any other sensible points.

144 Medical treatments

For example: Proton beam therapy
Source: http://www.nhs.uk/news/2014/09September/Pages/what-is-proton-beam-therapy.aspx
Source: http://scienceblog.cancerresearchuk.org/2015/07/16/proton-beam-therapy-where-are-we-now/
- New type of radiotherapy used in cancer treatment.
- Beams of accelerated protons focused on cancer cells.
- Proton beams stops when it hits the target tissue.
- Advantage: tight focus on cancer cells.
- Advantage: less damage to surrounding tissues.
- Used for treating cancer when it is especially important not to damage surrounding tissue, for example, in brain, eye cancers.
- Reduces side effects but not better at improving survival rates or cures.
- Disadvantage: may not be as good at destroying cancer cells as conventional radiotherapy.
- Economic issue: more expensive treatment of unproven value.
- Social issue: not widely available in the UK so have to travel abroad to get proton therapy – available to few patients.
- Ethical issue: relatively little research evidence for the value of proton beam therapy.
- Any other sensible points backed up from sources.

145 Pharmaceuticals

Example of the content you would include if you chose antibiotic resistance:
- Penicillin was the first antibiotic.
- Antibiotics made it possible to cure bacterial infections.
- Bacterial pathogens are becoming resistant to the antibiotics that are used to destroy them.
- Factors affecting the development of antibiotic resistance include overprescribing, patients not completing courses, overuse for livestock and in fish farming, poor infection control in hospitals etc, poor sanitation and hygiene, lack of new antibiotics.
- Scientists are racing to develop new antibiotics.
- Takes 10–12 years and over a billion pounds to discover and develop a successful new drug.

- Bacteria are developing antibiotic resistance faster than new antibiotics are being discovered.
- Fast and relatively cheap genome analysis of bacteria may help scientists get ahead.

146 Chemical developments

Some of the topics you need to cover:
- What are plastics?
- What are they made of?
- What are smart plastics?
- Why are plastics so useful?
- How do plastics affect the environment?
- How can we overcome the problems of plastic pollution?

You may use images to carry information, for example, graphs or other data, images of ocean pollution, etc.
Reference all the sources you use.

147 Nanotechnology

1 Need electron microscope to be able to see and manipulate objects at the nanoscale.

2

Current uses of nanotechnology	Planned uses of nanotechnology
• replacing solder • instant-on computers • wearable electronics to record vital signs for fitness training and medical use with 100 nm thick silicon patches • removing heavy metal particles from water	• tiny space probes and rovers • nanosensors that can detect poisons and pathogens will help diagnose many diseases accurately and almost immediately • nanoparticles may be engineered to recognise cancers for diagnosis • nanoparticles may be made to carry drugs to specific sites in the body, including the heart, the brain or tumours • lighter batteries • more efficient catalysts • more efficient solar panels • coatings for wind turbines to prevent water/ice build-up

148 Food technology

Benefits
- Can give increased yield: 2014 study showed GM soybeans, maize and cotton have 22% increased yields (www.geneticliteracyproject.org/2014/11/19/gmo-crops-increase-yields-benefit-the-environment/).
- Can give increased nutritional value, for example, added vitamins, more protein. GM Golden Rice gives increased vitamin A, this article.
- Can survive adverse conditions, for example, drought resistant, storm/wind resistant, survive flooding. Flood resistant rice can survive up to three weeks submersion and still give 80% yield instead of dying after days (http://irri.org/our-impact/increase-food-security/flood-tolerant-rice-saves-farmers-livelihoods).
- Can be disease resistant. British scientists develop potatoes resistant to potato blight – biggest threat to potato crops for centuries (http://www.bbc.co.uk/news/science-environment-26189722).
- Fruit can have longer shelf life so less wastage. GM tomatoes last up to 45 days on shelf (http://www.telegraph.co.uk/news/science/7128622/Scientists-create-GM-tomatoes-which-stay-fresh-for-a-month-longer-than-usual.html).

Concerns
- GM genes may pass into wild populations of plants, for example, genes from GM herbicide resistant common bentgrass found in wild grass (http://sitn.hms.harvard.edu/flash/2015/challenging-evolution-how-gmos-can-influence-genetic-diversity/).
- GM crops will be too expensive for farmers in developing countries which need them most. Suicides in African farmers who could not afford the cost of GM cotton seed (https://www.theguardian.com/global-development/poverty-matters/2013/jun/24/gm-crops-african-farmers).
- The inserted genes may affect people when they eat GM food (although we digest the DNA of every organism we eat anyway). Thousands of independent studies have found that GMOs are perfectly safe to eat (www.geneticliteracyproject.org/2014/11/19/gmo-crops-increase-yields-benefit-the-environment/).

149 Government and global organisations

1 Global organisations such as the WHO can collect data from all over the world. This can highlight areas where scientific research is particularly needed/direct global legislation which influences what scientific issues are seen as important. Global organisations such as the UN/WHO have money from many different countries and so can give financial support to research into some scientific issues and not to others.

2 Governmental organisations such as the FSA can influence which scientific issues are brought to the public notice through advertising and funding research/influence which scientific issues attract funding/pass legislation that impacts on different scientific issues.

150 NGOs

1 Professional bodies and associations are non-governmental organisations so they are not-for-profit and are independent of the state or government. They look after the interests of particular groups, for example, Royal Society of Biology, Medical Research Council.

2 They may advise government on priorities for research funding/they may allocate research funding directly to scientific issues they think are important.

151 Universities and research groups

Advantages: for example:
Top scientists are the best in their areas so it is good to have them researching the most important science issues.
The best universities will have a lot of very good scientists working for them, so they will attract lots of funding for the areas of science they think are important.
Disadvantages: for example:
Scientists who are less well known may come up with the better science but will find it hard to attract funding.
Top universities will get a lot of funding and use it in the areas where their top scientists are experts, which may not be the most important scientific issues of the day.

152 Private and multinational organisations

1 One similarity: for example, both multinational companies and private organisations can invest a lot of money in research into important scientific issues.
One difference: for example, multinational companies have to make a profit, so their funding will be into science that could help their company.
Private organisations are often philanthropic and fund research for the good of other people without trying to make a profit.

2 For example: rich people could fund research that would enable them to make more money without focusing on issues that affect many people globally. Rich individuals could fund unethical research, for example, human cloning, which the scientific community does not approve.

153 Voluntary pressure groups

For example:

	Cancer Research UK	Greenpeace
What they are?	UK charity	international pressure group
Where does their money come from?	charity donations, tax relief	donations, tax relief
What they fund?	• supporting five core institutes researching the causes of and ways to treat cancer • supporting research into cancer and cancer treatments at other institutions • providing information on cancer to the general public • policy development to advise the government on research funding	• lobby politicians globally over perceived environmental issues • campaign against environmentally damaging practices • work for peace • investigate environmental abuse and take action to stop it or get it into the news

154 Qualitative evidence?

The dates vary from 1939 through to 2005. This tells you that the paper is based on data collected over a long period of time. When commenting on the quality of the paper you could look at this in two ways:

You might comment that recording equipment and knowledge were very different in the 1930s and even in the 1960s than they are today and so this reduces the quality of the data and makes the paper less reliable.

However, you could equally say that measurements taken over a long period of time show consistency and allow us to look for changes over the long term – even allowing for improvements in technology this is still high quality content. Also lots of citations from a range of scientists over time are usually associated with good quality research

Both answers would be acceptable.

155 Quantitative evidence?

(a)

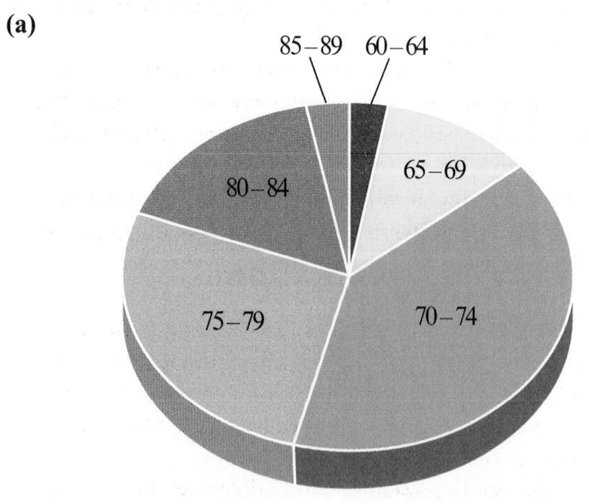

(b)

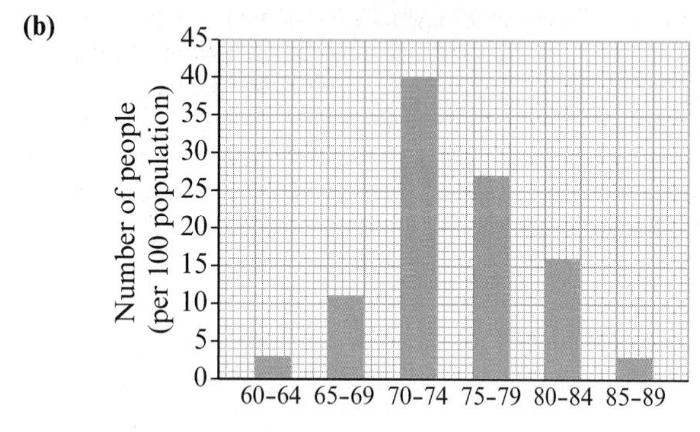

(c) Whichever answer you give is acceptable, if you justify your choice with clear reasons, e.g. 'I think the bar chart shows the patterns in this data most clearly, because it is very easy to see exactly how many people per hundred population have the different ranges of heart rate.'

156 The importance of statistics

Small error bars suggest there was little variation in the data, which can make it seem reliable. However, in a study with only 10 individuals, any variation is likely to be due to chance and the data cannot be reliable. So, although the data looks reliable, it is not and it could be used to mislead people. The data used in the second article has bigger error bars, which could suggest it is not so good because there is more variation in the data. However, it involves 30 000 people, which means it is much more reliable – this is the data I would trust because it is such a big sample and the variation is still within an acceptable range.

157 The validity and reliability of data

Read the Wikipedia article carefully.

Bullets should include:

- Whitehall I Study 18 000 male civil servants aged 20–64. Whitehall II Study 10 308 civil servants aged 35–55, $\frac{2}{3}$ male, $\frac{1}{3}$ female: large studies so reliable.
- Length of study: Whitehall I 10 years, Whitehall II started in 1985 and still continues: long studies so reliable.
- Scientists: led by Sir Michael Marmot, Department of Epidemiology and public health at UCL. He was also involved with WHO: very reputable scientist and institutions – adds to reliability.
- Only Whitehall II quoted – 73%: this is high.

158 Use and misuse of data

Choose a story that interests you.

- List the main results – you can get these from this page but you may be able to add more detail.
- List what is wrong with the results.
- The misuse of data can be deliberate fraud, or a genuine mistake. In most cases, other scientists will raise queries when they cannot replicate the results. In some of these examples it was deliberate fraud. Then, someone involved may admit the fraud or other scientists get together to raise concerns. Sometimes, financial benefits are discovered by journalists or scientific bodies. See what you can find.

159 Potential areas for R&D

Some or all of the following points:

Most of the factors that decide whether a project will get R&D funding involve either that it has a positive benefit for a lot of people or it has the potential to make a lot of money – or both. Blue skies research is not driven by either of these factors and – even if it discovers new science, this could have no direct impact on peoples' lives, be unpatentable and have no potential for making money.

As a result, philanthropists are less likely to invest in blue skies research as there is no guarantee of benefit to people, industries are less likely to invest as there is less chance of making money. When the financial situation in a country is difficult, fewer people are prepared to invest in blue skies research as they want results for their money.

160 Supporting evidence

(a) It is an advert and designed to make money, so it may be created more for selling a product than delivering good science. Look for what you would want to see in a valid scientific claim and see if it is present in the advert, for example, references to influential, reliable, valid scientific studies with large samples carried out over a long period of time by well-known scientists in good research institutions. Ideally, you want several scientists finding the same results. Published papers in influential journals are also good evidence. Data clearly displayed. If none of this is available, you would be suspicious of the claim.

(b) Look in scientific journals/articles in reliable magazines such as *New Scientist*/reputable websites to see if there is authentic, reliable evidence supporting the claim that these supplements help you to get fitter.

161 Target audiences

(a) This is a small font and is full of technical language. It refers to statistical analysis. Target audience: physicists/people who are scientifically very literate – probably a scientific journal.

(b) Large font, headline script, simple language, content that would appeal to lots of people: target audience probably ordinary people who do not have any specialist scientific training. Probably newspaper or popular magazine.

(c) This is about scientific findings and uses some technical language but it is generally interesting and a relevant topic. Target audience probably general population with a scientific interest, for example, *New Scientist* magazine.

162 Presentation of science

(a) Keep your target audience in mind all the time to make your piece fun yet retain the science.

(b) Think what information makes science reliable and valid. You would need evidence and to know the source of the research. So your list would include some or all of the following:

- the names of the scientists who did the work
- the university or research institution where the work was done
- how many patients the adhesive had been tried on
- where the research was published
- who funded the research
- the date of the research
- any other scientific groups which had tested the adhesive and got the same results.

163 Quantity, quality and bias

Person 1 wants freedom for people to use cars and doesn't accept the problems resulting from burning fossil fuels.
Person 2 wants all fossil fuels to be banned because of their impact on global warming.
Both are relatively extreme views.
They would use the data differently.
Person 1 would cast doubt on the validity of the data, undermine the scientists who did the research, the scientific institutions involved and the journal in which the data was published. They would find other data published by lobbying groups undermining the impact of fossil fuels on global warming.
Person 2 would accept the data and highlight its validity, along with status and influence of the scientists, scientific institution and journals involved. They would use supporting data from other influential institutions to support their arguments.

164 Print media

Count up the science articles. There will probably – but not necessarily – be more in one paper than the other.
Compare the way one science topic is covered in both newspapers using a table – for example:

	Article from paper 1	Article from paper 2
Title of article		
How many lines of text?		
How many photos?		
How many diagrams?		
Name of the scientist/s involved		
Where is the research done?		
Size of study		
References		

165 TV and digital media

When you are deciding whether a scientific website is reliable or not, you need to look for a number of things, for example:

- Who does the website belong to? If it belongs to a university, a major science charity, such as the Wellcome Trust, or an organisation, such as the BBC, this should mean it will be reputable.
- Where does the science reported come from? Check the reputations of the scientists, the organisations or the science journals involved to decide if the content is reliable and free from bias.
- Are there references? Check whether the site gives any references to the papers behind the stories – and check some of them to be sure they are authentic and not biased.
- Can you find other sites describing the same science in a similar way?
- Any other points you can think of.

Unit 7, Skills

167 Planning your research

Your plan should make sure you allow enough time for researching, writing and organising your notes.

168 Reading Article 1

Main individuals: Dame Karen Dunnell (chairman of panel), Richard Faragher (Professor of Biogerontology).
Organisations: Logevity Science Panel, University of Brighton, Cardiff University.

169 Analysing Article 1

1. Longevity Science Panel: very influential because it was set up by the Government with top experts. Research shows that, as well as the Chair Dame Karen Dunnell, these include:
   Professor Sir Colin Blakemore FRS, FMedSci, FRSB, FBPhS, who is a top neurobiologist specialising in vision and development of the brain. He has worked at Cambridge, Oxford and London universities. Hugely influential.
   Professor Sir John Pattison, studied at Oxford and Middlesex Medical School, worked at St Barts hospital, became Professor of Medical microbiology at UCL Medical school, then became head of research and Development in the NHS. Chaired the UK Stem Cell Initiative. Has massive influence and great understanding of NHS.
   Professor Steven Haberman is an expert in insurance and actuarial science, so he brings great expertise in data on life spans and disease profiles – he has published 140 refereed papers. Has also a lot of expertise and influence.
   Professor Kim McPherson is a public health specialist from the University of California. She is also emeritus professor of Public Health Epidemiology at Oxford University.
2. **Dame Karen Dunnell**: National Statistician and Chief Executive of the Office for National Statistics and head of the Government Statistical Service. Chief Executive of UK Statistics Authority. She is the Chair of the panel and very well placed to make sense of all the evidence presented. Widely respected and influential.
3. **Professor Richard Faragher**: leading UK expert on ageing, professor of Biogerontology at University of Brighton, and has received several honours. He has at least 40 published papers in the area and is widely respected in his field.
4. **Cardiff University**: Ranked 5th among UK universities in 2014 Research Excellence Framework for quality, and 2nd for impact of research, so it is an influential and well-regarded institution.

170 Reading Article 2

Evidence might include: Okinawa Centurian Study, with data showing very low plasma total homocysteine levels for Okinawa which is linked to reduced risk of coronary heart disease (see graph);
low levels of yearly cancer deaths and high average life expectancy (from WHO and Japanese Ministry of Health and Welfare data); Okinawa women have a high intake of soy oestrogens in the diet (shown in the table) which could be linked with slower bone loss during the menopause (shown in papers published in two journals: *Am J Clin Nutr* and *Obstet Gynecol*).

171 Analysing Article 2

Answer might include these examples:
Claim 1: Okinawans have great longevity.
Evidence 1: They have low levels of yearly cancer deaths and a high average life expectancy (from WHO and Japanese Ministry of Health and Welfare data).
Claim 2: Okinawans have reduced cardiovascular problems.
Evidence 2: Okinawa Centurian Study, with data showing very low plasma total homocysteine levels for Okinawa. This is linked to a reduced risk of coronary heart disease (see graph of page 170).
Claim 3: Okinawans have reduced risk of complications due to menopause.
Evidence 3: Okinawan women have a high intake of soy oestrogens in the diet (shown in the table on page 170), which could be linked with slower bone loss during the menopause (shown in papers published in two journals: *Am J Clin Nutr* and *Obstet Gynecol*).

172 Reading Article 3

Areas for development and research could include:

- impact of exercise levels on ageing
- impact of different diets, e.g. Mediterranean diet and calorie restriction on ageing
- development of drugs to reduce ageing.

173 Analysing Article 3

Implications might include:

- The lack of agreement among experts and the lack of clear-cut evidence means it will be very difficult to find a single, simple way to slow or halt the process of ageing.
- Animal models are important in scientific research, but there are lots of differences in the ageing in many different species, so it is not easy to find a useful animal model.
- The conflict between scientists wanting to find a drug that would be taken from an early age to help reduce or prevent the effects of ageing, and the feeling that ageing is a natural process and should not be medicalised.

174 Considering the scientific issues

(a) In the first paragraph the answer highlights social issues from Article 1 and Article 2. It also comments on Article 3, which demonstrates both ethical and economic aspects of anti-ageing innovations.

(b) The second paragraph highlights the links between the different issues. It shows that scientific developments have many different implications, from the environmental problems of needing more housing, the economic and social implications of an ageing and often ill population on the rest of the population and the ethical issues of choice over length of life and the inequality of richer people being able to afford more anti-ageing interventions.

175 Identifying organisations and individuals

Organisations/individuals	Reference
Longevity Science Panel	Article 1 line 6 and whole of Article 3
Dame Karen Dunnell	Article 1 line 11 and author of Article 3
Okinawa Centenarian Study	Whole of Article 2
Richard Faragher	Article 1
World Health Organization	Article 2 data in table
Japanese Ministry of Health and Welfare	Article 2 data in table

176 Considering the validity of judgements

Key points to include in your plan:

1. State what is meant by valid judgements. For example, for an article to make valid judgements, the data on which the judgements are made must be valid (effectively measure what is being investigated) and reliable (can be reproduced).
2. Evidence for the validity and reliability of the data given in the article. For example,
   - The data used comes from a wide number of reliable sources including:
   - A the World Health Organization, which collects data from scientists and doctors all over the world.
   - B the Okinawa study, which has been running for many years and involves scientists from several countries studying a population that has a high number of centenarians on the island of Okinawa in Japan.
   - C There are 62 references, all to well-cited papers. For example, *Cell* and *Ageing Cell* are well-respected peer-reviewed journals).

   You could add several examples, such as:
   - Scientists referenced are well-known experts in their field, for example, Professor Tom Kirkwood, who is Director of the Institute for Ageing and Health at Newcastle University, Dr Makoto Suzuki MD, PhD, Director of the Okinawa Research Centre for Longevity Studies and Principal Investigator in the Okanawa Centenarian Study
   - Reference to metadata studies that analyse results from a number of other studies into a given area shows that the data is reliable as it has been reproduced by many different teams, for example, Nakagawa et al.
3. Who made the judgements in the article: the panel involved includes many top scientists, including Dame Karen Dunnell (formerly National Statistician), Professor Sir Colin Blakemore (London, emeritus Oxford) Professor Sir John Pattison (previously Director of R&D at the Department of Health in England) and others – well capable of analysing the evidence and making sound judgements – plus many other experts in ageing specifically.
4. Summarise that the judgements made in the article are made on data that appears valid and reliable, by people who are appropriately qualified and experienced to make those judgements.

177 Suggest areas for development/ research

Any two examples from Articles 1–3. Could include: (Article 2) research which aspect of Okinawan lifestyle lowers rates of breast, prostrate, ovary and colon cancers and develop supplements/drugs to give benefits to other populations. Research into ways to help people stick to a healthy regime so they live longer more healthily. Develop ways to enable the whole population to make the same changes. (Article 3) research into ways of measuring biological age so people can have a realistic picture of how their lifestyle is affecting their aging, so they can make informed choices about how they live or any drugs they take.

178 Writing for a specific audience

A good way to structure this answer might be:

- Intro to attract the reader (for example, 'would you like to live until 100?' 'being fit and active 'til you are at least 100 years old!').
  There are a few places in the world where this is exactly what happens – and scientists are working hard to find the secret. The studies have been varied, and there don't seem to be any clear-cut answers (reference some examples from Articles 1, 2 and 3). However, it's beginning to look as if lifestyle factors can have a big impact on how long we live.
- Scientific evidence for (a) healthier eating, (b) more exercise, (c) not smoking (reference especially Articles 1 and 2).
- Conclusion – although some evidence exists, we need more research to establish exactly what we can all do to ensure longer, healthier lives – end with call for donations.